GLOBE
High School
MATHEMATICS

GLOBE BOOK COMPANY
Englewood Cliffs, New Jersey

Authors

Debra R. Clithero teaches mathematics at Southwest Missouri State University in Springfield, Missouri. She holds a bachelor's degree in mathematics education from Southeast Missouri State University and a master's degree in mathematics education from the University of Missouri in Columbia. She also serves as a mathematics curriculum consultant.

Ricardo Flores is a mathematics writer and consultant. He holds a bachelor's degree in education from Concordia Teachers College in Seward, Nebraska. He has taught mathematics at St. Paul's Lutheran School in Strassburg, Illinois, and he has taught mathematics methods at the University of Missouri at St. Louis.

Albert F. Kempf is an educational consultant and author. He has served as a teacher at John Marshall Senior High School in Rochester, Minnesota, as a mathematics curriculum writer for the Rochester school district, and as an editorial director of mathematics projects for a major educational publisher. He holds graduate degrees in mathematics from the University of Wyoming and the University of Illinois.

Leo Gafney is a mathematics curriculum adviser, author, and educational consultant. He has taught mathematics at Choate Rosemary Hall in Wallingford, Connecticut, and at the Ralph Young School in Baltimore, Maryland, where he received a Mayor's Citation. He holds a master's degree in mathematics from Fordham University, and a doctorate in education from the University of Pennsylvania.

Michael J. O'Neil is director of mathematics and science at Salem High School in Salem, New Hampshire. He holds a bachelor's degree in mathematics education from Keene State College and a master's degree in supervision and administration from Salem State College.

Consultants

Mary Elmore
Mathematics Coordinator
Greenville County
   School District
Greenville, South Carolina

Robert L. Hoburg
Mathematics Professor
Western Connecticut
   State University
Danbury, Connecticut

Marion M. Kelley
Mathematics Teacher
Evanston Township High School
Evanston, Illinois

Kaz Ogawa
Secondary Mathematics Advisor
Los Angeles Unified School
   District
Los Angeles, California

Developmental assistance by Ligature, Inc.

ISBN: 1-55675-904-5

 **GLOBE BOOK COMPANY**
A Division of Simon & Schuster

Printed in U.S.A.
9 8 7 6 5 4 3 2 1

Englewood Cliffs, New Jersey

Table of Contents

*Problem-solving lessons
are shown in blue.*

Chapter 3 Dividing Whole Numbers and Decimals

Chapter 4 Metric Measurement

*Problem-solving lessons
are shown in blue.*

Chapter 5   Introducing Fractions and Mixed Numbers

Chapter 6   Multiplying and Dividing Fractions and Mixed Numbers

Chapter 7   Adding and Subtracting Fractions and Mixed Numbers

Chapter 8   Customary Measurement

Chapter 9   Graphs

*Problem-solving lessons
are shown in blue.*

Chapter 10   Statistics

Chapter 11   Ratio and Proportion

Chapter 12   Percents

Chapter 13   Probability

Chapter 14   Geometry: Perimeter and Area

Chapter 15   Geometry: Surface Area and Volume

*Problem-solving lessons
are shown in blue.*

Chapter 16   Pre-Algebra: Expressions and Sentences

Chapter 17   Pre-Algebra: Integers and Equations

Be Comfortable with Math

Math is everywhere! You can hardly make it through a day without dealing with mathematics. So, the more thorough your understanding of math, the more comfortable you will be handling math situations in daily life. *Globe High School Mathematics* can help.

Each chapter is introduced by a photograph of students telling you how mathematics helped them with a project, pastime, or job.

"...$\frac{2}{3}$ cup shortening, $3\frac{1}{2}$ cups flour, $1\frac{1}{2}$ teaspoons cinnamon. To bake this cake, I need to know a lot about measurement. Would you believe that I have to measure the length and width of the baking pan to be sure that it's the right size?"

172 *Chapter 8*

1   2

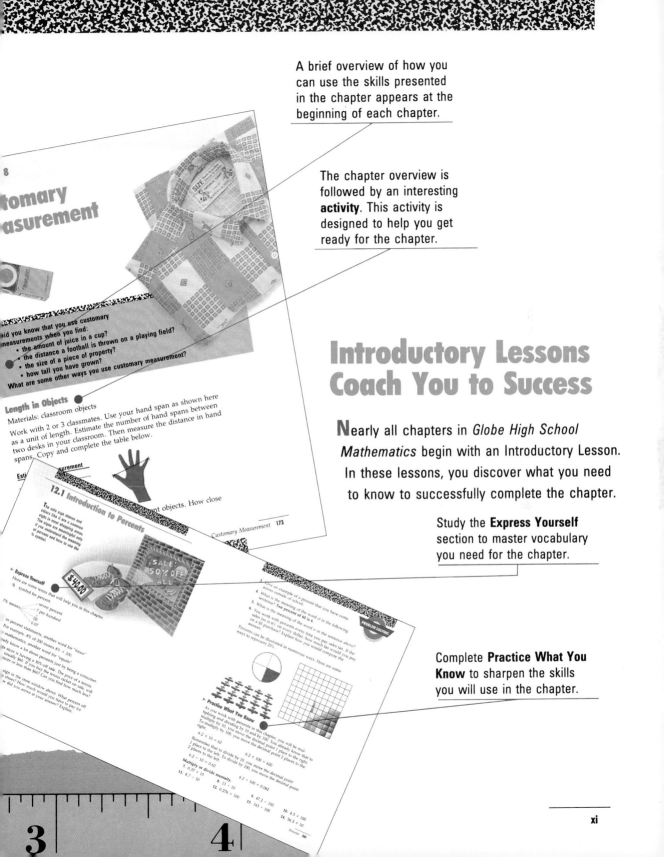

A brief overview of how you can use the skills presented in the chapter appears at the beginning of each chapter.

The chapter overview is followed by an interesting **activity**. This activity is designed to help you get ready for the chapter.

Introductory Lessons Coach You to Success

Nearly all chapters in *Globe High School Mathematics* begin with an Introductory Lesson. In these lessons, you discover what you need to know to successfully complete the chapter.

Study the **Express Yourself** section to master vocabulary you need for the chapter.

Complete **Practice What You Know** to sharpen the skills you will use in the chapter.

Skills Lessons Direct You Step-by-Step

About three-fourths of the lessons in *Globe High School Mathematics* are **Skills Lessons**. These lessons present step-by-step directions and practice problems.

Skills lessons have the Skills logo in the corner.

Each lesson starts with clear examples of the skills needed in the exercises. Important new terms are shown in heavy type.

Important steps in examples are shown in red type. The answers to examples are shown in heavy type.

Working through the **Think and Discuss** questions gives you a chance to check your understanding. This is the time to get more help if you need it.

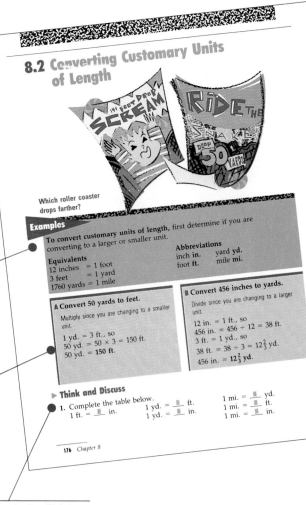

8.2 Converting Customary Units of Length

Which roller coaster drops farther?

Examples

To convert customary units of length, first determine if you are converting to a larger or smaller unit.

Equivalents
12 inches = 1 foot
3 feet = 1 yard
1760 yards = 1 mile

Abbreviations
inch **in.** yard **yd.**
foot **ft.** mile **mi.**

A Convert 50 yards to feet.

Multiply since you are changing to a smaller unit.

1 yd. = 3 ft., so
50 yd. = 50 × 3 = 150 ft.
50 yd. = **150 ft.**

B Convert 456 inches to yards.

Divide since you are changing to a larger unit.

12 in. = 1 ft., so
456 in. = 456 ÷ 12 = 38 ft.
3 ft. = 1 yd., so
38 ft. = 38 ÷ 3 = $12\frac{2}{3}$ yd.
456 in. = **$12\frac{2}{3}$ yd.**

▶ **Think and Discuss**

1. Complete the table below.
1 ft. = ▧ in.   1 yd. = ▧ ft.   1 mi. = ▧ yd.
                 1 yd. = ▧ in.   1 mi. = ▧ ft.
                                  1 mi. = ▧ in.

176 *Chapter 8*

The first exercise sets are matched with the examples. Because the lesson always appears on two facing pages, it's easy to look back to the examples for help.

Make sure you're on the right track by comparing some of your answers with those in the **Selected Answers** section beginning on page 479.

Mixed Practice is more challenging. You must decide which example to follow.

If you need **Extra Practice**, check the page reference for additional practice problems in the back of the book.

Applications give you a chance to apply what you've learned to real-life situations.

Every lesson ends with a **Review** of a past lesson or lessons. Often the skill reviewed will be needed in the next lesson.

SKILLS

2. Convert 5 miles to feet.

3. Refer to the introduction to this lesson. Which roller coaster drops farther?

4. It takes Jan 8 minutes to run a mile. How many feet does she run in 1 minute?

5. Convert 5.1 feet to inches in two ways. First, round to the nearest foot and then convert. Second, convert and then round to the nearest inch. Which answer is a better estimate? Why?

6. Explain how you would order 35 yards, 1392 inches, and 114 feet from shortest to longest.

Exercises

Convert each measure. (See Example A.)
7. 4 ft. to in.   8. 25 yd. to ft.   9. 2 mi. to ft.   10. 7 yd. to in.
11. 20 ft. to in.   12. 3 yd. to ft.   13. 10 yd. to in.   14. 9 mi. to ft.

Convert each measure. (See Example B.)
15. 561 ft. to yd.   16. 540 in. to yd.   17. 15,840 ft. to mi.
18. 228 in. to ft.   19. 7040 yd. to mi.   20. 36,960 ft. to mi.

▶ **Mixed Practice** (For more practice, see page 425.)

Convert each measure.
21. 4 ft. to in.   22. 108 in. to ft.   23. 297 ft. to yd.
24. 10 ft. to in.   25. 1800 in. to ft.   26. 21,600 in. to yd.

▶ **Applications**
27. A marathon can be 26 miles 385 yards long. How many yards is that?

28. Bill told his mother he was going to race in the 440 this weekend. "Don't you usually run the $\frac{1}{4}$-mile race?" she asked. Was Bill running in a different race than usual? Explain.

▶ **Review** (Lessons 7.1, 7.3, 7.4)

Add.
29. $\frac{1}{6}$
    $+ \frac{5}{6}$

30. $2\frac{1}{2}$
    $+ 1\frac{3}{4}$

31. $\frac{3}{10}$
    $+ \frac{1}{4}$

32. $\frac{4}{5} + \frac{1}{2}$

33. $2\frac{2}{3} + 6\frac{1}{3}$

Customary Measurement **177**

Problem-Solving Lessons Provide Winning Strategies

The main reason for learning math skills is to apply them to the mathematical problems that are a part of daily life. **Problem-Solving Lessons** introduce you to plans and strategies for dealing with those problems.

Problem-Solving Lessons have the Problem-Solving logo in the corner.

Four problem-solving lessons present information about using a **4-Step Problem-Solving Process**. Look for the **Read**, **Plan**, **Do**, and **Check** logo in these lessons.

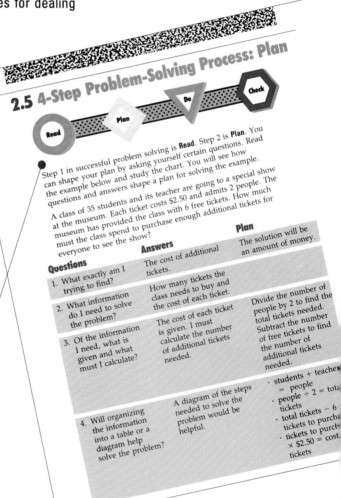

2.5 4-Step Problem-Solving Process: Plan

Read ▸ Plan ▸ Do ▸ Check

Step 1 in successful problem solving is **Read**. Step 2 is **Plan**. You can shape your plan by asking yourself certain questions. Read the example below and study the chart. You will see how questions and answers shape a plan for solving the example.

A class of 35 students and its teacher are going to a special show at the museum. Each ticket costs $2.50 and admits 2 people. The museum has provided the class with 6 free tickets. How much must the class spend to purchase enough additional tickets for everyone to see the show?

Questions	Answers	Plan
1. What exactly am I trying to find?	The cost of additional tickets.	The solution will be an amount of money.
2. What information do I need to solve the problem?	How many tickets the class needs to buy and the cost of each ticket.	
3. Of the information I need, what is given and what must I calculate?	The cost of each ticket is given. I must calculate the number of additional tickets needed.	Divide the number of people by 2 to find the total tickets needed. Subtract the number of free tickets to find the number of additional tickets needed.
4. Will organizing the information into a table or a diagram help solve the problem?	A diagram of the steps needed to solve the problem would be helpful.	· students + teacher = people · people ÷ 2 = total tickets · total tickets − 6 tickets to purchase · tickets to purchase × $2.50 = cost of tickets

Keep your knowledge of problem-solving strategies fresh by completing the **Review** section at the end of each lesson. The more you practice these strategies, the easier it becomes for you to use them.

PROBLEM SOLVING

Questions | **Answers**

5. What operations must I carry out? | Addition, division, subtraction, multiplication

6. In what order should I do the operations? | Follow the sequence under Plan for Question 4.

...ning how to solve a problem
Another point to co...
is whether you hav...
yourself if any of...
can be used to so...

▶ **Think and Disc...**
1. Solve the exa...
2. As you read...
   this problem...
   Part of...
   5 minut...
   that ne...
   also n...
   much...
3. Complet...
   problem...
   • Boo...
   exa...
   • Co...
   ex...
   ...
   • ...
4. Di...
   di...
5. ...

4.7 Using a Table to Find Information

"You may have already won a trip..."

AUSTRALIA
• Cairns
• Alice Springs
• Perth
• Adelaide
• Sydney
• Melbourne

	Adelaide	Alice Springs	Cairns	Melbourne	Perth	Sydney
	1693					
	2845	2435	3501			
	755	2488				
	2713	3772	4727	3468		
	1422	2960	2853	893	4135	

Before you even open the envelope, you could start mentally planning your trip with the help of a map and a kilometer table.

Use the kilometer table to answer the following questions.
1. How far is it from Alice Springs to Perth? Look down the column from Alice Springs until you come to the row that goes across to Perth.
2. Which city is 2853 kilometers from Sydney? Find 2853 in the row that goes across to Sydney, and see which city is at the head of that column.
3. Which two cities are 755 kilometers apart? Find 755 in the table, and read the cities for that column and row.

For part of your trip, you plan to take trains. Below is a train schedule for four cities. Each train is named after one of the special animals of Australia.

Train	Arrives at Sydney	Arrives at Wollongong	Arrives at Canberra	Arrives at Melbourne
Kangaroo	8:30 a.m.	10:00 a.m.	11:50 a.m.	4:30 p.m.
Wallaby	12:00 noon	1:40 p.m.	3:45 p.m.	8:45 p.m.
Koala	4:00 p.m.	5:30 p.m.	7:15 p.m.	11:30 p.m.

Use the train schedule to answer the following questions.
4. What time does the Wallaby arrive at Canberra?
5. About how long is the trip from Sydney to Wollongong? Is it the same for all three trains?

PROBLEM SOLVI...

6. It is 11:00 a.m. in Canberra. How long will it be until the next train to Melbourne leaves?
7. The trip from Canberra to Melbourne is about 500 kilometers. Use the schedule to find how long the trip on the Koala takes. Then estimate how fast the train travels.

For another part of your trip, you plan to rent a car. The table below shows how far you can go, in kilometers, on a single tank of gasoline, depending on the size of the car's gas tank and the car's kilometer-per-liter rating.

Size of Tank	Maximum Distance (per tank of gas)			
	5.9 km/L	7.8 km/L	9.9 km/L	13.7 km/L
32 L	188.8	249.6	316.8	438.4
40 L	236	312	396	548
50 L	295	390	495	685
65 L	383.5	507	643.5	890.5

Use the table above to answer the following questions.
8. What is the greatest distance you can travel on one tank of gasoline if your tank holds 40 liters and your car averages 7.8 kilometers per liter?

9. On one tank, how far could you go in a car with a 32-liter tank that averages 5.9 kilometers per liter? How much farther could you go in a car with a 65-liter tank that averages 13.7 kilometers per liter?

▶ **Review** (Lesson 3.8)
10. Each section of a theater seats 175 people. There are 8 sections. What operation would you use to find how many seats are in the theater? How many seats are there?

Measurem...

Make sure you're on the right track by comparing some of your answers with those in the **Selected Answers** section beginning on page 479.

Knowing the Score—
Reviews and Tests

On the job and in learning, checking your performance is important. It helps you learn where to concentrate your efforts.

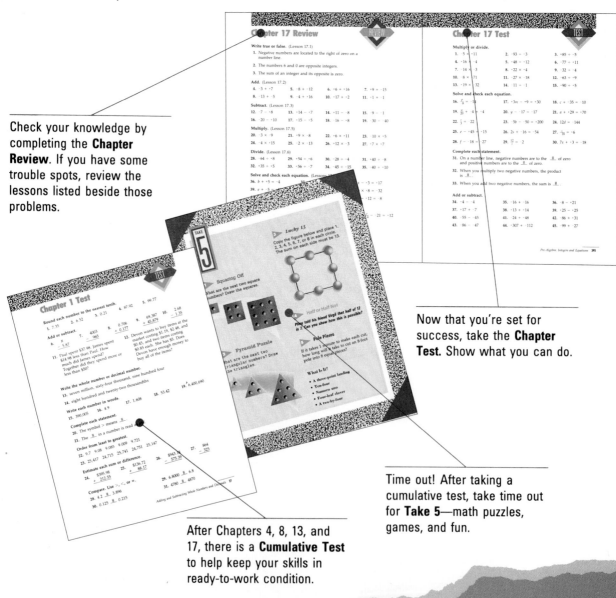

Check your knowledge by completing the **Chapter Review**. If you have some trouble spots, review the lessons listed beside those problems.

Now that you're set for success, take the **Chapter Test**. Show what you can do.

After Chapters 4, 8, 13, and 17, there is a **Cumulative Test** to help keep your skills in ready-to-work condition.

Time out! After taking a cumulative test, take time out for **Take 5**—math puzzles, games, and fun.

Call on the Reserves— Turn to the Appendix

When you need help, turn to the **Appendix**. You'll find a variety of information to help you succeed.

If you have access to a calculator, check out the techniques for getting the most out of it in **Calculator Applications**.

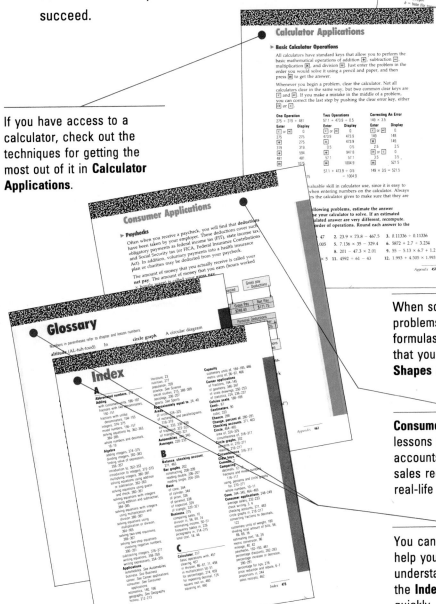

When solving geometry problems, you can find the formulas and abbreviations that you need in **Geometric Shapes and Formulas**.

Consumer Applications lessons on checking accounts, paychecks, and sales receipts help you gain real-life math skills.

You can use the **Glossary** to help you pronounce and understand math terms and the **Index** to find information quickly.

"We couldn't believe how many times we used addition and subtraction the day we ran a concession stand. All day long we were adding prices and subtracting to make change. Then, at the end of the day, we had to add together all the money we collected."

Adding and Subtracting Whole Numbers and Decimals

Adding and subtracting whole numbers and decimals are a part of everyday life. You add or subtract when you:
- count the amount of change from a purchase.
- balance a checking account.
- figure the number of hours worked in one week.
- count daily intake of calories.
- keep score in a game.

The following game is one way to learn more about whole numbers.

Maxi

Materials: number cube

Work with 1 or 2 classmates. Each player draws six blank spaces on a sheet of paper, as shown below.

—— —— —— —— —— ——

One player rolls a number cube.
Each player secretly writes the number that appears on the number cube in one of the blank spaces. The players take turns rolling the number cube until each blank is filled (6 rolls in all). The player with the greatest number after six rolls is the winner.

1.1 Reading and Writing Whole Numbers

17115233 SOLD EVERY MONTH

Jeremy glanced quickly at a billboard and saw this ad.
He stopped to look more closely. *"How* many sold?" he thought.

Jeremy had to put in the missing commas to read this large number.

Examples

To read and write whole numbers correctly, you can use a place value chart. Each group of three digits has a name. A comma separates each group.

Billions	Hundred Millions	Ten Millions	Millions	Hundred Thousands	Ten Thousands	Thousands	Hundreds	Tens	Ones
	Millions			**Thousands**			**Ones**		

A Find the place value of the 5 in the number 17,115,233.

Count the number of places from the right. Find the column in the place value chart that is fourth from the right.

The 5 is in the **thousands place**.

B Write 15 million, 37 as a standard numeral.

Write each digit in the appropriate place. Write a zero where the place value is empty.

The standard numeral for 15 million, 37 is **15,000,037**.

C Write 1,536,012 in words.

1,536,012 is written as **one million, five hundred thirty-six thousand, twelve**.

▶ Think and Discuss

1. Write two billion, three hundred forty-two thousand, sixty as a standard numeral.

2. Find the place value of the 5 in the number 256,006,902.

3. Write the largest standard numeral that can be made from the digits in the White House phone number, 202-456-1414.

4. How would you teach someone the correct way to write 3,065,000 in words?

Exercises

Find the place value of the 3 in each number. (See Example A.)

5. 947,342   6. 263,004   7. 31,872,961   8. 1,322,654

9. 53,000,021   10. 198,536,724   11. 57,231   12. 3,120,000,000

Write the standard numeral. (See Example B.)

13. 15 million, 312 thousand   14. 15 million, 312

15. 15 billion, 312 thousand   16. 402 thousand, 17

17. 402 million, 17   18. 402 billion, 17 million

19. 6 billion, 421 thousand, 92   20. 6 million, 421 thousand, 5

Write each number in words. (See Example C.)

21. 23,001   22. 30,001,412   23. 637,000,000   24. 3,791,060

25. 11,300,000,000   26. 4972   27. 3,433,012   28. 45,005

▶ Mixed Practice (For more practice, see page 400.)

29. Write the standard numeral for 63 thousand, 40.

30. What place value does the zero hold in 23,049?

31. Write 17,049,001 in words.

32. What digit is in the ten thousands place in 12,390,485?

▶ Applications

33. Georgia bought a used car for $999. When she wrote the check, she wrote the total amount in words. What did Georgia write?

34. Sal and Tom toss number cubes to see who can form the greater six-digit number. Sal rolls 1, 5, 6, 1, 2, and 5. Tom rolls 6, 3, 2, 4, 4, and 5. Who wins?

1.2 Reading and Writing Decimals

Helen was surprised when her September bank statement showed that her balance was $90 less than her check register balance. When she compared her canceled checks against her check register, she found her error. On check #1052, she had written $100.00, but in her check register, she had entered $10.00.

This error taught Helen the importance of placing the decimal point accurately.

To read and write decimals, you can use a place value chart. A decimal point separates the ones place and the tenths place.

Hundreds	Tens	Ones	Tenths	Hundredths	Thousandths	Ten-thousandths	Hundred-thousandths	Millionths

A **Find the place value of the 3 in the decimal 0.0013.**

Count the number of places to the right of the decimal point.
Compare with the place value chart.

The 3 is in the **ten-thousandths place**.

B **Write 65 thousandths as a decimal.**

Write the digits so that they end in the thousandths place.
Write a zero for empty place values. When no whole number precedes the decimal point, write a zero in the ones place.

The decimal for 65 thousandths is **0.065**.

C **Write 7.025 in words.**

When the decimal point follows a whole number, you use the word *and*.

7.025 is written as **seven and twenty-five thousandths**.

▶ Think and Discuss

1. What is the place value of 5 in 17.456?

2. Write 0.033 in words.

3. With ten dimes in a dollar, 1 dime = 0.1 dollar, or one-tenth of a dollar. What part of a dollar is a penny?

4. Which would you rather have, 0.8 dollar or 8.0 dimes? Explain.

5. Write one hundred two thousandths and one hundred and two thousandths as decimals. Explain how they differ.

Exercises

Find the place value of the 8 in each decimal. (See Example A.)

6. 0.08   **7.** 0.891   **8.** 0.1028   **9.** 176.8   **10.** 18.492

Write the decimal. (See Example B.)

11. 927 and 3 tenths   **12.** eleven thousandths

13. six hundred sixteen thousandths   **14.** 18 thousand and 18 thousandths

Write each decimal in words. (See Example C.)

15. 4.03   **16.** 0.019   **17.** 27.204   **18.** 2000.2   **19.** 729.07   **20.** 20.501

▶ Mixed Practice (For more practice, see page 400.)

21. What digit is in the hundredths place in 3768.425?

22. Write the decimal for 4 thousand 93 and 42 thousandths.

23. Write 500.025 in words.

▶ Applications

24. Airmail paper is one thousand seven hundred fifty-five millionths of an inch thick. Write as a decimal.

25. Manuel received a paycheck for $169.65. Write the amount of the check in words.

▶ Review (Lesson 1.1)

Find the place value of the underlined digit.

26. 5601   **27.** 18,964   **28.** 437,289   **29.** 99,754   **30.** 28,700

31. 302,401   **32.** 782,050   **33.** 4,062,893   **34.** 913,662   **35.** 256,238

1.3 A 4-Step Problem-Solving Process

Which maze would you choose if you wanted to reach the correct solution quickly?

The guidesigns in the second maze keep you on track. They help you move quickly toward the solution and avoid errors.

In much the same way, word problems also contain guidesigns that lead you to the correct solution. Following the 4-Step Problem-Solving Process will help you learn to recognize and understand the guidesigns in word problems.

Read the problem below.

It took Maria weeks to save enough money to buy the $128 10-speed bicycle she wanted. When she went to buy it, she found the price reduced by $15. As the clerk wrote up the sale, he told Maria that the bike had an $8 manufacturer's rebate. How much did the bike cost Maria?

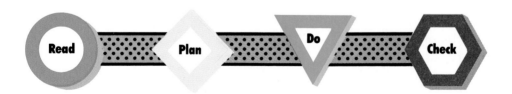

The 4-Step Problem-Solving Process is introduced briefly in the left column on the next page. See Lessons 1.4, 2.5, and 3.7 for more detailed information on each step. As you read each step, look in the right column to see how you can solve the problem about Maria's bicycle.

▶ 4-Step Problem-Solving Process

PROBLEM SOLVING

• Step 1:

Read the problem.

Decide what you need to find.

Examine the facts.

Look up any unfamiliar words.

Your Thinking

After reading the problem, think, *"I have to find what the final price of the bicycle is. I know 'price reduced' means Maria had to pay less money."* You might have to look up *rebate* in a dictionary. It means *return of part of a payment*. Thus, getting a rebate also means paying less money.

• Step 2:

Organize your information.

Determine if you need more information.

Choose a strategy to solve the problem.

You find that you have all the information you need. You plan to add the price reduction and the rebate together and then subtract the sum from the original price.

• Step 3:

Carry out your plan.

Compute.

Reduction + rebate = total reduction.
$$\$15 + \$8 = \$23$$
Price − total reduction =
final price of bicycle.
$$\$128 - \$23 = \$105$$

• Step 4:

Decide if your answer is reasonable.

Check your calculations—recompute if necessary.

The answer seems reasonable. It is less than the original price. Check the calculations.

▶ Think and Discuss

1. You used addition and subtraction in your plan. Can you think of another plan to solve the problem about Maria's bicycle?

2. How would you have to change your plan if, instead of a rebate of $8, the clerk told Maria she must pay a registration fee of $8? Would Maria pay the same amount?

3. How might using the 4-Step Problem-Solving Process help you solve problems that are not mathematical? Discuss.

Adding and Subtracting Whole Numbers and Decimals **7**

1.4 4-Step Problem-Solving Process: Read

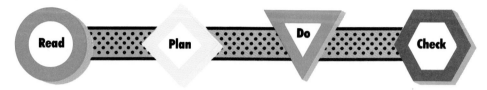

1. Take this test. Try to finish in 3 minutes or less.

```
(1) Record the time you are starting the test.

(2) Read through the whole test before computing.

(3) Add the number of pages in this book to the
    number of pages in Chapter 8.

(4) Subtract the number of pages in chapter 16
    from the answer to Step 3.

(5) Add the number of pages in Chapters 11 and 2.
    Subtract the sum from the answer to Step 4.

(6) Skip steps 3, 4, and 5.

(7) Record the time you finish the test.
```

2. How long did the test take? Did you compute anything?

Step 1 in the 4-step problem-solving process is **Read**. If you followed the directions in Step 2, you quickly realized that you needed to read the problems, but did not need to work them.

If you read through the problems and did not compute anything, you practiced the main rule of the **Read** step: Always keep reading until you fully understand the problem and have all the information you need to solve it. If you are not absolutely certain of a word's meaning, look it up in a glossary or a dictionary.

The following chart shows you how asking certain questions helps you decide when you are ready to move on to Step 2 in the problem-solving process, **Plan**.

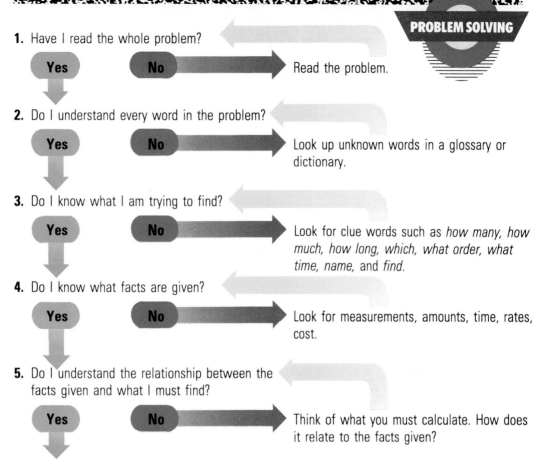

1. Have I read the whole problem?

Yes | No → Read the problem.

2. Do I understand every word in the problem?

Yes | No → Look up unknown words in a glossary or dictionary.

3. Do I know what I am trying to find?

Yes | No → Look for clue words such as *how many, how much, how long, which, what order, what time, name,* and *find.*

4. Do I know what facts are given?

Yes | No → Look for measurements, amounts, time, rates, cost.

5. Do I understand the relationship between the facts given and what I must find?

Yes | No → Think of what you must calculate. How does it relate to the facts given?

6. Go to Step 2 in the problem-solving process: **Plan**.

Read the problem and answer the questions.

How much fuel is burned in 8 hours by a motorcycle that consumes 2 liters of gasoline per hour?

1. What term is used to mean the same thing as *fuel*?

2. What term is used to mean the same thing as *burned*?

3. What other meanings does *consume* have?

4. What facts are given in the problem?

5. What are you supposed to find? Name some clue words.

Discuss how you would work through the chart to solve the following problem.

6. The toll to cross a certain bridge is 40¢. Denise drives across the bridge each day on her way to and from work. How much money does she spend on bridge tolls for 5 days?

1.5 Comparing and Ordering Whole Numbers

Four seniors were competing to see who could sell the most popcorn at the home basketball games. Randy sold 1260 bags, Todd sold 1206 bags, Marlee sold 1602 bags, and Antonia sold 620 bags.

Who sold the most popcorn?

Examples

To compare whole numbers, align the ones places and work from left to right. Compare digits within a column. Use > for "is greater than," < for "is less than," and = for "is equal to."

A Compare. 1260 ▨ 1206

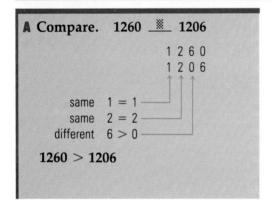

```
1 2 6 0
1 2 0 6
```

same   1 = 1
same   2 = 2
different 6 > 0

1260 > 1206

B Order 1260, 1206, 1602, and 620 from greatest to least.

In Example A we saw that 1260 > 1206. Now compare 1602 with 1260 and 1206, using the same procedure.

1602 > 1260 > 1206

Finally, compare 620 with these three numbers.

1602 > 1260 > 1206 > 620

▶ Think and Discuss

1. Compare the numbers 5702 and 5207. Use >, <, or =.

2. Order the numbers 396, 609, and 369 from greatest to least.

3. Refer to the introduction to this lesson. Who sold the most popcorn?

4. Describe 3 situations for which you might need to order numbers.

Exercises

Compare. Use >, <, or =. (See Example A.)

5. 78 ▒ 65   **6.** 88 ▒ 92

7. 230 ▒ 302   **8.** 637 ▒ 736

9. 76 ▒ 74   **10.** 802 ▒ 803

11. 4239 ▒ 5293   **12.** 6013 ▒ 6103

13. 7887 ▒ 7787   **14.** 53,722 ▒ 53,227

15. 362,109 ▒ 354,109 **16.** 188,889 ▒ 188,888

Order from greatest to least. (See Example B.)

17. 75,581   57,851   75,851   **18.** 246   2464   266

19. 828,331   882,313   282,313   **20.** 467   476   475   645

21. 409,320   49,320   490,230   409,230

22. 71,003   711,003   71,030   73,003   713,030

▶ **Mixed Practice** (For more practice, see page 401.)

Compare. Use >, <, or =.

23. 309 ▒ 409   **24.** 586 ▒ 585

25. 5421 ▒ 4421   **26.** 264 ▒ 245

Order from greatest to least.

27. 1,437,911,200   1,437,901,200   1,437,912,100

28. 4,693,233   4,393,233   493,633   469,333   4,693,333

29. 355,779   335,797   355,797   353,779   355,997

▶ **Applications**

30. Richard and Hideo just completed a community fund-raising drive. Richard raised $36,940. Hideo raised $34,247. Who raised more money?

31. The top basketball scorers are: Carr, 1106; McGill, 1009; Maravich, 1381; Selvy, 1209; Williams, 1010. List the players' names in order of points scored, from greatest to least.

▶ **Review** (Lessons 1.1, 1.2)

Write each number in words.

32. 4083   **33.** 28,607   **34.** 39.4   **35.** 7.052

36. 309,866   **37.** 0.974   **38.** 4.002   **39.** 5.67

1.6 Comparing and Ordering Decimals

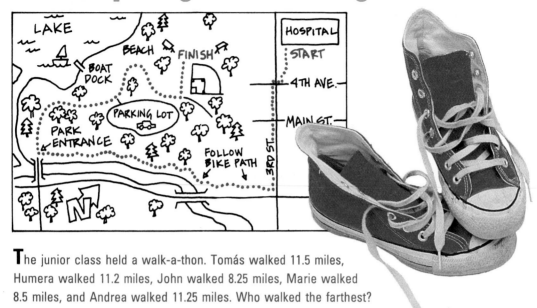

The junior class held a walk-a-thon. Tomás walked 11.5 miles, Humera walked 11.2 miles, John walked 8.25 miles, Marie walked 8.5 miles, and Andrea walked 11.25 miles. Who walked the farthest?

Examples

To compare decimals, align the ones places and work from left to right. Compare digits within a column. Use >, <, or =.

A Compare. 20.11 ▨ 20.12

$$2\ 0.1\ 1$$
$$2\ 0.1\ 2$$

same 2 = 2
same 0 = 0
same 1 = 1
different 1 < 2

20.11 < 20.12

B Order 11.5, 11.2, and 11.25 from least to greatest.

Use Example A as a model.
Compare 11.5 and 11.2.
11.2 < 11.5
Compare 11.2 and 11.25.
11.20 < 11.25
Compare 11.25 and 11.5.
11.25 < 11.5
11.2 < 11.25 < 11.5

▶ Think and Discuss

1. Compare 1462.86 and 1426.86. Use >, <, or =.

2. Order 200.95, 205.9, 20.95, and 205.95 from least to greatest.

3. Refer to the introduction to this lesson. Who walked the farthest?

4. Find two numbers between 8.01 and 8.02.

Exercises

Compare. Use >, <, or =. (See Example A.)

5. 0.3 ⬚ 0.4

6. 0.7 ⬚ 0.8

7. 0.12 ⬚ 0.21

8. 0.312 ⬚ 0.313

9. 1.4 ⬚ 1.7

10. 3.013 ⬚ 3.013

11. 5.0117 ⬚ 5.011

12. 20.080018 ⬚ 20.080008

Order from least to greatest. (See Example B.)

13. 0.1006   0.1060   0.6001   0.0601

14. 1.771107   1.770117   17.71017   1.771017   1.777101

15. 36.991132   36.901132   3.6911329   369.01132   36.091132

▶ **Mixed Practice** (For more practice, see page 401.)

Compare. Use >, <, or =.

16. 0.5 ⬚ 0.50

17. 43.56 ⬚ 44.65

18. 414.111 ⬚ 414.11

19. 10.03003 ⬚ 10.30303

20. 46.09119 ⬚ 460.9119

21. 371.371 ⬚ 371.371

Order from least to greatest.

22. 3.72   3.702   3.7   3.07

23. 0.98   0.098   0.908   9.09   0.89

24. 0.0025   0.0205   0.05

25. 5.73   5.75   5.37   5.57   5.35

▶ **Applications**

26. Claudia ran three hundred-meter heats at the Saturday track meet. Her times were 15.36 seconds, 15.32 seconds, and 15.26 seconds. In which heat did Claudia run fastest?

27. In baseball, ERA stands for Earned Run Average. In general, the lower an ERA is, the better that pitcher is. A team's starting pitchers had the ERAs shown here. Which pitcher had the best ERA?

TEAM ERA	
Pitchers	**ERA**
McMoore, J.	4.27
Satherson, W.	3.13
Winsom, T.	3.19
Adlers, G.	4.61
Dempke, P.	4.40
Andrews, A.	4.56
Dickens, R.	3.72
Stark, E.	4.10
Mathews, R.	3.70

▶ **Review** (Lesson 1.5)

Compare. Use >, <, or =.

28. 3792 ⬚ 2792

29. 92 ⬚ 920

30. 430 ⬚ 430

Order from least to greatest.

31. 79,433   79,344   74,337

32. 802   892   908   1802

33. 9008   8009   19,090   998

34. 243   422   244   234   423

1.7 Rounding Whole Numbers

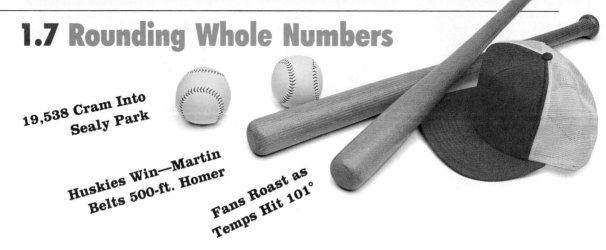

19,538 Cram Into Sealy Park

Huskies Win—Martin Belts 500-ft. Homer

Fans Roast as Temps Hit 101°

Read the numbers in the headlines. Which of the numbers do you think are actual counts? Which are close to the actual count? Numbers that are close to an actual count are called estimates. The examples below show how to estimate numbers by rounding.

Examples

To round whole numbers, determine the place value you will round to. Then examine the digit to the right.
Use the Rule of Rounding:
  Round down if the digit is less than 5.
  Round up if the digit is 5 or greater.

A Round to the nearest 10.	**B** Round to the nearest 100.	**C** Round to the nearest 1000.
19,538 └ tens place	19,538 └ hundreds place	19,538 └ thousands place
Since the digit to the right of the tens place is 5 or greater, 38 is closer to 40 than 30. Round up to **19,540**.	Since the digit to the right of the hundreds place is less than 5, 538 is closer to 500 than 600. Round down to **19,500**.	The digit to the right of the thousands place is 5. Round up to **20,000**.

▶ Think and Discuss

1. Round $68,869 to the nearest ten thousand dollars.

2. When rounding 869,500 to the nearest thousand, do you round up or down? Why? What is 869,500 to the nearest thousand?

3. Refer to the introduction to this lesson. Why did some of the headlines use an actual count? Why did some use an estimated number? Explain.

4. You are buying a number of items at the store and want to be sure that you have enough money. Should you round up or down before adding the prices? Explain.

Exercises

Round to the nearest ten. (See Example A.)

5. 125 6. 179 7. 384 8. 403 9. 798

10. 6215 11. 8491 12. 10,322 13. 17,008 14. 20,212

Round to the nearest hundred. (See Example B.)

15. 983 16. 829 17. 4055 18. 10,775 19. 13,228 20. 6314

21. 2864 22. 9008 23. 8330 24. 6741 25. 17,929 26. 4998

Round to the nearest thousand. (See Example C.)

27. 9261 28. 8590 29. 4801 30. 975 31. 12,334 32. 19,730

33. 17,512 34. 62,000 35. 67,750 36. 40,001 37. 39,808 38. 1499

▶ Mixed Practice (For more practice, see page 402.)

Round $64,508 to the nearest

39. ten dollars. 40. hundred dollars. 41. thousand dollars.

Round 72,287 to the nearest

42. ten. 43. thousand. 44. ten thousand.

Round 25,019 to the nearest

45. ten. 46. hundred. 47. thousand.

▶ Applications

48. The seating capacity of Wrigley Field in Chicago is 37,272. To the nearest thousand, how many seats are in Wrigley Field?

49. There were 17,000 fans at Monday's game, rounded to the nearest thousand. What is the least number of fans that might have been at the game?

▶ Review (Lessons 1.5, 1.6)

Order from least to greatest.

50. 875 80 807 87 857

51. 46.2 46.02 4.602 462.2

1.8 Rounding Decimals

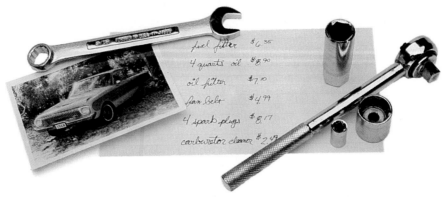

Kevin is going to work on his neighbor's car. He went to the auto supply store to buy the parts. As Kevin found each item, he wrote down its price. Kevin is in the checkout line and wants to quickly find if he has enough money with him to pay for the parts.

Kevin can quickly estimate the total amount of his bill by rounding the figures and adding.

Examples

To round decimals, determine the place value you will round to. Then examine the digit to the right. Use the Rule of Rounding:
Round down if the digit is less than 5.
Round up if the digit is 5 or greater.

A Round $6.35 to the nearest dollar.	**B** Round 12.86 to the nearest tenth.	**C** Round 7.9354 to the nearest hundredth.
The digit to the right of the decimal point is less than 5. 6.3 is closer to 6 than 7.	The digit to the right of the tenths place is 5 or greater. 12.86 is closer to 12.9 than 12.8.	The digit to the right of the hundredths place is 5.
Round down to **$6**.	Round up to **12.9**.	Round up to **7.94**.

▶ Think and Discuss

1. What digit is in the ten-thousandths place in 10.37185?

2. Round $15.59 to the nearest ten cents.

3. Refer to the introduction to this lesson. To estimate his total, Kevin rounded each amount to the nearest dollar and then added. What estimate did Kevin get?

4. When you round 3.999 to the nearest tenth, why do you write 4.0 and not 4?

5. If a price is rounded to $9, what is the highest amount it could be? What is the lowest amount?

Exercises

Round to the nearest whole number or dollar. (See Example A.)

6. 6.3 **7.** 9.5 **8.** $2.81 **9.** $14.36 **10.** 47.93

11. $218.52 **12.** 373.007 **13.** $99.89 **14.** 407.581 **15.** 8.099

Round to the nearest tenth or ten cents. (See Example B.)

16. 1.62 **17.** 3.47 **18.** $9.85 **19.** $41.54 **20.** 71.899

21. 40.02 **22.** 611.58 **23.** 18.215 **24.** $20.25 **25.** 9.519

Round to the nearest hundredth. (See Example C.)

26. 0.987 **27.** 1.426 **28.** 8.733 **29.** 6.9108 **30.** 43.0071

31. 0.0915 **32.** 14.8157 **33.** 0.2525 **34.** 2.3623 **35.** 8.909

▶ **Mixed Practice** (For more practice, see page 402.)

Round 473.1486 to the nearest

36. tenth. **37.** whole number. **38.** thousandth.

Round $837.527 to the nearest

39. 10 dollars. **40.** dollar. **41.** 10 cents. **42.** 100 dollars.

Round 12,932.7117 to the nearest

43. hundred. **44.** thousandth. **45.** whole number.

▶ **Applications**

46. On a recent bicycling trip, Hector rode 268.3 miles in two days. When he told his friends about his trip, he rounded the number to the nearest ten. What number did he report?

47. Fran finished first in a long jump contest. Rounded to the nearest centimeter, her jump was four hundred forty-seven centimeters. What is the shortest distance she could have jumped to receive her score?

▶ **Review** (Lessons 1.5, 1.6)

Compare. Use >, <, or =.

48. 5.8 _▨_ 8.123 **49.** 39.3 _▨_ 39.300 **50.** 0.0309 _▨_ 0.00489

51. 481 _▨_ 2026 **52.** 752,894 _▨_ 752,849 **53.** 49.35 _▨_ 4.935

1.9 Adding Whole Numbers and Decimals

Television sponsors spend great amounts of money for commercial time. On a recent sports special, two commercial spots each cost $28,329 and $40,397. How much did the two commercials cost in all?

You need to add the two amounts to figure out the total cost.

Examples

To add, align numbers by place value. Begin at the right and add each column. Regroup when a sum is 10 or greater.

A Add. $28,329 + $40,397

```
      1 1
  $2 8,3 2 9     Align the ones place.
+   4 0,3 9 7    Add and regroup.
  $6 8,7 2 6 ←—— 9 + 7 = 16 = 1 ten + 6 ones
            ↑——— 1 + 2 + 9 = 12 tens = 1 hundred + 2 tens
```

$28,329 + $40,397 = **$68,726**

B Add.   0.348 + 0.897

```
   1 1 1
  0.3 4 8      Align the ones place.
+ 0.8 9 7      Add and regroup.
  1.2 4 5      Place a decimal point in
               the sum.
```

C Add.   4.2 + 8 + 3.185 + 7.69

```
    2 1 1
   4.2 0 0      Place decimal points and
   8.0 0 0      zeros as needed.
   3.1 8 5      Add and regroup.
+  7.6 9 0      Place a decimal point in
  2 3.0 7 5      the sum.
```

▶ Think and Discuss

1. $9 + 7 = 16$. What digit is in the tens place in the sum 16?

2. Add. $4.362 + 17 + 2909.11 + 36.0303 + 5$

3. Add $32,798 and $41,599. How many times must you regroup?

4. Refer to the introduction to this lesson. What was the total cost of the two commercials?

5. Look at Example C. Explain why 8 and 8.000 are the same.

Exercises

Add. (See Example A.)

6. 5267 + 6924	**7.** 4328 + 3761	**8.** 2854 + 3657	**9.** 7649 + 8784	**10.** 24,372 + 81,989

Add. (See Example B.)

11. 0.77 + 0.19	**12.** 0.83 + 0.48	**13.** 0.845 + 0.194	**14.** 0.797 + 0.526	**15.** 1.78 + 5.56

Add. (See Example C.)

16. 4.3 + 0.5 + 2.22

17. 1.95 + 0.173 + 312

18. 0.9 + 0.99 + 0.999 + 9

19. 62 + 1.8 + 5.55 + 713

▶ Mixed Practice (For more practice, see page 403.)

Add.

20. 45,621 + 3,879	**21.** 0.78 + 0.53	**22.** 4387 + 5377	**23.** 25.25 + 7.93	**24.** 54,849 + 3,478

▶ Applications

25. Megan bought a new dress, some earrings, and a necklace. The cost for these items was $54.59, $4.98, and $12, respectively. The sales tax was $4.65. How much did Megan pay in all?

26. A thirty-minute TV show had three commercial breaks. If the show aired in segments of 6.25 minutes, 9.75 minutes, 7 minutes, and 1.5 minutes, what was the actual length of the program?

▶ Review (Lessons 1.7, 1.8)

Round each number to the nearest hundred.

27. 8449

28. 9511

29. 16,087

30. 27,499

31. 35,631

32. 209,276

Round each number to the nearest tenth.

33. 166.88   **34.** 20.639   **35.** 474.853   **36.** 9.947   **37.** 3.9727   **38.** 27.88

1.10 Subtracting Whole Numbers and Decimals

Simon saved $545 to buy a stereo system. In the Sunday paper he found the three ads shown here. "But wait," says Simon. "If I get the most expensive system, I won't have much money left for records and tapes."

Great sound from this new stereo system with 80 watt amplifier, turntable, and 3-way speakers. $500.00

Super stereo system offers AM/FM receiver, turntable, tape deck, and speakers. Now only $435.00

Stereo system with AM/FM receiver, manual turntable, and 2-way speakers. Only $375.00

Enjoy listening to a single-unit system with the quality of a multi-unit system. All for only $349.00

Brand new stereo with AM/FM receiver, turntable, and 2-way speakers. A great system at only $99.00

Quality and low cost: That is what you'll get when you invest in this great stereo system. $269.00

To figure out how much money he would have left with each system, Simon has to subtract each price from $545.

Examples

To subtract, align numbers by place value. Begin at the right and subtract each column. Regroup whenever the top digit in a column is smaller than the bottom digit.

A Subtract. $545 − $375

$$\begin{array}{r} {}^{4}\;{}^{14}\\ \cancel{5}\;\cancel{4}\;5 \\ -\;3\;7\;5 \\ \hline \$1\;7\;0 \end{array}$$

Tens column: 4 is smaller than 7, so regroup.
540 = **4** hundreds + **14** tens

$545 − $375 = **$170**

B Subtract.   21.525 − 13.78

$$\begin{array}{r} {}^{1}\;{}^{10}\;{}^{14}\;{}^{12}\\ 2\;1.\cancel{5}\;\cancel{2}\;5 \\ -\;1\;3.7\;8\;0 \\ \hline 7.7\;4\;5 \end{array}$$

Write a decimal point in the difference.

21.525 − 13.78 = **7.745**

▶ Think and Discuss

1. Subtract 165.9 from 241.23.

2. Refer to the introduction to this lesson. How much money would Simon have left if he bought the $375 stereo system?

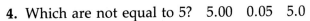

3. You can regroup 34 as 2 tens and 14 ones. How can you regroup 72?

4. Which are not equal to 5?   5.00   0.05   5.0

5. How can you check the result of a subtraction problem?

Exercises

Subtract. (See Example A.)

6.	6625	7.	3152	8.	9738	9.	5464	10.	2817
	− 437		− 874		− 3869		− 863		−1528

11.	9249	12.	6443	13.	7411	14.	14,869	15.	85,728
	− 637		− 4791		− 658		− 3,545		− 83,922

Subtract. (See Example B.)

16.	0.37	17.	5.42	18.	6.35	19.	7.21	20.	0.726
	− 0.29		− 3.61		− 4.21		− 5.83		− 0.363

21.	4.659	22.	8.943	23.	5.347	24.	23.13	25.	3.234
	− 0.271		− 5.786		− 1.998		− 7.47		−2.602

▶ **Mixed Practice** (For more practice, see page 403.)

Subtract.

26.	82,333	27.	9.34	28.	61.312	29.	18.763	30.	7482
	− 4,949		− 7.67		− 0.049		− 7.98		− 847

▶ **Applications**

31. Brian is buying his first car. He has narrowed down his choices to a blue car with an odometer reading of 65,721 miles and a red one with a reading of 98,035 miles. How much farther has the red car been driven?

32. Zina ran the one-hundred-meter sprint in 13.15 seconds. The women's world record is 10.76 seconds. How much more time did Zina take?

▶ **Review** (Lesson 1.9)

Add.

33.	4506	34.	62.9	35.	499.85	36.	72,111	37.	6.7125
	+ 893		+ 95.5		+ 23.99		+ 45,789		+ 0.8435

38. 4.607 + 82 + 0.11   39. 43,222 + 986 + 52   40. 0.115 + 0.87 + 0.6

1.11 Subtracting Whole Numbers and Decimals with Zeros

San Valley Daily Herald

702,007
−698,209

Sam is a sports writer covering the Terriers' rugby match against the Blazers. An announcement comes over the public address system: "Today's attendance is 15,275, bringing the season total to 698,209." Sam wonders, "If the league's record for a season is 702,007, how many fans must come to the next game to break the record?" Sam has to subtract.

Examples

To subtract from numbers with zeros, remember the rules for regrouping.

A Subtract.   702,007 − 698,209

$$\begin{array}{r} {}^{6}\ {}^{9}\ {}^{11}\ {}^{9}\ {}^{9}\ {}^{17} \\ 7\ \cancel{0}\ 2,\cancel{0}\ \cancel{0}\ 7 \\ -\ 6\ 9\ 8,2\ 0\ 9 \\ \hline 3,7\ 9\ 8 \end{array}$$ = 69 ten thousands + 11 thousands + 9 hundreds + 9 tens + 17 ones

B Subtract.   9 − 3.87

$$\begin{array}{r} {}^{8}\ {}^{9}\ {}^{10} \\ \cancel{9}.\cancel{0}\ \cancel{0} \\ -\ 3.8\ 7 \\ \hline 5.1\ 3 \end{array}$$ Align the ones place. Place decimal points and write zeros as needed. Subtract and regroup.
Write a decimal point in the difference.

C Subtract.   0.805 − 0.6774

$$\begin{array}{r} {}^{7}\ {}^{9}\ {}^{14}\ {}^{10} \\ 0.8\ \cancel{0}\ \cancel{5}\ \cancel{0} \\ -\ 0.6\ 7\ 7\ 4 \\ \hline 0.1\ 2\ 7\ 6 \end{array}$$ Align the ones place. Write zeros as needed.
Subtract and regroup.
Write a decimal point in the difference.

▶ Think and Discuss

1. Subtract 60.25 from 85.

2. Refer to the introduction to this lesson. How many fans must come to the Terriers' next game to break the record?

3. Show how you regroup 7 to solve the problem $7 - 3.25$.

4. Do you need to regroup to subtract 2043 from 3047? Explain.

Exercises

Subtract. (See Example A.)

5.	6.	7.	8.	9.
3008	12,004	16,038	28,140	46,004
− 1409	− 7,706	− 14,956	− 12,067	− 12,148

Subtract. (See Example B.)

10. $7 - 4.6$ 11. $402 - 11.8$ 12. $1300 - 99.4$ 13. $20 - 11.3$

14. $238 - 163.74$ 15. $109 - 40.38$ 16. $40 - 13.76$ 17. $632 - 209.74$

Subtract. (See Example C.)

18.	19.	20.	21.	22.
0.704	0.606	0.7	0.93	0.85
− 0.5993	− 0.4932	− 0.392	− 0.5432	− 0.693

▶ Mixed Practice (For more practice, see page 404.)

Subtract.

23.	24.	25.	26.	27.
34,000	48.007	66,090	0.7	3803
− 18,401	− 9.139	− 19,002	− 0.482	− 49

28. $8 - 4.7$ 29. $10 - 2.5$ 30. $16 - 2.48$ 31. $33 - 0.9$ 32. $5 - 0.39$

▶ Applications

33. Mark Twain was born in 1835. He died in 1910. Approximately how old was he when he died?

34. The stadium capacity for the cross-town football game was 14,007. 8987 attended the game. How many seats were empty?

▶ Review (Lessons 1.5, 1.6)

Order from least to greatest.

35. 2.0191   2.091   2.919

36. 41.73   41   403   437   43.72

37. 78,536   87,536   76,536

38. 602   6002   62.02   60.2   620

1.12 Estimating Sums and Differences

On their first date, Kirk and Tiffany went out for dinner. Everything was perfect until Tiffany ordered lobster for her main course. At $12.95, this was the most expensive item on the menu. Kirk knew he only had $25 to pay for the entire meal and the tip. Meanwhile, the waiter was waiting for Kirk to order.

Kirk quickly figured out what he could spend on his entree by estimating how much of his $25 he had left.

Examples

To estimate sums and differences, use rounding to establish numbers that you can compute mentally. The symbol ≈ means "is approximately equal to."

A Estimate the sum of $12.95, $0.80, and $1.10.

$$
\begin{array}{rl}
\$12.95 \approx \$13 & \text{Round each} \\
0.80 \approx 1 & \text{number. Then add.} \\
+\ \underline{\ 1.10 \approx \ 1} & \\
\$15 &
\end{array}
$$

The sum is approximately **$15**.

B Estimate. 67,312 − 43,897

$$
\begin{array}{rl}
67,312 \approx 70,000 & \text{Round to the} \\
-\ \underline{43,897 \approx 40,000} & \text{nearest ten} \\
30,000 & \text{thousand.}
\end{array}
$$

or

$$
\begin{array}{rl}
67,312 \approx 67,000 & \text{Round to the} \\
-\ \underline{43,897 \approx 44,000} & \text{nearest} \\
23,000 & \text{thousand.}
\end{array}
$$

Two possible estimates are **30,000** and **23,000**.

▶ Think and Discuss

1. Round 397.4 to the nearest whole number and the nearest ten.

2. Estimate the sum of $49.95, $7.29, and $24.49.

3. Refer to the introduction to this lesson. If Kirk orders a $7.45 dinner and two colas for $0.95 each, estimate how much he has left for tax and a tip. Explain how you estimated.

4. Estimate the sum of 3906 and 849 by rounding to the nearest ten and the nearest hundred.

5. Give three different estimates for the sum of 397.4 and 531.9. Which estimate is most exact?

Exercises

Estimate the sum. (See Example A.)

6. 387
  + 522

7. 27.3
  + 5.6

8. 1653
  + 967

9. 56,891
  + 13,555

10. 522.48
  + 749.83

11. $52.50
  37.75
  + 12.20

12. 98
  24
  + 122

13. 3,491
  643
  + 49,113

14. $22.81
  7.22
  + 41.99

15. 407.3
  22.7
  + 373.4

Estimate the difference. (See Example B.)

16. 270
  − 209

17. 489.7
  − 43.2

18. $95.30
  − 42.95

19. 4311
  − 2909

20. 624
  − 609

21. 23.599
  − 7.741

22. $21.75
  − 15.24

23. 72
  − 49

24. 707
  − 183

25. 4197
  − 989

26. 473.21
  − 58.74

27. $932.30
  − 769.99

▶ Mixed Practice (For more practice, see page 404.)

Estimate the sum or difference.

28. 49.7
  − 19.9

29. 403
  + 771

30. $35.86
  7.81
  + 29.99

31. 5211
  − 4817

▶ Applications

32. Emma is in a bookstore. She has found four books that she would like to buy. They cost $4.95, $3.75, $12.95, and $7.25, respectively. Estimate how much she will have to spend.

33. Spenser is treating his 3 younger sisters to lunch at a seafood restaurant. Spenser has a $20 bill. Estimate how much change he will receive if they order items costing $2.95, $3.85, $3.60, $4.25, and four colas at $0.85 each?

▶ Review (Lessons 1.9, 1.10, 1.11)

Add or subtract.

34. 78 + 499

35. 883 + 97

36. 1.9 + 71.1 + 5.5

37. 50.3 + 0.89

38. 800 − 123

39. 926 − 647

40. 475.24 − 89.5

41. 4.63 − 0.85

42. 70.03 − 27.985

Chapter 1 Review

Complete each statement. (Lessons 1.1, 1.2)

1. The 3 in the number 12,002.0031 is in the ___ place.

2. The decimal point is read as the word ___ .

3. The number 436,792.9183 has a(n) ___ in the hundreds place.

4. A penny is one ___ of a dollar.

5. 8,075,006.3 is written in words as ___ .

Write the whole number or decimal number. (Lessons 1.1, 1.2)

6. 2 million, 20 thousand, 93   7. 1 and 13 hundredths

8. 10 and 81 thousandths   9. five hundred eight thousandths

Order from least to greatest. (Lessons 1.5, 1.6)

10. 27,341   72,314   73,214   23,714   72,413

11. 6.03   2.9   6.294   6.87   2.694

Compare. Use >, <, or =. (Lessons 1.5, 1.6)

12. 7908 ___ 8097   13. 862,112 ___ 862,121

14. 5.2 ___ 5.200   15. 4.603 ___ 4603

Round each number to the nearest thousand. (Lesson 1.7)

16. 7455   17. 529   18. 3712   19. 19,937   20. 83,096

Round each decimal to the nearest whole number. (Lesson 1.8)

21. 4.62   22. 0.9   23. 25.08   24. 49.516   25. 78.267

Add or subtract. (Lessons 1.9, 1.10, 1.11)

26.
$$74,629 + 9,378$$

27.
$$106.43 - 27.59$$

28.
$$72,846 - 23,752$$

Estimate. (Lesson 1.12)

29. Kim and Marcus ordered two pizzas at $9.89 each. They have $25. About how much change will they receive?

30. Tracy ordered items costing $1.89, $3.75, $0.65, and $1.49. She has nine $1 bills. Does she have enough money?

Chapter 1 Test

Round each number to the nearest tenth.

1. 7.35   **2.** 6.52   **3.** 0.21   **4.** 87.92   **5.** 99.77

Add or subtract.

6.	**7.**	**8.**	**9.**	**10.**
8	4003	0.706	69,387	2.68
− 5.87	− 965	+ 0.177	+ 45,879	− 1.35

11. Paul spent $37.98. James spent $14.98 less than Paul. How much did James spend? Together did they spend more or less than $50?

12. Devon wants to buy items at the market costing $1.19, $2.48, and $0.45, and two items costing $0.85 each. She has $5. Does Devon have enough money to buy all of the items?

Write the whole number or decimal number.

13. seven million, sixty-four thousand, nine hundred four

14. eight hundred and twenty-two thousandths

Write each number in words.

15. 390,005   **16.** 4.9   **17.** 1.608   **18.** 53.42   **19.** 6,400,690

Complete each statement.

20. The symbol > means _⁂_ .

21. The _⁂_ in a number is read as *and*.

Order from least to greatest.

22. 9.7   9.08   9.085   9.009   9.721

23. 25,417   24,715   25,741   24,751   25,147

Estimate each sum or difference.

24.	**25.**	**26.**	**27.**
$395.98	$136.72	$943.18	864
+ 212.55	+ 88.17	− 575.39	− 325

Compare. Use >, <, or =.

28. 4.2 _⁂_ 3.896   **29.** 6.8000 _⁂_ 6.8

30. 0.125 _⁂_ 0.215   **31.** 4780 _⁂_ 4870

"Knowing how to multiply has really helped me at my new job. When a customer calls and orders 2 or 3 bunches of flowers, I can multiply to figure out the cost of the flowers. Then I multiply again to find the sales tax."

Multiplying Whole Numbers and Decimals

You need to multiply whole numbers or decimals
- to find the price of concert tickets for you and two friends.
- to estimate how much wallpaper you need for your bedroom.
- to project your yearly salary from one week's paycheck.
- to determine how many miles you ride a bike in a month if you ride the same distance each week.

What are some other situations that involve multiplication?

Supermarket Grouping

Choose 2 or 3 classmates to work with. Imagine that you work in a supermarket and the manager has asked you to arrange 36 boxes of cereal. You can arrange the boxes in one display or in several, but all of the boxes must be used and each display must contain the same number of boxes. For example, you could have 3 displays of 12 boxes, but you cannot have 1 display of 30 boxes and 1 display of 6 boxes.

On a sheet of paper, draw the different ways the boxes can be displayed. Compare your group's drawings with those of other groups.

2.1 Introduction to Multiplication

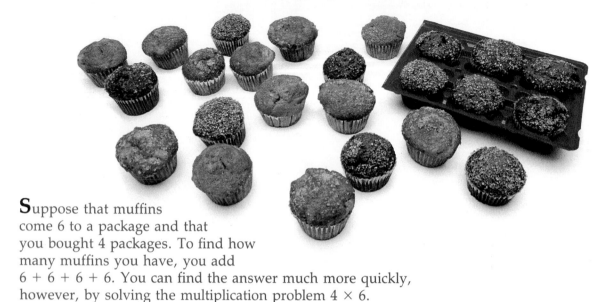

Suppose that muffins
come 6 to a package and that
you bought 4 packages. To find how
many muffins you have, you add
6 + 6 + 6 + 6. You can find the answer much more quickly,
however, by solving the multiplication problem 4 × 6.
For problems that can be solved by either addition or multiplication,
multiplication is often the quicker method.

▶ Think and Discuss

1. Describe two situations when you have used multiplication.

2. Why might the girl in the picture need to use multiplication?

For each problem below, decide whether using addition or multiplication is more appropriate.

3. Mrs. Lopez arranged her classroom to make 8 rows with 5 desks in each row. How many student desks are in her room?

4. Jennifer bought presents for her family costing $21, $15, $11, and $30. How much did Jennifer spend?

5. Rafael's teacher assigned 6 homework questions every day for 5 days. How many questions did Rafael have to do in all?

▶ Express Yourself

Here are some terms used in multiplication.

times   multiplied by

"3 times 5" means three multiplied by five.

product   the result of multiplication

The product of 3 and 5 is 15.

multipliers or **factors**   the numbers being multiplied

To get 15, you can multiply the factors 3 and 5.

To get 15, you can multiply the multipliers 3 and 5.

doubled   multiplied by 2

Seven doubled is the same as 2×7.

Knowing the following two characteristics, or **properties**, of multiplication will help you to solve multiplication problems.

identity property   The product of 1 and any number is that number.

$1 \times 6 = 6$ $32 \times 1 = 32$

zero property   The product of 0 and any number is 0.

$0 \times 3 = 0$ $5{,}987{,}732 \times 0 = 0$

Complete each statement using the terms defined in this lesson.

6. In words, 15×2 is 15 ___ 2.

7. 20 multiplied by 7 can be written as 20 ___ 7.

8. The problem $7 \times 0 = 0$ illustrates the ___ .

9. When you multiply the ___ or ___ 7 and 5, the result is 35.

10. 24 is the ___ of 6 and 4.

11. The problem $500 \times 1 = 500$ illustrates the ___ .

▶ Practice What You Know

When you multiply, you must regroup if a product is 10 or greater. Regrouping in multiplication is much like regrouping in addition. When adding, you must regroup numbers if a sum is 10 or greater.

Regroup for addition.   **18 + 18 + 18**

```
    2
    18       Add the ones column.
    18       8 + 8 + 8 = 24 ones = 2 tens + 4 ones
 +  18       Add the tens column.
    54
```

12. Write three addition problems in which you have to regroup to find the answer.

2.2 Multiplying by 1-Digit Multipliers

During orientation, a group of freshmen overheard a senior remark, "Hey, only 172 school days until we graduate!" The remark prompted a couple of freshmen to figure out how many school days were left before their graduation.

▸ The freshmen could multiply 172 by 4 to find their answer.

Examples

To multiply by a 1-digit number, begin at the ones places.
Multiply each digit in the top number by the 1-digit number.

A Multiply. 27 × 6

$$\overset{4}{2}\,7$$
$$\times\quad 6$$
$$\overline{2}$$

6 × 7 = 42
Regroup.
42 = **4** tens + **2** ones

$$\overset{4}{2}\,7$$
$$\times\quad 6$$
$$\overline{1\,6\,2}$$

6 × 2 tens = 12 tens
12 tens + 4 tens = **16** tens

27 × 6 = **162**

B Multiply. 172 × 4

$$1\,7\,2$$
$$\times\quad\quad 4$$
$$\overline{8}$$

4 × 2 = **8**

$$\overset{2}{1}\,7\,2$$
$$\times\quad\quad 4$$
$$\overline{8\,8}$$

4 × 7 tens = 28 tens
Regroup.
280 = **2** hundreds + **8** tens

$$\overset{2}{1}\,7\,2$$
$$\times\quad\quad 4$$
$$\overline{6\,8\,8}$$

4 × 1 hundred = 4 hundreds
4 hundreds + 2 hundreds =
6 hundreds

172 × 4 = **688**

▸ Think and Discuss

1. Find the product of 403 and 8. Explain each step.

2. Refer to the introduction to this lesson. How many school days do the freshmen have left before their graduation?

SKILLS

3. Rewrite 56 ones as tens and ones.

4. In Example A, it was necessary to regroup after the first multiplication step. In Example B, it was not necessary to regroup until the second step. Explain why.

Exercises

Multiply. (See Example A.)

5.	73	6.	31	7.	55	8.	32	9.	47	10.	86
	× 3		× 5		× 6		× 9		× 4		× 7

11. 92×6 12. 67×9 13. 53×5 14. 45×2 15. 79×7 16. 98×3

Multiply. (See Example B.)

17.	923	18.	965	19.	6513	20.	2869	21.	6024
	× 6		× 2		× 7		× 8		× 3

22. 289×3 23. 263×4 24. 4821×2 25. 7152×7 26. 725×5

▶ **Mixed Practice** (For more practice, see page 405.)

Multiply.

27.	3844	28.	603	29.	49	30.	908	31.	6756
	× 5		× 9		× 8		× 6		× 4

32. 84×3 33. 389×4 34. 7543×2 35. 825×7 36. 99×5

▶ **Applications**

37. Kevin drives fifty-five miles an hour to get to a college football game. It takes him four hours. About how far away is the game?

38. Three squads of 16 cheerleaders need new uniforms. Uniforms at All-Pro cost $33. Uniforms at Sports-Plus cost $36. How much money will the cheerleaders save in all if they buy from All-Pro?

▶ **Review** (Lesson 1.12)

Estimate each sum or difference.

39.	332	40.	994	41.	75	42.	559	43.	804
	+ 614		+ 766		+ 88		− 51		− 443

Multiplying Whole Numbers and Decimals **33**

2.3 Multiplying by Powers and Multiples of 10

Sue had resolved to cut down on junk food. She was doing fine until one evening she suddenly craved something crunchy. All she could find was potato chips. "Well, there are only 12 calories in a chip," she thought. "That means 10 chips are 120 calories." Before she realized what she was doing, Sue had eaten 100 chips! "So much for that resolution," she said as she popped the 101st chip into her mouth.

Sue had multiplied mentally to find the number of calories in 10 potato chips.

Examples

To multiply by powers or multiples of 10 mentally, first count the number of zeros at the end of the power or multiple.
Numbers like 10, 100, 1000, and 10,000 are **powers** of 10.
Numbers like 10, 20, 50, 200, and 250 are **multiples** of 10.

A Multiply mentally. 12×10

Add a zero to the 12 for each zero at the end of the power of 10.

$12 \times 10 = 120$

one zero   one place

$12 \times 10 = \mathbf{120}$

B Multiply mentally. 9×400

Mulitply 9 by 4. Add a zero to the product for each zero at the end of the multiple of 10.

$9 \times 400 = (9 \times 4) \times 100$
$\qquad\qquad 36 \times 100 = 3600$

$9 \times 400 = \mathbf{3600}$

▶ Think and Discuss

1. Multiply 126 by 1000. How many zeros did you add to 126?
2. Multiply 1260 by 10,000. How many zeros are in your answer?

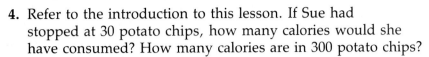
3. Describe how to solve the problem 71 × 300. Write the product.

4. Refer to the introduction to this lesson. If Sue had stopped at 30 potato chips, how many calories would she have consumed? How many calories are in 300 potato chips?

5. Describe how you would multiply 50 by 80 mentally. How many zeros are in the product?

Exercises

Multiply mentally. (See Example A.)

6. 9 × 10	7. 8920 × 100	8. 56 × 10,000	9. 72 × 100
10. 48 × 1000	11. 52 × 10,000	12. 8172 × 100	13. 8 × 10
14. 5063 × 1000	15. 47 × 100,000	16. 9 × 1000	17. 430 × 100

Multiply mentally. (See Example B.)

18. 4 × 30	19. 800 × 60	20. 9 × 500	21. 7 × 6000	22. 5 × 400
23. 3 × 8000	24. 120 × 20	25. 7 × 9000	26. 25 × 300	27. 13 × 200

▶ **Mixed Practice** (For more practice, see page 405.)

Multiply mentally.

28. 45 × 200	29. 137 × 10	30. 255 × 100	31. 4 × 400,000
32. 4 × 5000	33. 35 × 10,000	34. 7 × 700	35. 75 × 10
36. 17 × 100	37. 92 × 10,000	38. 50 × 300	39. 600 × 6000

▶ **Applications**

40. If sixty band members each invite six people to a concert, how many people will have been invited?

41. If each student in a class of 34 students sells 2 candles like the one below, how much money will be collected?

$5.00

▶ **Review** (Lessons 1.10, 1.11)

Subtract.

42.   6004
  − 2453
	43. 8.9
	  − 4.532
		44. 520.15
		  − 318.67
			45. 338,714
			  − 189,855

46. 10 − 0.446   47. 82,695 − 66,787   48. 4129 − 754

2.4 Multiplying by 2-Digit and 3-Digit Multipliers

Bernadette was so curious one day that she started to count the holes in the ceiling tiles. One tile had 159 holes. "It'll take me forever to count every hole in the ceiling," she thought.

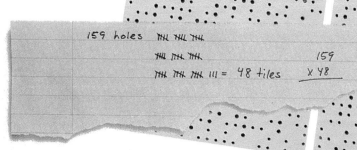

159 holes   ̶T̶H̶L̶ ̶T̶H̶L̶ ̶T̶H̶L̶
  ̶T̶H̶L̶ ̶T̶H̶L̶ ̶T̶H̶L̶   159
  ̶T̶H̶L̶ ̶T̶H̶L̶ ̶T̶H̶L̶ III = 48 tiles   × 48

Since the ceiling was so large, Bernadette decided to simply multiply 159 by 48, the number of tiles in the ceiling.

Examples

To multiply by a 2-digit or 3-digit number, begin at the ones places. Multiply each digit in the top number by each digit in the bottom number. Add the products.

A Multiply. 159 × 48

```
      1 5 9
    ×   4 8          Regroup: 48 = 40 + 8
    1 2 7 2      → 159 × 8
  + 6 3 6 0      → 159 × 40
    7 6 3 2
```

B Multiply. 824 × 604

```
        8 2 4          Regroup:
      × 6 0 4          604 = 600 + 4
      3,2 9 6      → 824 × 4
  + 4 9 4,4 0 0    → 824 × 600
    4 9 7,6 9 6
```

▶ Think and Discuss

1. Find the product of 573 and 19.

2. Find the product of 234 and 615. How is this problem different from the problem in Question 1?

3. Refer to the introduction to this lesson. How many holes does the ceiling have?

4. Which would be easier to find, 65 × 76 or 86 × 77? Explain.

5. How are the problems in Examples A and B like 1-digit multiplication? How are they different?

Exercises

Multiply. (See Example A.)

6. 74
× 42

7. 79
× 34

8. 155
× 18

9. 579
× 42

10. 203
× 49

11. 670
× 81

12. 943
× 71

13. 257
× 12

14. 336
× 25

15. 210
× 44

16. 425 × 36 **17.** 376 × 34 **18.** 15 × 99 **19.** 123 × 52

Multiply. (See Example B.)

20. 569
× 143

21. 3104
× 379

22. 368
× 244

23. 3007
× 403

24. 711
× 639

25. 750
× 179

26. 6348
× 205

27. 9002
× 791

28. 303
× 303

29. 4718
× 608

30. 204 × 99 **31.** 72 × 305 **32.** 601 × 207 **33.** 911 × 602

▶ **Mixed Practice** (For more practice, see page 406.)

Multiply.

34. 682
× 76

35. 5409
× 45

36. 82
× 49

37. 707
× 95

38. 49
× 94

39. 525
× 45

40. 4393
× 702

41. 7005
× 75

42. 823
× 44

43. 3909
× 64

44. 53 × 51 **45.** 416 × 25 **46.** 905 × 387 **47.** 301 × 500

▶ **Applications**

48. Coach Murphy told each player to bring twelve tennis balls to the first practice. How many tennis balls will he have if there are twenty-four players on the team?

49. To raise money for their school, the 162 seniors sponsored a rock concert. They sold 748 of the tickets below. How much money did the seniors earn from ticket sales?

CENTER CITY HIGH SCHOOL
presents SHARKSKIN
Appearing in the main auditorium
on Saturday at 8:00 PM
ROW A SEAT 20 $15.00

CENTER CITY HIGH SCHOOL
presents SHARKSKIN
Appearing in the main auditorium
on Saturday at 8:00 PM
ROW A SEAT 22 $15.00

▶ **Review** (Lesson 1.9)

Add.

50. 3.917
+ 4.48

51. 87,420
+ 29,658

52. 6956.3
+ 6975.7

53. 4,396,087
+ 5,857,317

2.5 4-Step Problem-Solving Process: Plan

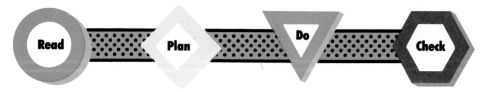

Step 1 in successful problem solving is **Read**. Step 2 is **Plan**. You can shape your plan by asking yourself certain questions. Read the example below and study the chart. You will see how questions and answers shape a plan for solving the example.

A class of 35 students and its teacher are going to a special show at the museum. Each ticket costs $2.50 and admits 2 people. The museum has provided the class with 6 free tickets. How much must the class spend to purchase enough additional tickets for everyone to see the show?

Questions	Answers	Plan
1. What exactly am I trying to find?	The cost of additional tickets.	The solution will be an amount of money.
2. What information do I need to solve the problem?	How many tickets the class needs to buy and the cost of each ticket.	
3. Of the information I need, what is given and what must I calculate?	The cost of each ticket is given. I must calculate the number of additional tickets needed.	Divide the number of people by 2 to find the total tickets needed. Subtract the number of free tickets to find the number of additional tickets needed.
4. Will organizing the information into a table or a diagram help solve the problem?	A diagram of the steps needed to solve the problem would be helpful.	· students + teacher = people · people ÷ 2 = total tickets · total tickets − 6 = tickets to purchase · tickets to purchase × $2.50 = cost of tickets

Questions	Answers
5. What operations must I carry out?	Addition, division, subtraction, multiplication
6. In what order should I do the operations?	Follow the sequence under Plan for Question 4.

Another point to consider when planning how to solve a problem is whether you have solved a similar problem before. Ask yourself if any of the steps that you used to solve that problem can be used to solve the new problem.

▶ Think and Discuss

1. Solve the example on page 38.

2. As you read the following problem, look for ways in which this problem is like the example on page 38.

   Part of Mary's job is making book covers. It takes Mary 5 minutes to make each cover. Ned brought Mary 15 books that need covers. Mary put them with 9 other books that also need covers. If Mary already has 8 covers ready, how much time must she spend making covers for these books?

3. Complete these sentences to show how elements in the above problem can be compared to those in the example.
   • <u>Books</u> in this problem can be compared to <u>people</u> in the example.
   • <u>Covers</u> in this problem can be compared to ▒ in the example.
   • ▒ in this problem can be compared to <u>cost of tickets</u> in the example.
   • <u>Covers already made</u> can be compared to ▒ in the example.

4. Division is not needed in the library problem. Where was division used in this example?

5. Make a chart like the one in the example to plan how to solve the following problem.

   To earn a jogging award, Tom must jog 90 miles. He jogs for $\frac{1}{2}$ hour a day, 5 days a week. If it takes Tom about 10 minutes to jog 1 mile, how many weeks will it take him to earn the award?

2.6 Estimating Products

Ken borrowed his parents' car to take his girlfriend to a concert 45 miles away. The car got 19 miles to the gallon. Ken knew by looking at the gas gauge that he had about 4 gallons of gas. He didn't want to run out of gas on the way home.

Ken needed quickly to figure out how much gas he would need.

Examples

To estimate a product, round one or both multipliers. Then multiply. The symbol ≈ means "is approximately equal to."

A Estimate. 19 × 4

19 rounds up to 20.
There is no need to round 4.

$20 \times 4 = 80$
$19 \times 4 \approx 80$

B Estimate. 227 × 48

227 rounds down to 200.
48 rounds up to 50.

$200 \times 50 = 10,000$
$227 \times 48 \approx \mathbf{10,000}$

▶ Think and Discuss

1. Refer to the introduction to this lesson. About how many miles can Ken go on 4 gallons of gas? Estimate how many miles he can go with a full tank, or 16 gallons of gas. Explain how you estimated.

2. Nancy earns $4 an hour, and she works 18 hours a week. Estimate how much she earns each month.

3. If both multipliers are rounded down, will the estimated product be higher or lower than the actual product? Explain.

4. When is it useful to estimate?

Exercises

Estimate the product. (See Example A.)

5.	6.	7.	8.	9.	10.
42	68	53	81	41	49
× 4	× 3	× 7	× 9	× 8	× 3

Estimate the product. (See Example B.)

11.	12.	13.	14.	15.	16.
15	172	39	478	25	93
× 11	× 23	× 32	× 68	× 13	× 11

17.	18.	19.	20.	21.	22.
98	142	283	17	407	874
× 32	× 21	× 46	× 92	× 81	× 19

▶ Mixed Practice (For more practice, see page 406.)

Estimate the product.

23.	24.	25.	26.	27.	28.
77	205	89	476	719	595
× 17	× 71	× 3	× 43	× 21	× 7

29.	30.	31.	32.	33.	34.
43	68	281	29	18	348
× 2	× 87	× 14	× 6	× 6	× 92

▶ Applications

35. Mr. Harris has just started a sales job that involves some driving. For accounting purposes, Mr. Harris must estimate his monthly mileage. He travels 185 miles 6 times a month. Estimate Mr. Harris's monthly mileage.

36. Rod, a speed reader, started reading a book one night and read eighty-five pages the first hour. He continued reading at the same pace for one hour each night. Did he finish the six-hundred-eighty-five-page book in seven days? Solve mentally using estimation.

▶ Review (Lessons 1.5, 1.6)

Compare. Use >, <, or = .

37. 6.581 ___ 6.481

38. 37,855 ___ 37,585

39. 9.877 ___ 98.77

40. 8.240 ___ 8.2400

41. 1.74 ___ 1.6985

42. 564,972 ___ 571,792

2.7 Multiplying Decimals by Powers and Multiples of 10

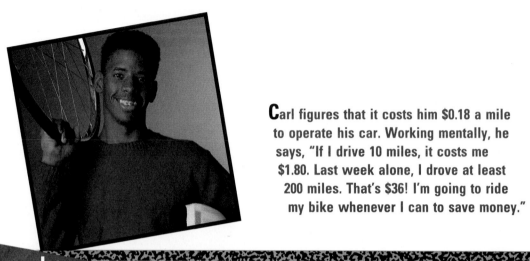

Carl figures that it costs him $0.18 a mile to operate his car. Working mentally, he says, "If I drive 10 miles, it costs me $1.80. Last week alone, I drove at least 200 miles. That's $36! I'm going to ride my bike whenever I can to save money."

Examples

To multiply a decimal by a multiple or power of 10 mentally, first count the number of zeros at the end of the power or multiple.

A Multiply. 0.18 × 10	**B Multiply. 4.5 × 100**	**C Multiply. 1.2 × 300**
Move the decimal point one place to the right.	Move the decimal point two places. Add a zero.	Multiply 1.2 by 3. Move the decimal point two places. Add a zero.
$0.18 \times 10 = 0.1.8$	$4.50 \times 100 = 4.50.$	$1.2 \times 300 =$
└ one zero	└ two zeros	$(1.2 \times 3) \times 100 =$
$0.18 \times 10 = \mathbf{1.80}$	$4.5 \times 100 = \mathbf{450}$	$3.6 \times 100 = 3.60.$
		$1.2 \times 300 = \mathbf{360}$

▶ Think and Discuss

1. Suppose you were solving a problem that involved multiplying a decimal by 10,000. How many places to the right would you move the decimal point?

2. Multiply 2.2 by four million. How many zeros did you add? How many places did you move the decimal point? Now multiply 2.2 by four billion. Did you add more or fewer zeros? Why?

3. What do you multiply 3.6 by to get 3600?

4. How is multiplying by a multiple of ten similar to multiplying by a power of ten? How is it different?

Exercises

Multiply. (See Example A.)

5. 1.29 × 10 **6.** 4.751 × 100 **7.** 37.95 × 10 **8.** 0.002 × 1000

9. 0.3511 × 1000 **10.** 4293.61 × 10 **11.** 9.3 × 10 **12.** 63.05 × 100

Multiply. (See Example B.)

13. 0.9 × 100 **14.** 32.71 × 1000 **15.** 0.6 × 100 **16.** 439.7 × 1000

17. 0.97 × 10,000 **18.** 113.3 × 1000 **19.** 3.01 × 100,000 **20.** 0.005 × 1000

Multiply. (See Example C.)

21. 0.95 × 400 **22.** $0.72 × 50 **23.** 27.1 × 200 **24.** $3.50 × 30

25. 0.25 × 4000 **26.** 0.009 × 70 **27.** $0.80 × 400 **28.** 4.50 × 20

▶ **Mixed Practice** (For more practice, see page 407.)

Multiply.

29. $8.50 × 100 **30.** 0.069 × 200 **31.** 321.1 × 1000 **32.** 0.9 × 10,000

33. $0.75 × 800 **34.** 0.07 × 7000 **35.** $0.69 × 30 **36.** 9.123 × 10

37. 9.305 × 1000 **38.** 4.1 × 20,000 **39.** $2.50 × 4000 **40.** 4201.3 × 100

▶ **Applications**

41. Rachel bought ten yards of fabric to make curtains. The fabric cost $4.29 a yard. How much did Rachel pay?

42. A ticket to a Polar Bears hockey game costs $8.15. Estimate the team's ticket revenues for last year, when more than two million tickets were sold.

43. How much does a set of four Rugged Rex Tires actually cost?

TIRE SALE
Drive 50,000 miles on a set of
RUGGED REX TIRES
and pay only $0.0024 a mile

▶ **Review** (Lessons 1.1, 1.2)

Write each number in words.

44. 8077 **45.** 1.005 **46.** 43,983 **47.** 26.9 **48.** 0.044 **49.** 731

50. 6.85 **51.** 57,060 **52.** 4714.1 **53.** 0.209 **54.** 8207 **55.** 25.25

56. 9111 **57.** 467.09 **58.** 10.002 **59.** 4652 **60.** 987,414 **61.** 56.343

2.8 Multiplying Decimals and Whole Numbers

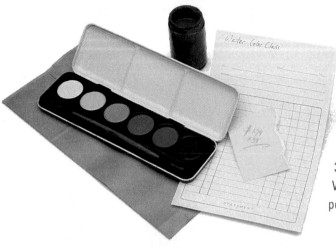

Winston teaches children's art classes at the Broomfield Arts Coalition. He has just bought supplies for several classes and must bill the Coalition for each class separately. For the children's watercolor class, he bought 39 paint boxes at $1.79 apiece. What was the total for this portion of his bill?

Examples

To multiply a whole number by a decimal, multiply as with whole numbers. The product has the same number of decimal places as the decimal factor.

A Multiply. 7 × 0.6

$$
\begin{array}{r}
7 \\
\times\ 0.6 \leftarrow \text{1 decimal place} \\
\hline
4.2 \leftarrow \text{1 decimal place}
\end{array}
$$

B Multiply. 1.79 × 39

$$
\begin{array}{r}
1.7\ 9 \leftarrow \text{2 decimal} \\
\times\quad 3\ 9 \quad \text{places} \\
\hline
1\ 6\ 1\ 1 \\
5\ 3\ 7\ 0 \\
\hline
6\ 9.8\ 1 \leftarrow \text{2 decimal} \\
\text{places}
\end{array}
$$

C Multiply.
4.815 × 235

$$
\begin{array}{r}
4.8\ 1\ 5 \leftarrow \\
\times\quad 2\ 3\ 5 \\
\hline
2\ 4\ 0\ 7\ 5 \\
1\ 4\ 4\ 4\ 5\ 0 \\
9\ 6\ 3\ 0\ 0\ 0 \\
\hline
1\ 1\ 3\ 1.5\ 2\ 5 \leftarrow
\end{array}
$$

3 decimal places

▶ Think and Discuss

1. How many decimal places are in the product of 67 and 1.038?

2. Refer to the introduction to this lesson. How much money did Winston spend on watercolors?

3. Which problem is easier to solve, 15 × 5 or 15 × 0.05? Why?

4. How many decimal places are in the product when you multiply whole numbers by dollars-and-cents amounts? Why?

Exercises

Multiply. (See Example A.)

5. 8×2.6 6. 15×1.5 7. 6×9.8 8. 3×3.1 9. 2×9.7

10. 200×0.6 11. 4×0.3 12. 29×3.3 13. 42×0.7 14. 1.9×68

15. 75×1.5 16. 50×7.9 17. 90×0.2 18. 7×4.6 19. 11×7.2

Multiply. (See Example B.)

20. 3×0.49 21. 17×0.35 22. 42×1.99 23. 7×7.75

24. 5.61×23 25. 3.03×33 26. 99×0.33 27. 50×0.25

28. 25×4.05 29. 9.23×7 30. 42×0.06 31. 150×0.75

Multiply. (See Example C.)

32. 14×12.725 33. 5×0.909 34. 134×5.557 35. 0.789×62

36. 42×4.007 37. 400×0.999 38. 3.325×7 39. 8.125×25

40. 8×0.123 41. 25×1.125 42. 6.001×30 43. 2.689×4

▶ Mixed Practice (For more practice, see page 407.)

Multiply.

44. 702×1.7 45. 1.93×4 46. 51×5.1 47. 0.405×9

48. 72×0.25 49. 63×0.333 50. 0.125×8 51. 49×0.08

52. 13.4×29 53. 0.909×555 54. $\$2.19 \times 32$ 55. 87×0.003

▶ Applications

56. Amy runs three and five-tenths miles five times a week. How far does she run in a week?

57. A sheet of Baltic plywood is made up of 16 layers of birch. If each layer is 0.03 inches thick, how thick is one sheet?

58. Alveta is making a poster for a school competition. She chose 17 colors of pastels. If each pastel crayon costs $0.87 and the total tax is $1.02, how much change will Alveta get back from $20?

▶ Review (Lesson 1.8)

Round each decimal to the nearest tenth and the nearest hundredth.

59. 6.581 60. 37.855 61. 9.877 62. 4.99999 63. 0.3695

2.9 Making Magic Squares

The figure shown here is a magic square. In a magic square, numbers are arranged so that each row, column, and diagonal has the same sum.

1. How many rows are there in the magic square shown above? How many columns? How many diagonals?

2. Show that each row, column, and diagonal has the same sum.

The magic square above is called a 4 × 4 ("four by four") magic square, since there are 4 rows and 4 columns. Each whole number from 1 to 16 occupies one of the 16 spaces.

Now the task is to create a 3 × 3 magic square. There are 362,880 3 × 3 squares and only 8 are magic. Rather than guessing, you can look for patterns when making magic squares. Use the following questions to help you discover patterns.

3. How many rows are there in a 3 × 3 magic square?

4. Find the sum. 1 + 2 + 3 + 4 + 5 + 6 + 7 + 8 + 9

5. Based on your answers to Questions 3 and 4, what is the sum of each row? Of each column? Of each diagonal?

The diagrams below are the key to finding a 3 × 3 magic square.

6. Look at the first diagram. How many sums is the number in a middle square a part of?

7. Look at the middle diagram. How many sums is the number in a corner square a part of?

8. Look at the last diagram. How many sums is the number in a side square a part of?

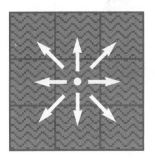 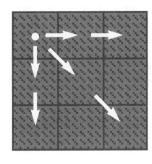 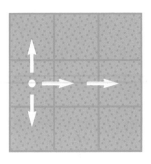

9. 1 + 6 + 8 = 15; List all the other ways 1 can be added to two different numbers from 2 to 9 to equal 15.

10. Use your answers to Questions 6–9 to decide whether 1 can fill a middle, corner, or side square. Explain your answer.

11. Which number fills the middle square? How did you find that number?

12. Which numbers fill the four corners?

13. Use your answers to Questions 6–12 to help you make a 3 × 3 magic square.

14. Compare all of the magic squares that your class created. How many different magic squares were made? Do you see a pattern in the answers? Describe the pattern.

▶ **Review** (Lessons 2.2, 2.4)

Multiply.

15.	16.	17.	18.	19.	20.
63	8895	408	45	179	59
× 4	× 9	× 52	× 76	× 81	× 23

2.10 Multiplying Decimals by Decimals

GERTRUDE'S
Fresh Fruit Market
• • • • • • • • • • • • • • • •
Broccoli/
Cauliflower @ $1.09 lb.

Total weight   3.5 lb.

TOTAL

Thank you

Suppose you are buying snacks for a party. They cost $1.09 a pound, and the scale registers 3.5 pounds. The total cost displayed on the scale is $3.82. How do you know whether the total cost shown on the scale is correct?

Examples

To multiply two decimals, multiply as with whole numbers. Add the number of decimal places in the multipliers to get the number of decimal places in the product.

A Multiply.
1.09 × 3.5

```
    1.0 9  ← 2 places
 ×    3.5  ← 1 place
    5 4 5
  3 2 7 0
  3.8 1 5  ← 3 places
```
1.09 × 3.5 = **3.815**

B Multiply.
0.49 × 0.05

```
    0.4 9  ← 2 places
 × 0.0 5  ← 2 places
  0.0 2 4 5  ← 4 places
```
Write zeros for placeholders
0.49 × 0.05 = **0.0245**

C Multiply.
0.832 × 6.01

```
    0.8 3 2  ← 3 places
 ×    6.0 1  ← 2 places
      8 3 2
  4 9 9 2 0 0
  5.0 0 0 3 2  ← 5 places
```
0.832 × 6.01
= **5.00032**

▶ Think and Discuss

1. Find the product of 0.13 and 2.05.

2. Refer to the introduction to this lesson. Was the reading on the scale correct?

3. How can estimation help when you multiply decimals?

4. When you multiply two decimals that are less than 1, do you expect the product to be greater or less than 1? Discuss.

Exercises

Multiply. (See Example A.)

5.	5.9	**6.**	1.401	**7.**	9.04
	× 3.5		× 8.2		× 7.18

8.	1.246	**9.**	9.11
	× 3.9		× 3.2

10. 2.119 × 4.303 **11.** 19.7 × 9.2 **12.** 401.5 × 50.75 **13.** 1.25 × 3.4

14. 6.003 × 5.12 **15.** 37.49 × 3.5 **16.** 25.5 × 5.8 **17.** 25.5 × 55.75

Multiply. (See Example B.)

18. 0.19 × 0.91 **19.** 0.528 × 0.7 **20.** 0.87 × 0.003 **21.** 0.473 × 0.473

22. 0.02 × 0.119 **23.** 0.05 × 0.73 **24.** 0.179 × 0.08 **25.** 0.49 × 0.12

Multiply. (See Example C.)

26.	0.308	**27.**	7.32	**28.**	0.776	**29.**	5.807	**30.**	0.951
	× 5.25		× 0.54		× 8.9		× 0.021		× 3.8

31. 0.4 × 4.4 **32.** 1.25 × 0.25 **33.** 33.33 × 0.33 **34.** 92.15 × 0.5

▶ Mixed Practice (For more practice, see page 408.)

Multiply.

35. 9.06 × 3.35 **36.** 0.2 × 0.0011 **37.** 5.71 × 0.246 **38.** 1.0931 × 2.9

39.	0.27	**40.**	6.3	**41.**	2.615	**42.**	0.707	**43.**	4.023
	× 0.95		× 9.8		× 0.83		× 0.15		× 1.351

▶ Applications

44. Miriam is enlarging a photo for the school paper. The photo is three and five-tenths inches wide. She decides to have it enlarged to one and five-tenths times its original width. How wide will the photo be after it is enlarged?

45. Carl owns Modesto Bakery. He overhears his assistant charge a customer $3.40 for a 1.89-pound loaf of rye bread. Carl knows that, at $1.29 a pound for rye bread, that amount is incorrect. How much should he refund to the customer?

▶ Review (Lesson 1.7)

Round each number to the nearest hundred and the nearest ten.

46. 18,609 **47.** 9499 **48.** 87,662 **49.** 59,527 **50.** 851,385

51. 729,717 **52.** 18,045 **53.** 442 **54.** 78 **55.** 19,709

Multiplying Whole Numbers and Decimals **49**

2.11 Choosing Ways to Compute

The student council at Humboldt High is sponsoring dances in the school gymnasium.

The students plan to sponsor 4 or 5 dances and to charge $2.00 to $3.00 admission. They expect between 300 and 400 people to attend each dance.

The students wanted to estimate the amount of money they would raise. First they used the smaller numbers; then they used the larger numbers.

1. What was the smallest amount of money they expected to raise? Use the smaller numbers:
   4 dances × $2.00 admission × 300 people = ▧ .

2. What was the largest amount they expected to raise? Use the larger numbers:
   5 dances × $3.00 admission × 400 people = ▧ .

3. Could you use mental math to answer Questions 1 or 2? Explain how.

4. What is the difference between the largest and smallest amounts of money they expect to raise?

Sometimes, when numbers are easy to work with, you can solve a problem quickly in your head. At other times, the numbers are large or difficult to work with. Then you want to use paper-and-pencil or a calculator.

5. At the first dance, the students charged $2.50 admission and 345 people attended. How much money did they raise?
   $2.50 admission × 345 people = ▧

6. Did you use a calculator, paper-and-pencil, or mental math to multiply 2.50 by 345? Explain why.

7. For the second dance, the students charged $3.00 admission and 300 attended. How much money did they raise? $3.00 admission × 300 people = ___ Which method did you use to multiply 3.00 × 300? Explain why.

Work with 2 or 3 classmates. For Questions 8–10, write out your solution to the problem. Tell whether you used mental math, paper-and-pencil, or a calculator. Explain why you chose your method.

8. The school gave the student council $22.00 for publicity for each dance. After 4 dances, how much money had the council received for publicity?

9. The student council sponsored 4 dances, and charged $2.25 admission at each dance. There were 325 people at the first dance, 250 at the second dance, 310 at the third dance, and 302 at the fourth dance. How much money did they raise in all?

10. The students knew that more people would attend dances if the admission charge were low. They estimated that an admission charge of $2.25 would result in about 375 people attending, and an admission charge of $2.75 would result in about 300 people attending. If the estimates are correct, which admission charge would raise more money?

▶ **Review** (Lesson 1.4)

Read the following problems. Write what you must find and the facts you need to solve each problem.

11. Jacklyn took a taxi from the airport to her home. She lives 23 miles from the airport. If the taxi meter registered 55¢ for each mile, and if Jacklyn tipped the driver $1.00, what was the cost of the taxi ride?

12. Paul works 8 hours a week at a cafeteria in a department store. He makes $3.75 an hour. Will he make enough money in 2 weeks to buy a camera that costs $79.95?

Chapter 2 Review

REVIEW

Complete each statement. (Lesson 2.1)

1. The answer to a multiplication problem is called the ▒ .

2. Numbers that are multiplied are called ▒ .

Multiply mentally. (Lesson 2.3)

3. 50×700 4. $100,000 \times 76$ 5. 519×1000 6. 90×30

7. 38×100 8. 600×800 9. 40×500 10. 2964×100

Multiply. (Lessons 2.2, 2.4)

11. 29   12. 6416   13.   55   14.   871   15.   125   16.   657
   $\times$ 8 $\times$ 9 $\times$ 25 $\times$ 43 $\times$ 230 $\times$ 801

17. 845   18.   237   19.   11   20. 7007
  $\times$ 2 $\times$ 49 $\times$ 14 $\times$ 8

21. A group of 22 people are going to a play. Tickets cost $12 each. How much must the playgoers spend on tickets in all?

22. A pear has about 63 calories. If you eat two pears a day for seven days, how many calories have you consumed?

Estimate the product. (Lesson 2.6)

23. 166×1018 24. 456×781 25. 6340×374 26. 909×862

Multiply. (Lesson 2.7)

27. 1000×42.89 28. 0.04×20 29. 6.3×100 30. 50×8.2

Multiply. (Lesson 2.8)

31.   4.5   32.   783   33.   15.25   34.   55   35.   8.1
   $\times$ 4 $\times$ 1.9 $\times$ 62 $\times$ 7.8 $\times$ 18

36. 49.013   37.   9.2   38.   607   39.   50.5   40.   28.69
  $\times$ 147 $\times$ 48 $\times$ 2.1 $\times$ 49 $\times$ 54

Multiply. (Lesson 2.10)

41.   1.2   42.   0.07   43.   43.6   44.   5.3   45.   0.208
   $\times$ 8.4 $\times$ 0.35 $\times$ 0.91 $\times$ 0.9 $\times$ 0.06

Chapter 2 Test

TEST

Multiply.

1. 3055×5

2. 99×44

3. 5.7×0.08

4. 404×2.3

5. 0.08×0.8

6. 600×90

7. 543×51

8. 205×55

9. 6263×290

10. 1000×43

11. 21.89×10

12. 0.004×0.01

13. 400×4.72

14. 80×70

15. 6.01×1.7

16. $3.6 \times 10,000$

17. 75.75×1.25

18. $8771 \times .43$

Estimate each product.

19. 626×551

20. 238×741

21. 97×139

Solve.

22. Jane works six hours on Saturdays. She earns $4.45 an hour. How much does she earn?

23. Gas costs $1.17 per gallon. If Nicholas buys 13.5 gallons, what is the total cost?

24. Samantha runs 3 miles a day during the week and 4 miles a day on Saturdays and Sundays. How far does she run in a week?

25. Karl makes $4.89 an hour at his summer job. If he works 27.5 hours a week for 13 weeks, how much will he earn?

Complete each sentence.

26. When 7 is multiplied by 28, the result is called the ░░ .

27. Factors are numbers that are ░░ .

28. To multiply a number by 1000, move the decimal point three places to the ░░ .

Multiplying Whole Numbers and Decimals **53**

"We were trying to figure out how we used division in sports and we came up with these ideas. Most playing fields are divided into sections. Games are divided into parts such as innings, sets, quarters, and halves. What ways can you think of?"

Dividing Whole Numbers and Decimals

Pro Basketball

Chicago Winds - Results to date	
Chicago vs. Washington	118 - 102
Chicago vs. Dallas	98 - 114
Chicago vs. Philadelphia	119 - 96
Chicago vs. Cleveland	108 - 105
Chicago vs. Atlanta	109 - 114
Chicago vs. Los Angeles	92 - 124
Chicago vs. Detroit	114 - 94
Chicago vs. Seattle	97 - 145
Chicago vs. Milwaukee	87 - 120
Chicago vs. Boston	143 - 136
Chicago vs. New York	130 - 106
Chicago vs. Houston	105 - 108
Chicago vs. San Antonio	84 - 97
Chicago vs. Denver	138 - 102
Chicago vs. Raleigh	101 - 108
Chicago vs. Sacramento	78 - 83
Chicago vs. Phoenix	159 - 78
Chicago vs. Nashville	94 - 120
Chicago vs. Portland	117 - 114
Chicago vs. Salt Lake City	138 - 92
Chicago vs. El Paso	88 - 130
Chicago vs. New Jersey	104 - 120

Examples of situations involving division include:
- finding your average test grade or homework grade.
- determining the cost of one item when you know the cost of a dozen.
- finding the average points scored per game when you know a team's total points for the season.

What are other situations in which you would use division?

Division in Sports

Work with 2 or 3 classmates to solve the following problem. Mrs. Acosta has 35 students in her physical education class. She wants to divide the class into teams so that every student is participating in one of the sports shown below. Assume that there is enough room and equipment to accommodate any number of students at all sports. List at least 6 ways that Mrs. Acosta can divide the class. Compare your results with the results of other groups.

Gym Class

Teacher: Mrs. Acosta

track relay   3 runners per team

table tennis 2 players per team

badminton   4 players per team

basketball   5 players per team

volleyball   6 players per team

3.1 Introduction to Division

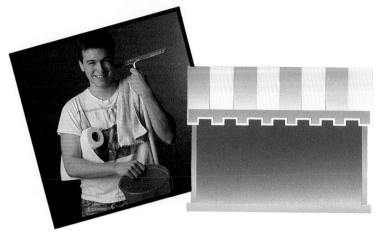
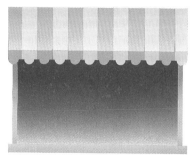

Robert can wash 9 windows in an hour. At this pace, how long would it take him to wash 81 windows? If he and 2 friends worked at the same pace, how long would it take them to complete the job? You can find the answers to these questions by dividing. But if you think about your multiplication facts, you will realize that you already know the answers.

Division is the inverse, or opposite, of multiplication. If you can multiply, then you can divide.

▶ Think and Discuss

Read the following problems.

150 freshmen are assigned to 5 homerooms. If all homerooms have the same number of students, how many are in each one?

Twenty presents are distributed evenly among four people. How many gifts did each person receive?

1. What is similar about the two problems?

2. Draw a diagram or picture that illustrates the second problem.

3. Write a division fact and a multiplication fact that describes the first problem.

4. How could you check your solution to the first problem?

5. Write a division fact and a multiplication fact that describes the second problem.

6. Write two multiplication facts. Then change each to a division fact.

7. Write two word problems that require multiplication to find an answer. Then change each to a division problem.

▶ Express Yourself

The parts of a division problem have names.
Look at the numbers in the following problems.

$$0.4\overline{)2.4}$$ with quotient 6

The **dividend** is 2.4.
The **divisor** is 0.4.
The **quotient** is 6.

$10 \div 2 = 5$

The **dividend** is 10.
The **divisor** is 2.
The **quotient** is 5.

Identify the dividend, the divisor, and the quotient in each problem.

8. $5\overline{)45}$ with quotient 9

9. $4.8\overline{)38.88}$ with quotient 8.1

10. $49.2 \div 8.2 = 6$

11. $63 \div 7 = 9$

▶ Practice What You Know

A factor is missing in the multiplication statement below.

$6 \times \underline{\text{\ }} = 18$

You can rewrite this multiplication statement as a division problem.

$18 \div 6 = \underline{\text{\ }}$

Complete each statement.

12. $7 \times \underline{\text{\ }} = 21$, so $21 \div 7 = \underline{\text{\ }}$.

13. $8 \times \underline{\text{\ }} = 40$, so $40 \div 8 = \underline{\text{\ }}$.

14. $9 \times \underline{\text{\ }} = 36$, so $36 \div 9 = \underline{\text{\ }}$.

15. $5 \times \underline{\text{\ }} = 45$, so $45 \div 5 = \underline{\text{\ }}$.

Solve.

16. Suppose you were asked to arrange 30 desks so that each row had the same number of desks. List all the possible arrangements. Write the division and multiplication facts involved.

Dividing Whole Numbers and Decimals **57**

3.2 Dividing by 1-Digit Divisors

June saw a leather jacket in the store window. "I must have that jacket," she thought. "I wonder how much it costs." When she went in to ask, she was told that the jacket cost $136. It could be hers in four equal monthly payments.

EASY
PAYMENT
PLAN!

To determine the amount of each payment, June divided $136 by 4.

Examples

To divide by a 1-digit divisor, remember: divide, multiply, subtract, bring down.

A Divide.   136 ÷ 4

```
     3 4
  4)1 3 6      How many 4s in 1? 0 In 13? 3
    1 2        Multiply. Subtract. Bring down
    1 6        the 6.
    1 6        How many 4s in 16? 4
       0       Multiply. Subtract.
               The remainder is 0.
```

136 ÷ 4 = **34**

B Divide.   535 ÷ 7

```
     7 6 R3
  7)5 3 5      How many 7s in 5? 0 In 53? 7
    4 9        Multiply. Subtract. Bring down
    4 5        the 5.
    4 2        How many 7s in 45? 6
       3       Multiply. Subtract. The
               remainder is 3.
               Write the remainder in the
               quotient.
```

535 ÷ 7 = 76 remainder 3
This is written **76 R3.**

▶ Think and Discuss

1. What is the first step in the problem 2)‾730? In the problem 9)‾116?

2. Refer to the introduction to this lesson. How much was June's monthly payment?

3. Identify the divisor, the dividend, and the quotient in the problem 75 divided by 3.

4. Divide. 683 ÷ 9

5. Describe three situations in which you might use division.

Exercises

Divide. (See Example A.)

6. $3\overline{)81}$
7. $6\overline{)72}$
8. $7\overline{)350}$
9. $7\overline{)714}$

10. $3\overline{)690}$
11. $4\overline{)836}$
12. $4\overline{)8388}$
13. $9\overline{)9513}$

14. $8040 \div 8$
15. $960 \div 2$
16. $48 \div 4$
17. $95 \div 5$

Divide. (See Example B.)

18. $2\overline{)33}$
19. $3\overline{)92}$
20. $9\overline{)60}$
21. $6\overline{)64}$

22. $9\overline{)39}$
23. $8\overline{)59}$
24. $7\overline{)79}$
25. $6\overline{)98}$

26. $6699 \div 4$
27. $791 \div 4$
28. $435 \div 6$
29. $3401 \div 7$

▶ Mixed Practice (For more practice, see page 408.)

Divide.

30. $4\overline{)404}$
31. $3\overline{)57}$
32. $8\overline{)6993}$
33. $2\overline{)510}$

34. $270 \div 9$
35. $3\overline{)711}$
36. $6\overline{)194}$
37. $391 \div 7$

38. $3\overline{)960}$
39. $4\overline{)63}$
40. $8200 \div 8$
41. $2\overline{)948}$

42. $7\overline{)781}$
43. $6\overline{)3640}$
44. $9\overline{)897}$
45. $6432 \div 2$

▶ Applications

46. Christiane had forty-eight cuttings from her plants that she was going to give to eight friends. If she gave the same number to each, how many cuttings did each receive?

47. Kevin is a waiter. He must pay taxes every three months on his tips. This year, he expects to owe a total of $300 in taxes. How much will each of his payments be?

48. Look at the ad for a radio. What is the sale price? What is the monthly payment if the payments are equal? How much would you owe after one payment? After two?

Was $48
Now $42

Only 6 Payments

▶ Review (Lesson 1.9)

Add.

49. $8.9 + 335.68$
50. $8924 + 36{,}951$
51. $0.053 + 9.87$

52. $375{,}209 + 617{,}959$
53. $5.5 + 4.7 + 0.51$
54. $0.07 + 0.008$

3.3 Dividing by Powers and Multiples of 10

When Lucky Jim won $3,000,000, he announced, "I don't need the money. I'm splitting it evenly among all my relatives." Within an hour, 10 relatives called, claiming their shares. The next day, the total number of relatives had increased to 60. By the third day, Lucky Jim had 100 relatives.

To figure out how much money each relative would receive, divide Lucky Jim's winnings by the number of relatives.

Examples

To divide by powers or multiples of 10 mentally, begin by counting the number of zeros at the end of the power or multiple.

A Divide. 3,000,000 ÷ 100

$3{,}000{,}000 \div 100 = 30{,}000.00 = \mathbf{30{,}000}$ Move the decimal point two places to the left.

B Divide. 625 ÷ 1000

$625 \div 1000 = 0.625 = \mathbf{0.625}$ Move the decimal point three places to the left. Add a zero as a placeholder.

C Divide. 350 ÷ 50

$350 \div 50 = 35.0 \div 5.0$ Move the decimal points one place to the left.
$= 35 \div 5 = \mathbf{7}$ Then divide.

▶ Think and Discuss

1. Divide. $15 \div 10{,}000$

2. Refer to the introduction to this lesson. How much money would each of Lucky Jim's 100 relatives receive?

3. Describe how dividing by 100 is similar to multiplying by 100. How is it different?

4. When might you divide by a power or other multiple of 10?

5. Describe what you do when you divide by 50. By 7000.

Exercises

Divide. (See Example A.)

6. 92,000 ÷ 10
7. 10,600 ÷ 100
8. 8300 ÷ 10
9. 5000 ÷ 1000
10. 250,000 ÷ 100
11. 390 ÷ 10

Divide. (See Example B.)

12. 47 ÷ 1000
13. 7 ÷ 10
14. 2725 ÷ 100
15. 1185 ÷ 10
16. 1457 ÷ 1000
17. 110 ÷ 10,000
18. 9527 ÷ 1000
19. 495 ÷ 100

Divide. (See Example C.)

20. 4900 ÷ 70
21. 1500 ÷ 300
22. 36,000 ÷ 600
23. 760 ÷ 40
24. 444 ÷ 200
25. 960,000 ÷ 80
26. 87,500 ÷ 50
27. 805 ÷ 50

▶ **Mixed Practice** (For more practice, see page 409.)

Divide.

28. 4000 ÷ 10
29. 3600 ÷ 60
30. 45,000 ÷ 100
31. 1525 ÷ 100
32. 5000 ÷ 50
33. 95 ÷ 1000
34. 9113 ÷ 10
35. 14,700 ÷ 70
36. 105 ÷ 10,000

▶ **Applications**

37. Find the price of one tea bag from each of the boxes shown in the picture. Which box is the best value?

$4.00 $3.00 $2.00

100 Tea Bags   70 Tea Bags   50 Tea Bags

38. Mei spent four dollars on twenty party favors. How much did each one cost?

39. Craig works 10 hours a week. He earns $1100 in 20 weeks. What is his hourly pay?

▶ **Review** (Lessons 2.8, 2.10)

Multiply.

40. 62.4
     × 0.08

41. 3.38
     × 2.21

42. 107
     × 6.5

43. 0.009
     × 0.006

44. $4.85
     ×   8

3.4 Estimating Quotients

Kieron wants to organize his tapes. He needs to know if he already has enough boxes or if he needs to buy some more. He has 115 tapes and can fit 24 tapes into each tape box.

To quickly figure out about how many boxes he needs, Kieron estimates the quotient of 115 ÷ 24.

Examples

To estimate a quotient, first round the divisor and the dividend. Then divide.

A Estimate. 115 ÷ 24

$115 \approx 120$ Round up to the nearest ten.
$24 \approx 20$ Round down to the nearest ten.
$120 \div 20 = 6$
$115 \div 24 \approx 6$

B Estimate. 64,305 ÷ 783

$64,305 \approx 64,000$ Round down to the nearest thousand.
$783 \approx 800$ Round up to the nearest hundred.
$64,000 \div 800 = 80$
$64,305 \div 783 \approx 80$

▶ Think and Discuss

1. Estimate.   14,080 ÷ 689

2. Estimate the quotient of 124 ÷ 5. Why is it not a good idea to round the divisor to the nearest 10? Find the actual answer.

3. Try Example A again, rounding the divisor to 25. Why is it easy to divide by 25?

4. Refer to the introduction to this lesson. How many boxes does Kieron really need?

Exercises

Estimate. (See Example A.)

5. $51 \div 9$ **6.** $138 \div 19$ **7.** $356 \div 88$ **8.** $922 \div 38$

9. $49 \div 11$ **10.** $178 \div 18$ **11.** $596 \div 41$ **12.** $54 \div 7$

13. $87 \div 6$ **14.** $539 \div 92$ **15.** $218 \div 14$ **16.** $634 \div 68$

Estimate. (See Example B.)

17. $1083 \div 23$ **18.** $5579 \div 73$ **19.** $3182 \div 49$

20. $6255 \div 98$ **21.** $4077 \div 79$ **22.** $35{,}911 \div 632$

23. $6432 \div 281$ **24.** $8591 \div 97$ **25.** $64{,}120 \div 807$

▶ Mixed Practice (For more practice, see page 409.)

Estimate.

26. $403 \div 7$ **27.** $97 \div 23$ **28.** $5217 \div 110$

29. $682 \div 7$ **30.** $124 \div 5$ **31.** $359 \div 59$

32. $477 \div 124$ **33.** $23{,}811 \div 62$ **34.** $99 \div 18$

35. $271 \div 32$ **36.** $3816 \div 79$ **37.** $9189 \div 576$

38. $253 \div 48$ **39.** $119{,}751 \div 19$ **40.** $16{,}327 \div 54$

▶ Applications

41. One thousand seven hundred fifty-nine students attend Carver High School. If fifty-eight classes are scheduled each period, estimate the number of students in each class.

42. A minor league team had a season attendance of 78,422 people. If 42 home games were played, estimate the attendance per game.

43. Peter is buying special paper for a science project. Estimate the cost of a single sheet of paper from each pad shown. Which pad is the better value?

50 Sheets $6.29

20 Sheets $4.39

▶ Review (Lessons 1.1, 1.2)

Write the whole number or decimal number.

44. eight and nine hundredths

45. six hundred thousand, twelve

46. seven million, forty

47. five thousand and five tenths

3.5 Dividing by 2-Digit Divisors

When Mrs. Hubbard had to divide a jar of jelly beans equally among her 31 children, she became quite concerned. You can imagine her relief when she discovered that dividing by 2-digit divisors was easier than she expected. After all, she already knew how to divide by 1-digit divisors.

Examples

To divide by a 2-digit divisor, remember: divide, multiply, subtract, bring down.

A Divide.   384 ÷ 52

```
      0 7 R20
52)3 8 4
   3 6 4
     2 0
```
52 does not divide 38. Write a **0** in the quotient.
How many times does 52 divide 384? To estimate, think: how many 5s in 38? **7** Write the remainder in the quotient.

384 ÷ 52 = 7 R20

B Divide.   7862 ÷ 39

```
    2 0 1 R23
39)7 8 6 2
   7 8
     0 6 2
       3 9
       2 3
```
How many times does 39 divide 78? To estimate, think: how many 3s in 7? **2** Multiply, subtract, and bring down the 6. Since 39 does not divide 6, write a **0** in the quotient. Bring down the 2. How many times does 39 divide 62? To estimate, think: how many 3s in 6? **2** Multiply. You cannot subtract 78 from 62. Try **1**.

7862 ÷ 39 = **201 R23**

▶ Think and Discuss

1. Divide. 683 ÷ 13

2. Is 587 ÷ 45 < 10? Why?

3. Is 587 ÷ 45 < 100? Why?

4. What does it mean to have a remainder of zero?

Exercises

Divide. (See Example A.)

5. $45\overline{)892}$ 6. $10\overline{)372}$ 7. $52\overline{)777}$ 8. $19\overline{)943}$

9. $65\overline{)731}$ 10. $94\overline{)391}$ 11. $39\overline{)303}$ 12. $40\overline{)950}$

13. $715 \div 11$ 14. $216 \div 88$ 15. $529 \div 23$ 16. $666 \div 73$

Divide. (See Example B.)

17. $43\overline{)4003}$ 18. $90\overline{)5850}$ 19. $12\overline{)7100}$ 20. $22\overline{)2012}$

21. $48\overline{)2687}$ 22. $61\overline{)1342}$ 23. $84\overline{)2687}$ 24. $57\overline{)6399}$

25. $1805 \div 25$ 26. $3030 \div 55$ 27. $1980 \div 30$ 28. $7171 \div 76$

▶ Mixed Practice (For more practice, see page 410.)

Divide.

29. $12\overline{)6660}$ 30. $3723 \div 62$ 31. $25\overline{)555}$ 32. $49\overline{)3000}$

33. $30\overline{)9999}$ 34. $8442 \div 21$ 35. $909 \div 90$ 36. $11\overline{)730}$

37. $41\overline{)1681}$ 38. $22\overline{)4444}$ 39. $4999 \div 50$ 40. $73\overline{)737}$

41. $43\overline{)4343}$ 42. $1255 \div 25$ 43. $44\overline{)1800}$ 44. $77\overline{)777}$

▶ Applications

45. Roger is planning a trip from Dallas to Boston. His car gets 28 miles to the gallon on the highway. The total distance is 1805 miles. At $1.09 a gallon for gas, estimate how much Roger can expect to pay for gasoline one way.

46. Shizuko is giving a party. She expects forty-five people to come. She has ordered fifteen large pizzas, which are cut into eight slices each. How many slices of pizza can each person have if all are to get the same number? How many pieces will there be left over?

▶ Review (Lessons 1.10, 1.11)

Subtract.

47. $\begin{array}{r} 8.03 \\ -\ 0.66 \end{array}$ 48. $\begin{array}{r} 52.31 \\ -\ 46.55 \end{array}$ 49. $\begin{array}{r} 6 \\ -\ 1.04 \end{array}$ 50. $\begin{array}{r} 19.40 \\ -\ 11.872 \end{array}$ 51. $\begin{array}{r} 7.8 \\ -\ 0.98 \end{array}$

52. $\begin{array}{r} 9.3 \\ -\ 0.457 \end{array}$ 53. $\begin{array}{r} 4.7 \\ -\ 2.63 \end{array}$ 54. $\begin{array}{r} 0.67 \\ -\ 0.28 \end{array}$ 55. $\begin{array}{r} 0.581 \\ -\ 0.39 \end{array}$ 56. $\begin{array}{r} 7.8 \\ -\ 0.9 \end{array}$

3.6 Dividing by 3-Digit Divisors

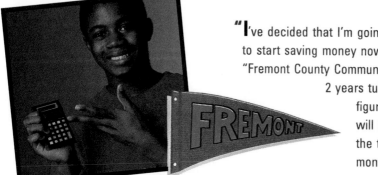

"I've decided that I'm going to college and I want to start saving money now," Lonnie told his parents. "Fremont County Community College costs $2800 for 2 years tuition. I need to figure out how soon I will have enough money to pay the tuition if I save $125 a month."

To find out how long it would take, Lonnie divided $2800 by $125.

Examples

To divide by a 3-digit number, it is helpful to use a calculator. Otherwise, use the method taught in Lesson 3.5.

A Divide.   2800 ÷ 125

Calculator Method

Enter 2800 ÷ 125 = 22.4.
To find the remainder,
multiply 22 by 125.
Enter 22 × 125 = 2750.
Subtract the product from 2800.
Enter 2800 − 2750 = 50.

2800 ÷ 125 = **22 R50**
Check to be sure your answer is reasonable by estimating.
2800 ÷ 125 ≈ 2800 ÷ 100 = 28, so 22 R50 is reasonable.

Paper-and-Pencil Method

$$
\begin{array}{r}
2\;2\ \text{R50} \\
125)\overline{2\;8\;0\;0} \\
2\;5\;0\ \downarrow \\
\hline
3\;0\;0 \\
2\;5\;0 \\
\hline
5\;0
\end{array}
$$

B Divide. $14,500 ÷ $325

Calculator Method

Enter 14500 ÷ 325 = 44.615385.
Enter 44 × 325 = 14300.
Enter 14500 − 14300 = 200.

$14,500 ÷ $325 = **44 R200**
Check your answer.

Paper-and-Pencil Method

$$
\begin{array}{r}
4\;4\ \text{R200} \\
325)\overline{1\;4\;5\;0\;0} \\
1\;3\;0\;0\ \downarrow \\
\hline
1\;5\;0\;0 \\
1\;3\;0\;0 \\
\hline
2\;0\;0
\end{array}
$$

▶ Think and Discuss

1. Estimate the quotient of 7572 ÷ 247. Then use a calculator to find the exact quotient. Compare your estimate with those of your classmates.

2. Refer to the introduction to this lesson. How long would it take Lonnie to save $2800?

3. Solve this mystery. The divisor was 789. The quotient was 80 R91. What was the dividend?

4. Sara used a calculator to find 37,920 ÷ 632. Her answer was 600. What might she have done wrong?

Exercises

Divide. (See Example A.)

5. 1411 ÷ 234   6. 4003 ÷ 511   7. 7316 ÷ 491   8. 3113 ÷ 369

9. 2789 ÷ 405   10. 5250 ÷ 735   11. 6090 ÷ 801   12. 1999 ÷ 101

Divide. (See Example B.)

13. 419,377 ÷ 572   14. 516,168 ÷ 856   15. 42,905 ÷ 285   16. 37,737 ÷ 212

17. 579,811 ÷ 666   18. 79,008 ÷ 374   19. 831,111 ÷ 909   20. 207,702 ÷ 431

▶ Mixed Practice (For more practice, see page 410.)

Divide.

21. 2970 ÷ 495   22. 121,112 ÷ 345   23. 400,707 ÷ 289   24. 4300 ÷ 826

25. 83,113 ÷ 491   26. 9190 ÷ 609   27. 79,097 ÷ 336   28. 5725 ÷ 225

▶ Applications

29. A rock group sold 137,528 tickets on their tour. If they held 41 concerts, estimate the number of tickets sold for each concert.

30. Fifty-six thousand three hundred twenty-four dollars was donated to South High to buy computers. A computer costs four hundred thirty-nine dollars. How many computers can the students buy? How much money will they have left over?

▶ Review (Lesson 2.3)

Multiply mentally.

31. 600 × 90   32. 700 × 1000   33. 5000 × 50   34. 100 × 40

3.7 4-Step Problem-Solving Process: Do and Check

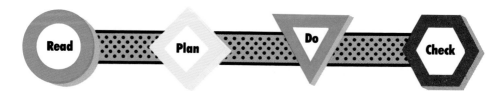

As you learned in Lessons 1.4 and 2.5, the first steps in successful problem solving are **Read** and **Plan**. You are now ready for the last two steps, **Do** and **Check**.

- ## Step 3: Do

In the **Do** step, you carry out the plan that you decided on in Step 2. Before you begin to compute, answer the following:

A Does your plan seem reasonable?

    Yes Write a mathematical sentence for solving the problem. Go to **B**.

    No Rethink your plan. Repeat **A**.

B Are the numbers in your mathematical sentence correct?

    Yes Compute.

    No Write the correct numbers. Then compute.

- ## Step 4: Check

In the **Check** step, determine if your answer is correct. If your answer is not correct, ask the following questions:

Questions	Hints
A **Did you check your answer correctly?** · Have you used the correct numbers? · Does your answer seem reasonable? · Have you computed correctly?	· Check that you have copied the numbers correctly. · Estimate the solution to the problem.
B **Did you follow your plan?** · Did you calculate the necessary information? · Did you follow the steps? · Is your answer in the correct form?	· Check your work against your plan. · Remember what you are computing: time, money, miles, etc.

▶ Think and Discuss

Use Read, Plan, Do, and Check to solve the problem below.

1. Cecil earns $80.00 per week. After payroll deductions, his take-home pay is $64.00. He saves half of his take-home pay and keeps the rest for expenses. His weekly expenses are $8.00 for bus fare and $12.00 for lunches. How much spending money does Cecil have left each week?

2. Three students solved the above problem. They were surprised when they realized that their answers were different. Their work is shown below. Study their plans. What errors did each student make?

Student A	Student B	Student C
Plan:	**Plan:**	**Plan:**
bus fare + lunch = expenses	take-home pay ÷ 2 = savings	take-home pay ÷ 2 = savings
take-home pay − expenses = savings and spending money	bus fare + lunches = expenses	bus fare + lunches = expenses
savings and spending money ÷ 2 = spending money	savings − expenses = spending money	savings − expenses = spending money
Computation:	**Computation:**	**Computation:**
8 + 12 = 20	80 ÷ 2 = 40	64 ÷ 2 = 32
64 − 20 = 44	8 + 12 = 20	8 + 12 = 16
44 ÷ 2 = 22	40 − 20 = 20	32 − 16 = 16
Solution: $22.00	Solution: $20.00	Solution: $16.00

Read the following problem. Write a Plan. Do the problem. Check your work.

3. Eli can carry 40 pounds. Mr. Smith hired him to carry some bricks and building equipment up from his cellar. One piece of equipment weighs 40 pounds and three other pieces each weigh 13 pounds. There are 100 2-pound bricks. Mr. Smith has offered Eli two rates: either a $40.00 flat rate or else $5.00 per trip, provided that Eli carries all he can each trip. At which rate would Eli make more money?

3.8 Choosing the Operation

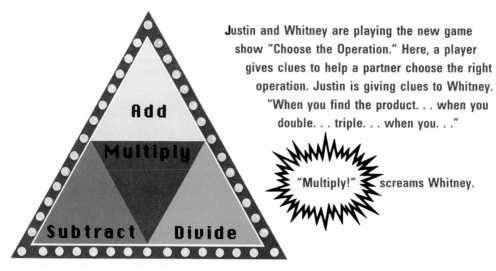

Justin and Whitney are playing the new game show "Choose the Operation." Here, a player gives clues to help a partner choose the right operation. Justin is giving clues to Whitney. "When you find the product. . . when you double. . . triple. . . when you. . ."

"Multiply!" screams Whitney.

1. Is Whitney correct?

   Continue playing the game with Justin and Whitney. "When you want to find out how much money you'll have left after you spend some. . . or how many eggs will remain if you take 2 out of a dozen. . . sometimes called finding the difference."

2. What is the correct response?

   Continue playing. "How to find the cost of one item if you know the cost of a dozen. . . or if you separate a crowd into small groups. . . sometimes called finding the quotient."

3. What is the correct response?

   The last operation is addition.

4. What clues would you give to describe addition?

   Justin and Whitney advanced to the next game: "Choose It and Use It." Whitney showed Justin 10 piles of baseball cards. Each pile contained a dozen cards. Whitney asked the questions. Help Justin by answering the questions.

5. How many piles of cards does Whitney have?

6. How many cards are in each pile?

7. What operation would you use to find how many cards are in all the piles?

8. How many cards are there in all?

Then Justin and Whitney played another game. This time Justin asked the questions. "I have 124 albums and 18 cassettes. Which operation would you use to find how many albums and cassettes I have in all?"

"Put them together. . . join. . . Add!" Whitney said. "So, Whitney, how many albums and cassettes do I have in all?"

"Add albums and cassettes. . . 124 plus 18 is 142." exclaimed Whitney.

9. Be the judge for the game. Did Whitney use the correct operation? Was Whitney's answer correct?

Work with 2 or more classmates to answer the following questions. Take turns asking the questions, answering the questions, and being the judge.

10. There are 8 granola bars in every box. Which operation would you use to find how many bars there are in 7 boxes? How many bars are there?

11. There are 1700 students enrolled in school. Which operation would you use to find how many girls are enrolled if there are 800 boys? How many girls are enrolled?

12. Each bus holds 50 passengers. Which operation would you use to find how many buses will be needed to transport 1500 students to school? How many buses will be needed?

13. Which operation would you use to find how many quarters there are in $6? How many quarters are in $6?

14. Which operation would you use to find how much money Tyrone spent if he bought a poster for $7, a shirt for $15, socks for $3, and a paperback for $4? How much did he spend in all?

15. Which operation would you use to find how much change you should receive if you give the cashier a $20 bill for a $13.50 purchase? How much change should you receive?

▶ **Review** (Lesson 2.11)

Solve. Did you use mental math, paper-and-pencil, or a calculator?

16. Find the cost of 6 greeting cards at $0.50 each.

17. What would you be paid for working 4.5 hours at a rate of $4.80 per hour?

3.9 Using the Order of Operations

The calendar shows the number of hours Jill worked her first week at her summer job. To compute the total number of hours for the week, she punched $3 + 7 \times 4 =$ into her calculator. To her amazement, Jill's calculator displayed 40. How many hours did Jill really work?

We need rules for working with expressions like $3 + 7 \times 4$. Otherwise, 2 people might read the same expression in 2 very different ways. The rules are called the *order of operations*.

Examples

To use the order of operations correctly, follow the steps below.

> 1. Do all operations within parentheses.
> 2. Do all multiplications and divisions from left to right.
> 3. Do all additions and subtractions from left to right.

A Simplify. $3 + 7 \times 4$

$$3 + 7 \times 4 = \qquad \text{Multiply.}$$
$$3 + \quad 28 \quad = \qquad \text{Add.}$$
$$31$$
$$3 + 7 \times 4 = \mathbf{31}$$

B Simplify. $(2 + 7) \times 4 - 6$

$$(2 + 7) \times 4 - 6 = \qquad \text{Work within}$$
$$\text{parentheses.}$$
$$9 \quad \times 4 - 6 = \qquad \text{Multiply.}$$
$$36 \quad - 6 = \qquad \text{Subtract.}$$
$$30$$
$$(2 + 7) \times 4 - 6 = \mathbf{30}$$

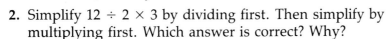

▶ Think and Discuss

1. Refer to the introduction to this lesson. How many hours did Jill really work?

2. Simplify 12 ÷ 2 × 3 by dividing first. Then simplify by multiplying first. Which answer is correct? Why?

3. Rewrite Example B without parentheses. Simplify and compare the two solutions.

4. One way to remember the rules for order of operations is as a set of symbols: () × ÷ + −. Explain what the sequence of symbols represents.

Exercises

Simplify. (See Example A.)

5. $2 \times 4 + 8$
6. $8 - 4 \div 2 - 2$
7. $8 \div 4 - 2 \div 2$

8. $12 + 6 \div 3$
9. $12 \div 6 + 3$
10. $1 + 2 \times 3 + 4$

Simplify. (See Example B.)

11. $(8 - 4) \div 2$
12. $7 + (3 \times 4) - 2$
13. $(7 + 3) \times (4 - 2)$

14. $(8 - 3) \div (2 + 3)$
15. $(21 + 14) \div 7$
16. $(35 \div 5) - (2 \times 3)$

▶ Mixed Practice (For more practice, see page 411.)

Simplify.

17. $7 - 2 \times 3$
18. $(7 - 2) \times 3$
19. $(8 \div 2) + 3$

20. $8 \div 2 + 3$
21. $(1 + 2) \times 3 + 4$
22. $(1 + 2) \times (3 + 4)$

23. $1 \times 2 + 3 \times 4$
24. $6 \div 6 \times 6 - 6 + 6$
25. $(6 + 6) \times (6 \div 6)$

▶ Applications

Solve. Add parentheses and operations signs to the numbers that follow each problem.

26. Andrea eats 3 bagels each Sunday and 2 bagels every other day of the week. How many does she eat in 52 weeks?
    52   3   2   6

27. Felicia bought six postcards at 15¢ apiece and ten 14¢ stamps. How much did she spend in all?
    6   15   10   14

▶ Review (Lessons 3.2, 3.3, 3.5)

Divide.

28. $695 \div 5$
29. $480 \div 40$
30. $8976 \div 25$
31. $9257 \div 3$

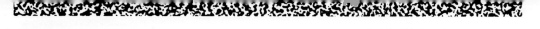

3.10 Dividing Decimals by Whole Numbers

Vicky & Rosa Go Fishing

"The Catch"

$ 19. 56 !

Vickie and Rosa received $19.56 for the fish they caught. They decided to split the profit equally.

To determine the amount of money each girl would receive, divide $19.56 by 2.

Examples

To divide a decimal by a whole number, place the decimal point in the quotient directly above the dividend's decimal point. Then divide as with whole numbers.

A Divide. 19.56 ÷ 2

```
      0 9.7 8
  2) 1 9.5 6
     1 8
     ─────
       1 5
       1 4
     ─────
         1 6
         1 6
     ─────
           0
```

2 does not divide 1. Write a zero in the quotient. Divide as with whole numbers.

19.56 ÷ 2 = **9.78**

B Divide. 1.08 ÷ 12

```
      0.0 9
  12) 1.0 8
      1 0 8
     ──────
          0
```

12 does not divide 1 or 10. Write zeros in the quotient. Divide as with whole numbers.

1.08 ÷ 12 = **0.09**

▶ Think and Discuss

1. Refer to the introduction to this lesson. How much money did Vickie earn?

2. Divide. $27.5 \div 11$

3. If a friend of Vicki and Rosa had helped them catch the fish and they had split the money three ways, how much money would each person have received?

4. Ramon claims that $\$638.35 \div 17 = \$37.55$. What are two ways to check Ramon's work? Which check is easier to perform?

Exercises

Divide. (See Example A.)

5. $4\overline{)39.2}$
6. $3\overline{)38.4}$
7. $9\overline{)97.2}$
8. $4\overline{)5.2}$

9. $3.96 \div 6$
10. $0.42 \div 7$
11. $0.432 \div 9$
12. $0.234 \div 3$

Divide. (See Example B.)

13. $38\overline{)5.51}$
14. $17\overline{)1.19}$
15. $88\overline{)0.792}$
16. $45\overline{)0.585}$

17. $7.90 \div 10$
18. $9.45 \div 63$
19. $0.72 \div 12$
20. $0.54 \div 18$

▶ Mixed Practice (For more practice, see page 411.)

Divide.

21. $8.68 \div 7$
22. $11\overline{)38.5}$
23. $5\overline{)0.745}$
24. $4.68 \div 12$

25. $22\overline{)18.48}$
26. $7\overline{)0.161}$
27. $0.144 \div 4$
28. $3.64 \div 52$

▶ Applications

29. Ken, Jay, and Juan hiked forty-three and six-tenths miles in four days, covering the same distance each day. How many miles did they hike per day?

30. Hannah and her two sisters bought their father a set of lures at the sale price shown. If each girl paid the same amount, how much did each girl spend?

31. Otis received a paycheck for $\$56.25$ one week. If he worked 15 hours, how much did he earn per hour?

Was $39.95

Now $34.95

▶ Review (Lesson 1.7)

Round to the nearest thousand.

32. 11,852
33. 65,395
34. 79,514
35. 492,186
36. 287,651

3.11 Rounding Quotients

Peter was selected by the 7 other members of the tennis team to buy a gift for their coach. He found a gift that cost $85.96. "So how much do we owe you?" asked Todd, one of the team members. After quickly dividing, Peter announced that each person owed $10.745. "You've got to be kidding!" responded Todd. "How are we supposed to pay half a cent?"

8)85.96

Perhaps Peter should have rounded $10.745. What would you have charged each person?

Examples

To round quotients, first divide. Then round.

A Divide. Round the quotient to the nearest tenth.   136.1 ÷ 56

```
        2.4 3
   56)1 3 6.1 0
      1 1 2
      ─────
        2 4 1
        2 2 4
        ─────
          1 7 0
          1 6 8
```
Since you are rounding to the nearest tenth, divide to the hundredths place. Then round 2.43 to 2.4.

136.1 ÷ 56 is about **2.4**.

B Divide. Round the quotient to the nearest cent. $85.96 ÷ 8

```
       1 0.7 4 5
   8)8 5.9 6 0
     8
     ─────
     0 5 9
       5 6
       ───
         3 6
         3 2
         ───
           4 0
           4 0
```
Divide to the thousandths place. Round to the hundredths place.

$85.96 ÷ 8 is about **$10.75**.

▶ Think and Discuss

1. Divide. Round the quotient to the nearest tenth.   86.4 ÷ 7
2. Divide. Round the quotient to the nearest cent. $58.95 ÷ 6
3. Is 156.98 closer to 156.9 or 157.0?
4. Round 289.95 to the nearest tenth.
5. Refer to the introduction to this lesson. If the 8 team members each contributed $10.75, how much money was collected?
6. Name a situation when rounding quotients might be useful.

Exercises

Divide. Round the quotient to the nearest tenth. (See Example A.)

7. $8\overline{)107.6}$ **8.** $19\overline{)5.63}$ **9.** $65\overline{)517}$ **10.** $9\overline{)47.35}$

11. $21\overline{)6.32}$ **12.** $7\overline{)10}$ **13.** $28\overline{)50.1}$ **14.** $19\overline{)4.87}$

15. $42\overline{)93.7}$ **16.** $61\overline{)102.8}$ **17.** $23.5 \div 41$ **18.** $1.6 \div 3$

19. $98 \div 52$ **20.** $485 \div 7$ **21.** $41\overline{)52}$ **22.** $38\overline{)98}$

Divide. Round the quotient to the nearest cent. (See Example B.)

23. $9\overline{)\$126.43}$ **24.** $24\overline{)\$2575}$ **25.** $5\overline{)\$229.49}$ **26.** $18\overline{)\$500}$

27. $2.6\overline{)\$58}$ **28.** $7\overline{)\$47.52}$ **29.** $55\overline{)\$189}$ **30.** $35\overline{)\$637.95}$

31. $43\overline{)\$92.25}$ **32.** $27\overline{)\$100}$ **33.** $68\overline{)\$215.52}$ **34.** $99\overline{)\$870.50}$

35. $\$1.98 \div 8$ **36.** $\$2.70 \div 16$ **37.** $\$383.50 \div 4$ **38.** $\$84.50 \div 9$

▶ Mixed Practice (For more practice, see page 412.)

Divide. Round the quotient to the nearest tenth or cent.

39. $\$22 \div 3$ **40.** $6\overline{)38}$ **41.** $4\overline{)\$63.50}$ **42.** $\$456.18 \div 5$

43. $11\overline{)381}$ **44.** $8\overline{)6.47}$ **45.** $3\overline{)\$49.49}$ **46.** $6.042 \div 7$

47. $87\overline{)11.53}$ **48.** $29\overline{)608}$ **49.** $56\overline{)42.2}$ **50.** $\$6.45 \div 8$

51. $294\overline{)205.7}$ **52.** $36\overline{)\$1200}$ **53.** $50\overline{)5.381}$ **54.** $\$405.75 \div 10$

▶ Applications

55. Monique saw an automatic-focus camera in the catalog shown here. She plans to save the same amount of money each month for six months. How much should she save each month?

$119.99

56. Suppose you wanted to buy the camera and a $40 lens kit. If you can save $20 each week, how many months will it take you to save the money for both items?

57. Aram wants to buy a stereo that costs $539.98. If he can pay for it in 8 equal payments, estimate how much he will owe after 2 payments.

▶ Review (Lessons 2.2, 2.3, 2.4)

Multiply.

58. $\begin{array}{r} 8995 \\ \times\ \ \ 9 \\ \hline \end{array}$ **59.** $\begin{array}{r} 22 \\ \times\ 47 \\ \hline \end{array}$ **60.** $\begin{array}{r} 907 \\ \times\ 206 \\ \hline \end{array}$ **61.** $\begin{array}{r} 274 \\ \times\ 10 \\ \hline \end{array}$ **62.** $\begin{array}{r} 15{,}012 \\ \times\ \ \ \ \ \ 5 \\ \hline \end{array}$

3.12 Dividing Decimals by Decimals

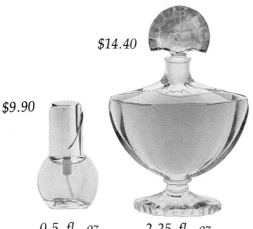

$14.40

$9.90

0.5 fl. oz.   2.25 fl. oz.

Richard wanted to buy his girlfriend some perfume. The salesperson suggested the bottles shown here.

How much does an ounce of perfume cost? You can find cost per ounce by dividing the total price by number of ounces.

Examples

To divide decimals, move the decimal point in the divisor to make a whole number. Then move the decimal point in the dividend the same number of places. Write zeros as needed. Write a decimal point in the quotient and divide.

A Divide.   9.90 ÷ 0.5

```
        1 9.8
0.5)9.9 0
    5
    4 9
    4 5
      4 0
      4 0
        0
```
Move each decimal point one place. Then place the decimal point in the quotient.

9.90 ÷ 0.5 = **19.8**

B Divide. Round to the nearest tenth.   17.1 ÷ 1.75

```
              9.7 7
1.7 5)1 7.1 0 0 0
      1 5 7 5
      1 3 5 0
      1 2 2 5
        1 2 5 0
        1 2 2 5
```
Move each decimal point two places. Write zeros. Then place the decimal point in the quotient.

17.1 ÷ 1.75 ≈ **9.8**

▶ Think and Discuss

1. Divide. 0.588 ÷ 2.1

2. Refer to the introduction to this lesson. What did each perfume cost per fluid ounce?

3. Write a division problem in which you need to move the decimal point 2 places.

4. In Example B, why was it necessary to write 3 zeros?

Exercises

Divide. (See Example A.)

5. $0.7\overline{)4.06}$ 6. $0.9\overline{)15.3}$ 7. $1.5\overline{)4.95}$ 8. $3.4\overline{)8.84}$

9. $11.6 \div 0.4$ 10. $31.2 \div 0.6$ 11. $0.96 \div 0.2$ 12. $2.58 \div 4.3$

Divide. Round to the nearest tenth. (See Example B.)

13. $1.43\overline{)0.322}$ 14. $2.87\overline{)4.79}$ 15. $1.32\overline{)11.99}$ 16. $2.16\overline{)0.1732}$

17. $1.283 \div 8.5$ 18. $8.7 \div 7.29$ 19. $8.643 \div 5.08$ 20. $10.376 \div 4.51$

▶ **Mixed Practice** (For more practice, see page 412.)

Divide. Round to the nearest tenth, if needed.

21. $1.134 \div 0.18$ 22. $0.3\overline{)2.089}$ 23. $5.12 \div 0.32$ 24. $9.2\overline{)29.47}$

25. $4.83 \div 0.23$ 26. $1.98\overline{)5.94}$ 27. $7.2\overline{)25.92}$ 28. $7.06\overline{)5.718}$

29. $12.7\overline{)80.01}$ 30. $9.6\overline{)5.27}$ 31. $70.1 \div 8.5$ 32. $0.45\overline{)3.555}$

▶ **Applications**

33. Kevin wants to know if all of his albums will fit into the record stand he bought. Ten albums are approximately two and eight tenths centimeters wide. Estimate the number of albums that fit in the record stand.

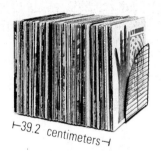

├39.2 centimeters┤

34. A tin of popcorn that weighs 9.375 pounds costs $23.99. A 5.5-pound tin of caramel corn costs $14.49. Which tin is cheaper per pound?

35. Mrs. Russell has a stack of homework papers on her desk 1.308 inches thick. If each sheet of paper is 0.003 inches thick, how many pages of homework does she have to grade?

▶ **Review** (Lesson 3.3)

Divide.

36. $6921 \div 100$ 37. $57.89 \div 10$ 38. $800 \div 20$ 39. $600 \div 30$

40. $192.4 \div 100$ 41. $63,000 \div 70$ 42. $3000 \div 60$ 43. $3.7 \div 100$

3.13 Dividing Whole Numbers by Decimals

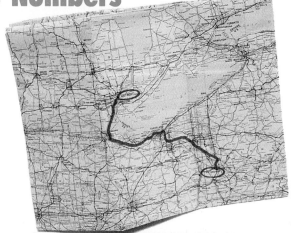

Imagine a car trip covering 304 miles and using 9.5 gallons of gas. To find the number of miles per gallon, you would divide 304 miles by 9.5 gallons.

Examples

To divide whole numbers by decimals, remember that the whole number has a decimal point to the right of the ones place. Then use the method taught in Lesson 3.12.

A Divide.   304 ÷ 9.5

```
        3 2.
  9.5)3 0 4.0
      2 8 5
        1 9 0
        1 9 0
            0
```
Move the decimal points one place. Write one zero.

304 ÷ 9.5 = **32**

B Divide. Round to the nearest tenth.   923 ÷ 3.06

```
            3 0 1.6 3
  3.0 6)9 2 3.0 0 0 0
        9 1 8
            5 0 0
            3 0 6
          1 9 4 0
          1 8 3 6
            1 0 4 0
              9 1 8
```
Move the decimal points two places. Place the decimal point in the quotient.

923 ÷ 3.06 ≈ **301.6**

▶ Think and Discuss

1. Refer to the introduction to this lesson. How many miles per gallon did the car get on the trip?

2. In Example A, how many places did you move the decimal point in the divisor? In the dividend? How many zeros did you write after the 4?

3. On another trip you travel 308 miles on 8.8 gallons of gas. Estimate the number of miles per gallon.

4. What do the division problems $0.008\overline{)10}$, $0.08\overline{)100}$, and $0.8\overline{)1000}$ have in common? Solve and see.

5. Why is it a good idea to estimate an answer when you begin a division problem? If you are using a calculator, is it still a good idea? Discuss.

Exercises

Divide. (See Example A.)

6. $0.4\overline{)6}$ 7. $0.6\overline{)9}$ 8. $2.5\overline{)16}$ 9. $5.6\overline{)14}$

10. $4.5\overline{)18}$ 11. $0.7\overline{)280}$ 12. $3.5\overline{)182}$ 13. $0.5\overline{)475}$

14. $3 \div 0.6$ 15. $75 \div 1.2$ 16. $221 \div 3.4$ 17. $16 \div 6.4$

Divide. Round to the nearest tenth. (See Example B.)

18. $0.06\overline{)20}$ 19. $0.15\overline{)7}$ 20. $4.25\overline{)69}$ 21. $3.82\overline{)765}$

22. $6.09\overline{)583}$ 23. $0.95\overline{)813}$ 24. $7.13\overline{)607}$ 25. $5.42\overline{)738}$

26. $5065 \div 2.11$ 27. $13 \div 0.75$ 28. $7470 \div 6.22$ 29. $1690 \div 8.43$

▶ Mixed Practice (For more practice, see page 413.)

Divide. Round to the nearest tenth, if needed.

30. $21 \div 0.5$ 31. $0.07\overline{)16}$ 32. $44 \div 2.5$ 33. $0.45\overline{)55}$

34. $0.7\overline{)41}$ 35. $65 \div 2.58$ 36. $23.2\overline{)187}$ 37. $998 \div 0.16$

38. $872 \div 1.35$ 39. $7.6\overline{)702}$ 40. $3.6\overline{)5}$ 41. $6.3\overline{)957}$

42. $0.92\overline{)1020}$ 43. $527 \div 0.31$ 44. $1018 \div 12.6$ 45. $7.6\overline{)689}$

46. $0.48\overline{)3456}$ 47. $0.2\overline{)21,828}$ 48. $9.27\overline{)33,380}$ 49. $1188 \div 9.85$

▶ Applications

50. A salesperson drove a rental car 725.2 miles and used 19.6 gallons of gasoline. What was his mileage per gallon?

51. Lockers are being installed along a wall that is 40 feet long. If each locker is 1.25 feet wide, how many lockers will fit?

▶ Review (Lesson 3.9)

Simplify.

52. $9 + 6 \times 5 - 2$ 53. $(8 \times 8) + (49 \div 7)$ 54. $28 \div 7 \times 6$

55. $25 \div 5 + 3 + 4 \times 5$ 56. $18 \div 3 \div 3$ 57. $8 - 36 \div 6 + 9$

Chapter 3 Review

If the answer is true, write T. If the answer is false, write F and explain why the answer is incorrect. (Lesson 3.1)

1. The number inside the division bracket is called the divisor.

2. The answer to a division problem is called the product.

Divide. (Lessons 3.2, 3.3)

3. $6\overline{)96}$	**4.** $703 \div 3$	**5.** $40\overline{)160}$	**6.** $10\overline{)960}$
7. $9000 \div 100$	**8.** $872 \div 5$	**9.** $8\overline{)93}$	**10.** $50\overline{)3500}$

Estimate. (Lesson 3.4)

11. $79 \div 17$	**12.** $31\overline{)963}$	**13.** $11\overline{)456}$	**14.** $52\overline{)1004}$
15. $502\overline{)2498}$	**16.** $1222 \div 38$	**17.** $73\overline{)492}$	**18.** $6587 \div 99$

Divide. (Lessons 3.5, 3.6)

19. $49\overline{)653}$	**20.** $491 \div 12$	**21.** $98\overline{)7625}$	**22.** $109\overline{)986}$
23. $533\overline{)872}$	**24.** $37\overline{)560}$	**25.** $387 \div 32$	**26.** $254\overline{)3670}$

Write which operation you would use to solve each problem. Then solve. (Lesson 3.8)

27. Pens come in packages of 18. How many packages can be made from 1158 pens?

28. Jolene spends an average of 45 minutes a night studying. About how many minutes does she spend each week studying?

29. An earthquake took place in China in 1556. In 1964 an earthquake took place in Alaska. How many years separated these events?

30. For assembly programs, all 962 students must sit in the 26 equal rows of chairs in the auditorium. How many chairs are there in each row?

Simplify. (Lesson 3.9)

31. $10 + 7 \times 7 - 8$	**32.** $39 - 90 \div 10$	**33.** $7 \times 8 + 9 \times 9$

Divide. Round the quotient to the nearest tenth or cent, if needed. (Lessons 3.10, 3.11, 3.12, 3.13)

34. $6\overline{)\$5.94}$	**35.** $24.80 \div 40$	**36.** $0.3\overline{)78}$	**37.** $0.50\overline{)0.625}$

Chapter 3 Test

Divide.

1. $60\overline{)960}$ 2. $19\overline{)494}$ 3. $14.7 \div 0.07$

4. $0.025\overline{)0.975}$ 5. $1000\overline{)5680}$ 6. $81\overline{)737.1}$ 7. $7500 \div 50$

8. $113\overline{)904}$ 9. $400\overline{)84,000}$ 10. $10.71 \div 6.3$ 11. $1.43\overline{)7.15}$

Complete each statement.

12. Order of operations tells you to ___ and ___ before you add or subtract.

13. The answer to a division problem is called the ___ .

14. When you divide a number by 100, you move the decimal point ___ places to the ___ .

Estimate.

15. $33\overline{)271}$ 16. $604 \div 99$ 17. $75\overline{)638}$ 18. $5\overline{)786}$

19. $408 \div 52$ 20. $68\overline{)489}$ 21. $118 \div 17$ 22. $9\overline{)987}$

Simplify.

23. $3 \times 9 + 8 \times 4$ 24. $16 - 72 \div 9 + 5$ 25. $45 + 6 \times 6 \div 2$

26. $18 \div 2 - 5 \times 1$ 27. $14 + 49 \div 7 - 7$ 28. $100 \div 5 \times 4 + 7$

Divide. Round the quotient to the nearest tenth or cent.

29. $7\overline{)\$51.49}$ 30. $30\overline{)90.6}$ 31. $\$984.73 \div 9$ 32. $4\overline{)73.93}$

Write which operation you would use to solve each problem. Then solve.

33. Kim rode 68 miles on his bike. Shishin rode eight times as far. How far did Shishin ride?

34. In 1876 Alexander Graham Bell patented the telephone. The first American landed on the moon 93 years later. In what year did that event occur?

"These days it really helps to know about the metric system. I went grocery shopping the other day and I couldn't believe the number of items that had metric measurements on them. But once I learned the system, I realized it was easy."

Metric Measurement

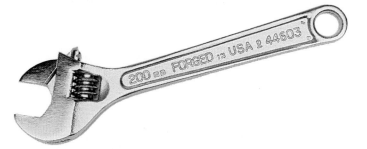

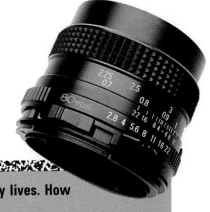

Metric measurements are very much a part of our daily lives. How many of the following examples have you noticed?
- milk, juice, and soft drink cartons measured in liters and milliliters
- track and road races marked off in meters and kilometers
- wrench sizes indicated in centimeters

Where else have you seen metric measurements?

Roadrunner Club

5-km *run today*

When Are Metric Measures Used?

Materials: newspaper or catalog

Work with 2 or 3 classmates. Look through your newspaper or catalog, and write a list of items that use metric measurements. Find as many objects and measurements as you can in 10 minutes. Compare your list with the lists of other groups.

Which type of items using metric measurements did you find most often? Why do you think this type of item is measured in metric units? In doing this activity, did you learn of items measured in metric units of which you were not aware?

4.1 Introduction to the Metric System

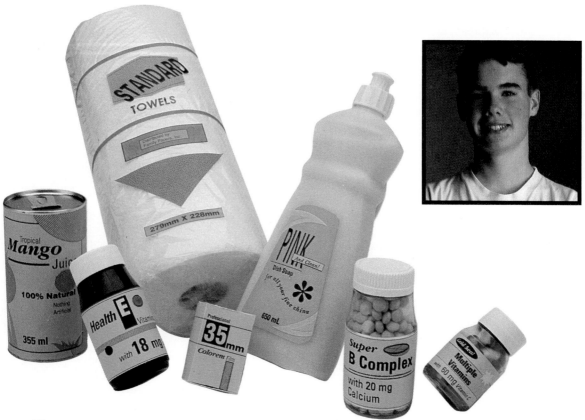

Todd works in a grocery store. One of his jobs is stocking shelves. When he first started working, Todd was surprised to find that many items in the store are labeled using metric units.

1. List the abbreviations shown above that represent metric units of measurement.
2. Determine whether each item shown above would be measured according to its length, mass, or liquid volume.

▶ Express Yourself

You are already familiar with many metric units of measurement. The basic units of the metric system are:

meters   used to measure length or distance

grams   used to measure mass

liters   used to measure liquid volume, or capacity

The prefixes below are used in the metric system.

kilo-	1000.	(times one thousand)
centi-	0.01	(times one hundredth)
milli-	0.001	(times one thousandth)

3. The basic unit of each of the following terms is underlined. Tell what each term means.

kilo*byte*　　　　milli*gram*　　　　milli*volt*　　　　centi*meter*

4. Look up the following terms in a dictionary: *decimal, decibel, decimate, decile*. What common meaning do these have?

▶ **Practice What You Know**

The metric system is organized by powers of ten. U.S. currency is another system that is organized by powers of ten.

100　　　　　　　　10　　　　　　　　1　　　　　　　0.10　0.01

If the dollar is the basic unit, what is a cent?

1¢ = $0.01 = one hundredth of a dollar

5. What would be a more reasonable way to say, "I earn 2500 cents a week?"

6. Suppose you want to exchange 2 one-dollar bills for dimes. You are changing a larger unit (dollars) to a smaller unit (dimes). How many dimes will you receive? Do you multiply or divide to get your answer?

7. Suppose you want to exchange 1800 pennies for dollar bills. You are changing a smaller unit (pennies) to a larger unit (dollars). How many dollars will you receive? Do you multiply or divide? By what number?

Remember that when you multiply by powers of ten, you move the decimal point to the right. When you divide by powers of ten, you move the decimal point to the left.

Multiply or divide.

8. 3×1000　　　**9.** $120 \div 100$　　　**10.** $650 \div 1000$　　　**11.** 5.9×1000

12. $9.7 \div 100$　　　**13.** $0.83 \div 1000$　　　**14.** 40.2×1000　　　**15.** $0.072 \div 100$

4.2 Converting Metric Units of Length

Jolene and Amy both claimed to have the longest fingernails in their class. To settle the dispute, they decided to measure the nail on their right index fingers. Jolene's nail was 3 centimeters long and Amy's nail was 26 millimeters long.

Who wins? To find out, you must convert metric units.

Examples

To convert metric units, first determine if you are converting to a larger or smaller unit.

× 1000　　× 100　　　× 10

kilometer (km)　meter (m)　centimeter (cm)　millimeter (mm)

÷ 1000　　÷ 100　　　÷ 10

A Convert 26 mm to cm.	**B Convert 4.7 km to m.**	**C Convert 8 km to cm.**
Divide since you are changing to a larger unit. 10 mm = 1 cm, so 26 mm = **2.6 cm**.	Multiply since you are changing to a smaller unit. 1 km = 1000 m, so 4.7 km = **4700 m**.	First convert to meters. Then convert to centimeters. 8 km = 8000 m, and 8000 m = 800000 cm, so 8 km = **800,000 cm**.

▶ Think and Discuss

1. Convert 63 centimeters to kilometers.

2. Convert 6 kilometers to centimeters.

3. Describe how to multiply and divide by powers of 10.

4. How do you convert measurements from millimeters to kilometers?

5. Refer to the introduction to this lesson. Did Jolene or Amy have the longer nail?

6. When converting metric units, how do you decide whether to multiply or divide?

Exercises

Convert each measure. (See Example A.)

7. 53 mm to cm

8. 77 mm to cm

9. 134 cm to m

10. 280 cm to m

11. 6384 m to km

12. 915 m to km

Convert each measure. (See Example B.)

13. 2.9 km to m

14. 6.1 km to m

15. 42 m to cm

16. 0.15 m to cm

17. 8.4 cm to mm

18. 27 cm to mm

Convert each measure. (See Example C.)

19. 9 mm to m

20. 431 mm to m

21. 5364 cm to km

22. 0.8 km to cm

23. 0.0052 m to mm

24. 12 km to mm

▶ Mixed Practice (For more practice, see page 413.)

Convert each measure.

25. 3100 m to km

26. 3.6 cm to mm

27. 5.8 m to cm

28. 3470 mm to m

29. 1.9 m to mm

30. 115 mm to cm

31. 3.2 km to m

32. 0.73 km to cm

33. 629 cm to km

▶ Applications

34. List the trails shown in the diagram at the right in order from shortest to longest.

35. The Blue Heron Trail is how many meters longer than the Mirror Lake Trail?

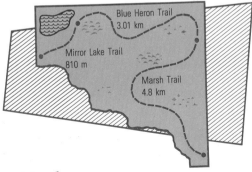

Blue Heron Trail 3.01 km

Mirror Lake Trail 810 m

Marsh Trail 4.8 km

▶ Review (Lessons 3.10, 3.11, 3.12)

Divide. Round the quotient to the nearest tenth.

36. $6.8 \div 3.2$

37. $44.731 \div 9$

38. $0.2695 \div 0.05$

39. $684.5 \div 4$

40. $3.91 \div 17$

41. $3.575 \div 6.5$

42. $87.55 \div 25$

43. $0.897 \div 0.3$

4.3 Measuring Length with Metric Units

James was in charge of getting trophies engraved for a sports banquet. When Coach Walker looked at the trophies, he exclaimed, "Let me see the list you gave the engraver. Somebody made a big mistake!"

What was the mistake? Were the measurements reasonable?

Examples

To measure length, line up one end of a measuring stick with one end of an object. Depending on the precision you need, you can round to the nearest millimeter, centimeter, or meter.

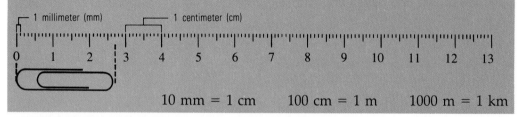

10 mm = 1 cm   100 cm = 1 m   1000 m = 1 km

A Measure the paper clip to the nearest centimeter.

The paper clip is closer to **3 centimeters** than 2 centimeters.

B Choose the more reasonable measure.

A man is 2 ___ tall. m km
The more reasonable measure is **meters**.

Two kilometers would be a distance between towns.

▶ Think and Discuss

1. Measure the length of your notebook to the nearest centimeter.

2. Name objects that can be measured in mm, cm, m, and km.

3. To the nearest cm, a bolt is 8 cm long. 8 cm = ___ mm. The bolt is at least ___ mm long but not as long as ___ mm.

Exercises

Measure the following to the nearest centimeter and the nearest millimeter. (See Example A.)

4. the length of the line segment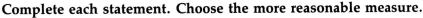

5. the length of your shoe

6. the distance from your wrist to your elbow

Complete each statement. Choose the more reasonable measure. (See Example B.)

7. The length of a public swimming pool is 25 ___.   cm   m

8. The width of your palm is 6 ___.   mm   cm

9. The length of a classroom is 10 ___.   m   km

▶ **Mixed Practice** (For more practice, see page 414.)

Measure the following to the nearest centimeter and the nearest millimeter.

10. the height of the frame at the right

11. the width of the frame at the right

12. the length of a sheet of notebook paper

Complete each statement. Choose the more reasonable measure.

13. A cat's nose is about 2 ___ wide.   mm   cm

14. A marathon is about 40 ___.   m   km

▶ **Applications**

15. Measure the length, width, and height of your desk to the nearest centimeter and the nearest millimeter.

16. Juan's goal is to swim a kilometer every day. If the length of the pool is 25 meters, how many lengths must he swim?

▶ **Review** (Lessons 1.9, 1.11)

Add or subtract.

17.   6.88
      + 3.29

18.   567,564
      + 78,946

19.   0.876
      + 2.45

20.   7
      − 3.459

21.   8007
      − 414

22.   25
      + 5.06

23.   3.004
      − 1.927

24.   97.03
      − 8.174

4.4 Measuring in Science

Suppose you hold a heavy stone and a light stone at eye level and drop them both. Which will hit the ground first? The scientist Galileo discovered that they will both hit the ground at the same time. But just how long would it take for the stones to reach the ground? More than 1 second? Less than 1 second? Estimate.

Mrs. Gravity's physical science class tried an experiment. They wanted to know how high above the ground a marble should be so that it would take exactly 1 second to reach the ground.

1. How might you conduct such an experiment?

The students were surprised by the results. They discovered that the answer is about 5 meters.

2. Look at a meter stick in your classroom. Are you 5 meters tall? Is your school building? How many meters above the ground is a basketball hoop?

3. Is anyone so tall that a marble dropped at eye level would take a full second to reach the ground?

4. Estimate the height, in meters, of the tallest student in the class. How many centimeters is this? How many kilometers?

5. If it takes 1 second for an object to fall 5 meters, estimate how long a drop of 10 meters would take.

The students in Mrs. Gravity's science class wondered about the answer to Question 5. "A drop of 5 meters takes 1 second, so for 10 meters, it's 2 seconds. It's obvious!" stated Rhonda.

Mrs. Gravity didn't agree. "Let's test it out. Let's take a field trip tomorrow. There's a well in my backyard that's about 10 meters deep. Rhonda, bring your stopwatch."

The next day, they dropped a stone down the well and listened for the splash. "Less than 2 seconds? Do it again!" Rhonda protested. They conducted the experiment several times, with both small rocks and large ones. Each time, it took about $1\frac{1}{2}$ seconds.

6. Explain why an object falling 10 meters took *less* than twice the time it takes for the object to fall 5 meters.

A science book stated that from the top of the Empire State Building, a stone would take about 8 seconds to reach the ground. The Empire State Building has a height of about 300 meters.

7. Give the height of the Empire State Building in centimeters and then in kilometers.

The class made a table like the one below. They used the table to look for a pattern.

Height in meters	5	10	300
Time of fall in seconds	1	1.5	8

8. Estimate the height of a cliff if a stone, dropped from the edge, took 4 seconds to reach the bottom.

9. A tree is about 30 meters tall. Estimate how long it would take an acorn to fall from the top to the ground.

10. Complete the table below. Work with 2 or 3 classmates. Discuss your estimation.

Height in meters	5	※	※	80	※	※	※	300
Time of fall in seconds	1	2	3	4	5	6	7	8

Compare your group's answers with those of other groups.

11. A football is kicked about 25 yards straight up into the air. About how long is the ball in the air?

▶ Review (Lesson 2.6)

Estimate.

12. 19×58 13. 72×103 14. 194×83 15. 312×47 16. 986×68

17. $\begin{array}{r} 87 \\ \times\ 63 \end{array}$ 18. $\begin{array}{r} 78 \\ \times\ 62 \end{array}$ 19. $\begin{array}{r} 367 \\ \times\ 431 \end{array}$ 20. $\begin{array}{r} 627 \\ \times\ 832 \end{array}$ 21. $\begin{array}{r} 782 \\ \times\ 918 \end{array}$

4.5 Metric Units of Mass

"**D**id you know," Pat asked Eileen, "that if I moved to Europe I'd weigh only 47 kilograms? Doesn't that sound great?" "No thanks," said Eileen, who wants to gain weight. "I'll stick with pounds." Pat thought for a moment. "Well then, how does 47,000 grams sound?" "Now you're talking," laughed Eileen.

Examples

To convert from one unit of mass to another, first determine if you are converting to a larger or smaller unit.

Equivalents
1 gram = 1000 milligrams
1 kilogram = 1000 grams
1 metric ton = 1000 kilograms

Abbreviations
milligram **mg**   kilogram **kg**
gram **g**   metric ton **T**

A Convert 47,000 grams to kilograms.

Divide since you are changing to a larger unit.

1000 g = 1 kg, so 47,000 g = 47.000 kg or **47 kg**.

B Complete the statement. Choose the more reasonable measure.

A frozen turkey has the mass of about 9 __▒__ . g   kg

A large paper clip has the mass of about 1 gram.
A hammer has the mass of about 1 kilogram.

A turkey is fairly heavy.
The more reasonable measure is **kilograms**.

▶ Think and Discuss

1. Convert 6 kilograms to grams.

2. Convert 96,802 milligrams to grams.

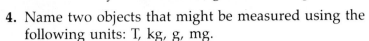

3. How would you convert kilograms to milligrams?

4. Name two objects that might be measured using the following units: T, kg, g, mg.

Exercises

Convert each measure. (See Example A.)

5. 3700 mg to g

6. 4.2 kg to g

7. 900 mg to g

8. 430 g to kg

9. 750 kg to g

10. 0.7 kg to g

11. 8725 g to kg

12. 36 mg to g

13. 45 kg to g

Complete each statement. Choose the more reasonable measure.
(See Example B.)

14. The mass of a canary is about 220 ___.   g   kg

15. The mass of a newborn baby is about 3.5 ___.   g   kg

▶ Mixed Practice (For more practice, see page 414.)

Convert each measure.

16. 8 kg to g

17. 500 g to kg

18. 37.5 kg to g

19. 0.04 mg to g

20. 51 g to mg

21. 9257 mg to g

Complete each statement. Choose the more reasonable measure.

22. The mass of a sewing needle is about 380 ___.   mg   g

23. The mass of a quarterback is about 80 ___.   g   kg

24. The mass of a pair of scissors is about 150 ___.   g   kg

▶ Applications

25. The Great Pyramid is built of 2 million blocks of sandstone. The total mass of the pyramid is 4.6 million metric tons. Estimate the mass of each block.

26. Mr. Guzman brought eight kilograms of peanuts to be divided among his 32 students. How many grams of peanuts did each student get?

▶ Review (Lessons 2.8, 2.10)

Multiply.

27. 8.1
    × 9.4

28. 0.75
    × 5.32

29. 0.004
    × 0.008

30. 327
    × 4.7

4.6 Metric Units of Capacity

The owner's manual for Jean's car states that she needs 4.75 liters of oil for each oil change. At the store she found containers of oil as shown here. How many containers should she buy?

Jean needs to convert metric units before she can determine the number of containers to buy.

Examples

To convert from one metric unit of capacity to another, first determine if you are converting to a larger or smaller unit.

Equivalent
1 liter = 1000 milliliters

Abbreviations
milliliter **mL**   liter **L**

A **Convert 4.75 L to mL.**

Multiply since you are changing to a smaller unit.
1 L = 1000 mL, so 4.75 L = 4750 mL.

B **Complete the statement. Choose the more reasonable measure.**

A sink holds about 20 ___.   L   mL

A large carton of milk could be measured in liters. A spoonful of vanilla or a small glass of juice could be measured in milliliters.

A sink holds a large amount of water, so the more reasonable unit is a **liter**.

▶ Think and Discuss

1. Convert 6.21 liters to milliliters.

2. Convert 0.3 milliliter to liters.

3. Refer to the introduction to this lesson. How many containers of oil does Jean need?

4. A kiloliter (kL) = 1000 L. What might be measured in kiloliters?

5. Which is larger, a 750-milliliter bottle or a 1-liter bottle?

Exercises

Convert each measure. (See Example A.)

6. 10.5 L to mL

7. 43,000 mL to L

8. 0.75 L to mL

9. 200 mL to L

10. 3 L to mL

11. 3750 mL to L

Complete each statement. Choose the more reasonable measure.
(See Example B.)

12. A thimble holds about 5 ___ of liquid.   L   mL

13. A gas tank has a capacity of about 60 ___ .   L   mL

14. A vase holds about 0.5 ___ of water.   L   mL

15. A can of soup contains about 500 ___ of liquid.   L   mL

▶ **Mixed Practice** (For more practice, see page 415.)

Convert each measure.

16. 50 mL to L

17. 45 L to mL

18. 5500 mL to L

19. 9.3 L to mL

20. 483 mL to L

21. 0.8 L to mL

Complete each statement. Choose the more reasonable measure.

22. An eyedropper can hold 10 ___ of fluid.   L   mL

23. A glass contains 250 ___ of milk.   L   mL

24. A large pitcher holds 2 ___ of water.   L   mL

25. A bottle of shampoo holds about 0.535 ___ .   L   mL

▶ **Applications**

26. A can of soup holds 350 milliliters. If Hank adds 3 cans of water to 3 cans of soup, how many liters of soup will he make?

27. Mike made 6 liters of punch for a party. If 82 ladles were served, how much punch was left?

72 mL

▶ **Review** (Lessons 1.5, 1.6)

Order from least to greatest.

28. 53,462   5482   52,642   53,624   5842

29. 4.065   4.605   4.506   4.650   4.056

4.7 Using a Table to Find Information

	Adelaide	Alice Springs	Cairns	Melbourne	Perth	Sydney
	1693					
	2845	2435				
	755	2488	3501			
	2713	3772	4727	3468		
	1422	2960	2853	893	4135	

"you may have already won a trip..."

Before you even open the envelope, you could start mentally planning your trip with the help of a map and a kilometer table.

Use the kilometer table to answer the following questions.

1. How far is it from Alice Springs to Perth? Look down the column from Alice Springs until you come to the row that goes across to Perth.

2. Which city is 2853 kilometers from Sydney? Find 2853 in the row that goes across to Sydney, and see which city is at the head of that column.

3. Which two cities are 755 kilometers apart? Find 755 in the table, and read the cities for that column and row.

For part of your trip, you plan to take trains. Below is a train schedule for four cities. Each train is named after one of the special animals of Australia.

Train	Arrives at Sydney	Arrives at Wollongong	Arrives at Canberra	Arrives at Melbourne
Kangaroo	8:30 a.m.	10:00 a.m.	11:50 a.m.	4:30 p.m.
Wallaby	12:00 noon	1:40 p.m.	3:45 p.m.	8:45 p.m.
Koala	4:00 p.m.	5:30 p.m.	7:15 p.m.	11:30 p.m.

Use the train schedule to answer the following questions.

4. What time does the Wallaby arrive at Canberra?

5. About how long is the trip from Sydney to Wollongong? Is it the same for all three trains?

6. It is 11:00 a.m. in Canberra. How long will it be until the next train to Melbourne leaves?

7. The trip from Canberra to Melbourne is about 500 kilometers. Use the schedule to find how long the trip on the Koala takes. Then estimate how fast the train travels.

For another part of your trip, you plan to rent a car. The table below shows how far you can go, in kilometers, on a single tank of gasoline, depending on the size of the car's gas tank and the car's kilometer-per-liter rating.

Size of Tank	Maximum Distance (per tank of gas)			
	5.9 km/L	7.8 km/L	9.9 km/L	13.7 km/L
32 L	188.8	249.6	316.8	438.4
40 L	236	312	396	548
50 L	295	390	495	685
65 L	383.5	507	643.5	890.5

Use the table above to answer the following questions.

8. What is the greatest distance you can travel on one tank of gasoline if your tank holds 40 liters and your car averages 7.8 kilometers per liter?

9. On one tank, how far could you go in a car with a 32-liter tank that averages 5.9 kilometers per liter? How much farther could you go in a car with a 65-liter tank that averages 13.7 kilometers per liter?

▶ **Review** (Lesson 3.8)

10. Each section of a theater seats 175 people. There are 8 sections. What operation would you use to find how many seats are in the theater? How many seats are there?

4.8 Making a Table to Organize Information

Imagine that your free trip to Australia gives you the opportunity to visit the nearby island of Tasmania at your own expense. Since you only have one week to see the entire island, you decide to rent a jeep. The local Tasmanian newspaper contains the following ads.

9   Classifieds-Car Rentals   TASMANIAN TIMES

ABC Jeep Rental	Landrover Jeep Rental	Deluxe Minijeep
$20 per day	$70 per day	$175 per week
$0.15 per km	No kilometer charge	First 1000 km free then $0.15 per km

You plan to rent the jeep for 7 days and to drive about 3000 kilometers. You can make a table to help figure out the costs.

Company	Rental Cost for 7 days	Km Charges	Total 7-day Cost
ABC	7 × $20 = $140	3000 × $0.15 = $450	$590
Landrover	7 × $70 = ※	no charge	※
Deluxe Minijeep	$175	1000 km free 2000 × $0.15 = ※	※

1. Copy the table above and fill in the missing costs.

2. What would the total cost be with Landrover Jeep Rental? What would the total cost be with Deluxe Minijeep?

3. Which jeep has the lowest cost for 7 days and 3000 kilometers?

4. Suppose you plan to drive 4000 kilometers in the 7 days. Make a table that shows the rental costs, kilometer charges, and costs for the week with each company.

5. Based on your table from Question 4, which company should have the lowest cost for 7 days and 4000 kilometers?

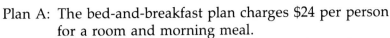

There are three vacation packages for your trip through Tasmania.

Plan A: The bed-and-breakfast plan charges $24 per person for a room and morning meal.

Plan B: The motel-only plan charges $20 per person for a room.

Plan C: The vacation-entertainment plan charges $36 per person for a room, an evening meal, and evening entertainment.

You have budgeted the following amounts to spend when an item is not covered by a package: $6 for breakfasts, $8 for lunches, $12 for dinners, and $8 for each evening's entertainment.

6. Copy the table below, and fill in the missing amounts.

Plan	Room	Breakfast	Lunch	Evening Meal	Entertainment	Total Cost
Plan A	24	(included)	8	12	8	※
Plan B	※	6	※	※	※	※
Plan C	36	※	※	(included)	(included)	※

7. Which plan gives the lowest total cost per day?

8. If you use Plan A for 2 days, Plan B for 3 days, and Plan C for 2 days, what will be your total cost?

9. If you make a reservation early, Plan A charges $20, Plan B charges $16, and Plan C charges $35. Make another table with the costs of rooms, meals, and entertainment using the advance registration costs.

10. Using the table for Question 9, decide which plan gives you the lowest total cost per day.

▶ **Review** (Lesson 2.11)

Solve. Did you use mental math, paper-and-pencil, or a calculator?

11. Find the cost per camera if 12 cameras cost $2880.

12. Divide 5800 by 100.

13. Find the sum 17 + 83 + 60 + 40.

Chapter 4 Review

Complete each statement. (Lesson 4.1)

1. The metric system is based on powers of ▒ .

2. The three basic units in the metric system are ▒ , ▒ , and ▒ .

Convert each measure. (Lesson 4.2)

3. 24 cm to mm

4. 7 m to cm

5. 4.8 km to m

6. 9.3 m to mm

7. 36.3 cm to m

8. 82.5 m to km

9. 0.54 m to cm

10. 1.98 cm to mm

11. 0.36 km to cm

Measure the line segments below to the nearest centimeter and the nearest millimeter. (Lesson 4.3)

12. _____

13. _____

Complete each statement. Choose the more reasonable measure. (Lessons 4.3, 4.5, 4.6)

14. The distance across a small town is about 1 ▒ .   m   km

15. The mass of a button is about 1 ▒ .   g   kg

16. The capacity of a large glass jar is about 1 ▒ .   mL   L

Convert each measure. (Lessons 4.5, 4.6)

17. 39 mL to L

18. 71 kg to g

19. 9.76 g to kg

20. 0.8 kg to g

21. 4.21 mL to L

22. 5 kg to mg

23. 0.41 g to kg

24. 32,840 mg to g

25. 4751 mg to kg

Make a table of the following information. Then use the table to answer the questions below. (Lessons 4.7, 4.8)

The areas in square miles of five island nations are as follows:
Fiji   7056   Australia   2,968,125   New Zealand   103,744
Madagascar   226,674   Japan   143,761.

26. Which country has the greatest area? The smallest?

27. Which two countries are closest in area?

28. Which country is about twice the size of New Zealand?

Chapter 4 Test

Complete each statement. Choose the more reasonable measure.

1. A large milk carton contains about 4 ___.   mL   L

2. The mass of a book is about 1 ___.   g   kg

3. The length of a glove is about 23 ___.   mm   cm

4. A sheet of paper is about 0.1 ___ thick.   mm   m

Make a table of the following information. Then use the table to answer the questions below.

The kangaroo is about 1.8 meters tall and has a mass of about 45 kilograms.
A koala is about 70 centimeters long and has a mass between 7 and 14 kilograms.
A platypus has a mass of about 2.3 kilograms. It is between 40 and 55 centimeters long.

5. Which mammal has the greatest mass?

6. How tall is the kangaroo in centimeters?

7. What is the mass of the platypus in grams?

8. How long is the koala in meters?

9. Four large koalas would have about the same mass as what other mammal?

Measure these lines to the nearest centimeter and the nearest millimeter.

10. _____   11. _____

12. _____

Convert each measure.

13. 98 L to mL	14. 11.6 g to kg	15. 492 mm to cm
16. 3.8 kg to g	17. 5.7 mL to L	18. 91 cm to mm
19. 64 mg to g	20. 225 m to km	21. 177 km to m
22. 4.2 L to mL	23. 53.9 km to m	24. 37 mm to m

Cumulative Test   Chapters 1–4

TEST

▶ **Choose the letter that shows the correct answer. Round each number to the nearest**

1. tenth.
7.35
 a. 7.3
 b. 7.4
 c. 7.5
 d. not given

2. hundredth.
99.7928
 a. 99.80
 b. 100
 c. 99.79
 d. not given

3. thousand.
39,622
 a. 40,000
 b. 39,000
 c. 39,600
 d. not given

4. dollar.
$25.49
 a. $25.50
 b. $26
 c. $24
 d. not given

▶ **Compute.**

5. 900 × 800
 a. 72,000   b. 7200   c. 720,000   d. not given

6. 679.4 ÷ 100
 a. 67.94   b. 67,940   c. 6.794   d. not given

7. 8.5 × 1000
 a. 85,000   b. 0.085   c. 8500   d. not given

8. 9 ÷ 1000
 a. 0.009   b. 0.09   c. 0.9   d. not given

9. 6.7 × 0.3
 a. 2.01   b. 201   c. 20.1   d. not given

10. 4515 ÷ 60
 a. 75   b. 85   c. 71 R15   d. not given

11. 903 − 265
 a. 748   b. 638   c. 648   d. not given

12. 69,387
 + 45,879
 a. 115,266
 b. 114,266
 c. 115,366
 d. not given

13.   8
 − 5.87
 a. 3.87
 b. 2.23
 c. 2.13
 d. not given

14.   29
 × 94
 a. 2728
 b. 2027
 c. 2726
 d. not given

15. 0.49 + 5 +
 3.9 + 7.672
 a. 7.765
 b. 77.65
 c. 17.62
 d. not given

▶ **Write the standard numeral or decimal.**

16. thirty-five thousand, seventeen
    a. 350,017   **b.** 3517   **c.** 35,017   **d.** not given

17. two thousand and nine tenths
    a. 0.209   **b.** 2,009.9   **c.** 2,000.9   **d.** not given

18. eight hundred forty-four and twenty-two thousandths
    a. 800.4422   **b.** 844.22   **c.** 844.202   **d.** not given

▶ **Compare. Use <, >, or =.**

19. 4.2 ▨ 3.896   **20.** 573,652 ▨ 572,978   **21.** 6.60 ▨ 6.6
    a. <   **a.** <   **a.** <
    b. >   **b.** >   **b.** >
    c. =   **c.** =   **c.** =
    d. not given   **d.** not given   **d.** not given

▶ **Simplify.**

22. $9 \times 2 + 56 \div 8$
    a. 9.25   **b.** 25   **c.** 126   **d.** not given

23. $19 + 21 \div 3 - 6$
    a. 21   **b.** 7.33   **c.** 20   **d.** not given

▶ **Complete each statement. Choose the most reasonable measure.**

24. A high school student is about 1.75 ▨ tall.
    a. cm   **b.** m   **c.** km

25. A small dog has a mass of about 5 ▨ .
    a. kg   **b.** mg   **c.** g

26. A gas tank of a car contains about 60 ▨ .
    a. mL   **b.** mg   **c.** L

▶ **Convert each measure.**

27. 91 mm to cm
    a. 9100 cm   **b.** 0.91 cm   **c.** 910 cm   **d.** not given

28. 0.765 kg to g
    a. 765 g   **b.** 7.65 g   **c.** 765,000 g   **d.** not given

▶ **Choose the letter that shows the operation you would use to solve each problem.**

29. Alfonso works 9 hours a week and earns $4.95 an hour. How much does he earn in one week?

  a. division
  b. addition
  c. multiplication
  d. not given

30. Kiri ordered a cheeseburger for $1.15, a salad for $2.45, and two milkshakes for $0.85 each. What is the total cost?

  a. addition
  b. multiplication
  c. subtraction
  d. not given

31. Jonita and her family drove 1278 miles in four days. They drove the same number of miles each day. How many miles did they drive each day?

  a. multiplication
  b. division
  c. subtraction
  d. not given

32. Juan saw an ad for a car that gets 30.5 miles per gallon on the road. His family car gets only 22.9 miles per gallon. How many more miles per gallon does the car in the ad get?

  a. subtraction
  b. division
  c. addition
  d. not given

▶ **Solve.**

Apples sell for $0.89 a pound. Oranges cost $1.77 for three pounds. Bananas are being sold for four pounds for $1.

33. What is the cost of 6 pounds of apples?

  a. $4.34
  b. $5.34
  c. $0.15
  d. not given

34. What is the cost of 2 pounds of bananas?

  a. $2
  b. $0.75
  c. $4
  d. not given

35. What is the cost of 6 pounds of oranges, 3 pounds of apples, and 6 pounds of bananas?

  a. $7.71   **b.** $3.65   **c.** $8.19   **d.** not given

TAKE 5

 ### 4 ▶ Coin Challenge

I am holding two coins whose total value is 30 cents. One of the coins is not a nickel. What are the coins?

1 ▶ Triple Threat

Make the numbers 3 to 10 using 5 threes.

$0 = (3 - 3) \times 3 \times 3 \times 3$
$1 = (3 \times 3) \div (3 + 3 + 3)$
$2 = (3 + 3 + 3 - 3) \div 3$

33333

5 ▶ Fast Money

Mrs. Puddles and Mr. Buddles each has the same amount of money. How much should Puddles give Buddles so that Buddles has $20 more than Puddles?

How Many?

- A baker's dozen
- A googol
- A hat trick
- Mark Twain
- "Four score and seven years ago..."
  A. Lincoln, Gettysburg Address

2 ▶ Number Pattern

Find the next three numbers in this pattern.

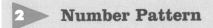

1 1 2 3 5 8 13

3 ▶ How Many Wholes?

How many two-digit whole numbers are there? (Time limit—30 seconds)

"It's amazing, but when I'm reading sheet music, I'm really working with fractions. A quarter note is actually one-quarter of a whole note. If a whole note lasts 16 beats, then a quarter note lasts 4 beats."

Chapter 5

Introducing Fractions and Mixed Numbers

Would you believe that you use fractions every time you
- **cut a pizza?**
- **read the gas gauge in a car?**
- **double a recipe?**
- **figure the price of a pair of jeans marked $\frac{1}{3}$ off?**
- **split a restaurant bill four ways?**

What are some other times that you use fractions?

Stained Glass Windows

Suppose you are designing two sets of stained glass windows. Each window is square. In the first set, each window must be divided into four panes of colored glass. All of the panes in a window must have straight edges and be the same size, but they need not be the same shape. Design as many windows as you can for this set. In the second set, each window must be divided into eight panes of glass. Design as many windows as you can for this set.

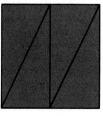

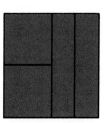

5.1 Understanding Fractions

What might have caused the confusion in the situation above?

When you use the fraction $\frac{1}{2}$, you know that something is divided into two equal parts. But $\frac{1}{2}$ of a 12-inch sandwich and $\frac{1}{2}$ of a 6-foot sandwich are quite different in size.

Numbers such as $\frac{1}{2}$, $\frac{1}{4}$, $\frac{3}{8}$, and $\frac{99}{100}$ are fractions.

▶ Express Yourself

fraction   a number that names part of a whole or part of a group. A fraction consists of a numerator and a denominator.

numerator   → 3   This fraction is read as "three-fourths"
denominator → 4   or "three-quarters."

A fraction can be used to describe how much of a picture is shaded. Use the numerator to tell how many parts are shaded. Use the denominator to tell the number of parts that make up the whole.

Write a fraction for the shaded part in each picture.

1.

2.

3.

4.

Knowing the following terms will help you when you work with fractions.

proper fraction   a fraction with a numerator less than the denominator; $\frac{2}{5}$, $\frac{14}{15}$, and $\frac{90}{100}$ are proper fractions.

improper fraction   a fraction with a numerator greater than or equal to the denominator; $\frac{11}{7}$, $\frac{50}{25}$, and $\frac{8}{8}$ are improper fractions.

mixed number   a number with a whole-number part and a fractional part; $4\frac{3}{5}$ and $12\frac{7}{12}$ are mixed numbers.

Identify each number below as either a proper fraction, an improper fraction, or a mixed number.

5. $\frac{1}{3}$ 6. $4\frac{2}{5}$ 7. $\frac{4}{7}$ 8. $\frac{11}{6}$ 9. $\frac{10}{10}$ 10. $1\frac{1}{4}$ 11. $\frac{6}{11}$

12. $\frac{2}{3}$ is read as "two-thirds." $\frac{3}{4}$ is read as "three-fourths." How is $\frac{4}{5}$ read? How is $\frac{4}{15}$ read?

13. Name five things, each of which is made up of equal parts. How might you use a fraction to describe the parts?

▶ **Practice What You Know**

When you work with fractions, you use multiplication facts to find the factors of numerators and denominators.

factors   numbers that are multiplied

$1 \times 6 = 6$ $2 \times 3 = 6$

$6 \times 1 = 6$ $3 \times 2 = 6$

The factors of 6 are 1, 2, 3, and 6.

14. Find the factors of 12.   15. Find the factors of 28.

5.2 Finding Equivalent Fractions

Everyone thought Francine was an excellent chef, until the day she made chili for her family. Suddenly everyone turned bright red and started coughing. "This stuff is *hot*!" her brother managed to choke out. Francine immediately knew what was wrong. The recipe called for $\frac{3}{8}$ teaspoon of hot sauce and she had added $\frac{3}{4}$ teaspoon.

To figure out how much extra hot pepper sauce she had added, Francine could find an equivalent fraction for $\frac{3}{4}$.

Examples

To find an equivalent fraction, multiply or divide the numerator and denominator by the same number.

A **Find 3 fractions equivalent to $\frac{3}{4}$.**

$$\frac{3 \times 2}{4 \times 2} = \frac{6}{8} \qquad \frac{3 \times 3}{4 \times 3} = \frac{9}{12} \qquad \frac{3 \times 4}{4 \times 4} = \frac{12}{16} \qquad \frac{3}{4} = \frac{6}{8} = \frac{9}{12} = \frac{12}{16}$$

$\frac{3}{4}$, $\frac{6}{8}$, $\frac{9}{12}$, and $\frac{12}{16}$ are equivalent fractions.

B **Use division to find a fraction equivalent to $\frac{20}{25}$.**

$$\frac{20 \div 5}{25 \div 5} = \frac{4}{5} \qquad \frac{20}{25} = \frac{4}{5}$$

$\frac{4}{5}$ is equivalent to $\frac{20}{25}$.

▶ Think and Discuss

1. Use division to find a fraction equivalent to $\frac{9}{21}$.

2. Refer to the introduction to this lesson. How many $\frac{1}{8}$ teaspoonfuls did Francine actually add?

3. What number do you multiply the numerator and the denominator of $\frac{3}{8}$ by to get $\frac{12}{32}$?

4. How many fractions are equivalent to $\frac{3}{8}$?

Exercises

Find 3 fractions equivalent to each fraction. (See Example A.)

5. $\frac{1}{2}$ **6.** $\frac{1}{4}$ **7.** $\frac{5}{4}$ **8.** $\frac{4}{9}$ **9.** $\frac{6}{7}$ **10.** $\frac{5}{8}$

11. $\frac{1}{6}$ **12.** $\frac{5}{9}$ **13.** $\frac{3}{5}$ **14.** $\frac{2}{3}$ **15.** $\frac{1}{8}$ **16.** $\frac{6}{5}$

Use division to find a fraction equivalent to each fraction.
(See Example B.)

17. $\frac{3}{6}$ **18.** $\frac{4}{8}$ **19.** $\frac{12}{16}$ **20.** $\frac{10}{15}$

21. $\frac{6}{27}$ **22.** $\frac{32}{24}$ **23.** $\frac{4}{10}$ **24.** $\frac{6}{18}$

25. $\frac{25}{20}$ **26.** $\frac{28}{38}$ **27.** $\frac{60}{72}$ **28.** $\frac{25}{45}$

▶ Mixed Practice (For more practice, see page 416.)

29. Find 3 fractions equivalent to $\frac{1}{3}$.   **30.** Find 3 fractions equivalent to $\frac{7}{8}$.

31. Find a fraction equivalent to $\frac{36}{24}$.   **32.** Find a fraction equivalent to $\frac{9}{21}$.

33. Find 3 fractions equivalent to $\frac{1}{10}$.   **34.** Find a fraction equivalent to $\frac{16}{20}$.

▶ Applications

35. Charles bought 18 cans from the case shown here. Write a fraction equivalent to $\frac{18}{24}$.

36. There are 16 ounces in a pound. Using equivalent fractions, find how many ounces are in $\frac{1}{4}$ pound.

37. A yard is divided into 36 inches. Using equivalent fractions, find how many inches are in $\frac{2}{3}$ yard.

▶ Review (Lesson 1.9)

Add.

38. $\begin{array}{r} 67 \\ + 29 \\ \hline \end{array}$ **39.** $\begin{array}{r} 583 \\ + 32 \\ \hline \end{array}$ **40.** $\begin{array}{r} 79 \\ + 8 \\ \hline \end{array}$ **41.** $\begin{array}{r} 494 \\ + 344 \\ \hline \end{array}$ **42.** $\begin{array}{r} 91 \\ + 36 \\ \hline \end{array}$

5.3 Writing Fractions and Mixed Numbers

A man went into a pizza parlor. "I'm starved!" he called to the chef. "Give me a large pizza to go." As the chef pulled the pizza from the oven, he asked, "Do you want that cut into six pieces or eight pieces?" "Six is plenty," replied the man. "I couldn't possibly eat eight!"

Did the man understand fractions? Which is more, $\frac{6}{6}$ or $\frac{8}{8}$ pizza?

Examples

To write fractions and mixed numbers, identify the parts and the whole.

A Write 2 as an improper fraction with a denominator of 8.

$2 = \frac{2}{1}$ Write 2 over a denominator of 1.

$2 = \frac{2 \times 8}{1 \times 8} = \frac{16}{8}$ Multiply the numerator and denominator by 8.

$2 = \frac{16}{8}$

B Write $2\frac{1}{6}$ as an improper fraction.

$2\frac{1}{6} = 2 + \frac{1}{6}$ Write the whole number as an improper fraction.

$2\frac{1}{6} = \frac{12}{6} + \frac{1}{6} = \frac{13}{6}$ Add the numerators.

Shortcut: $2\overset{+}{\underset{\times}{\frac{1}{6}}}$ $6 \times 2 = 12$

$12 + 1 = 13$ Write 13 over the original denominator.

$2\frac{1}{6} = \frac{13}{6}$

C Write $\frac{23}{7}$ as a mixed number.

$\frac{23}{7} = 23 \div 7$

$\begin{array}{r} 3\frac{2}{7} \\ 7\overline{)2\ 3} \\ \underline{-2\ 1} \\ 2 \end{array}$ Divide the numerator by the denominator.

$\frac{23}{7} = 3\frac{2}{7}$

▶ **Think and Discuss**

1. Write $\frac{14}{5}$ as a mixed number.

2. Write 5 as 3 different improper fractions.

3. Refer to the introduction to this lesson. Explain why the man in the pizza shop might have been confused.

Exercises

Write each whole number as an improper fraction.
(See Example A.)

4. $4 = \frac{⬚}{4}$ 5. $10 = \frac{⬚}{5}$ 6. $16 = \frac{⬚}{8}$ 7. $24 = \frac{⬚}{2}$ 8. $100 = \frac{⬚}{9}$

Write each mixed number as an improper fraction.
(See Example B.)

9. $4\frac{5}{6}$ 10. $10\frac{3}{4}$ 11. $2\frac{7}{8}$ 12. $5\frac{3}{8}$ 13. $3\frac{3}{9}$

Write each quotient as a whole or mixed number. (See Example C.)

14. $\frac{7}{3}$ 15. $9\overline{)33}$ 16. $\frac{45}{5}$ 17. $\frac{55}{4}$ 18. $12\overline{)65}$ 19. $\frac{72}{10}$

▶ **Mixed Practice** (For more practice, see page 416.)

Write each improper fraction as a whole or mixed number.

20. $\frac{19}{6}$ 21. $\frac{10}{2}$ 22. $\frac{31}{5}$ 23. $\frac{35}{10}$ 24. $\frac{7}{7}$

Write each mixed or whole number as an improper fraction.

25. $4\frac{3}{7}$ 26. $1\frac{3}{4}$ 27. 3 28. $9\frac{2}{3}$ 29. $1\frac{15}{16}$

▶ **Applications**

30. Juice comes in packages of 6 cartons. If Quentin has 7 full packages and one that is $\frac{5}{6}$ full, how many cartons does he have?

31. The maximum capacity of an elevator is shown here. If ninety-two people are waiting, what is the least number of trips needed to get everyone to the top?

EMPIRE·STATE
BUILDING
ELEVATOR CAPACITY
12 PEOPLE

▶ **Review** (Lessons 4.2, 4.5, 4.6)

Convert each measure.

32. 43 g to kg   33. 8.9 m to cm   34. 319 mL to L

5.4 Comparing Fractions and Mixed Numbers

Korinne was about to make banana bread when she discovered that she was almost out of honey. She found one recipe that called for $\frac{1}{3}$ cup of honey, and another that called for $\frac{3}{8}$ cup of honey. Which recipe required less honey?

Examples

To compare fractions, rewrite the fractions with a common denominator and compare their numerators. Use $<$, $>$, or $=$.

A Compare. $\frac{5}{9}$ ▧ $\frac{7}{9}$

Compare the numerators.

$\frac{5}{9}$ ▧ $\frac{7}{9}$ $5 < 7$

$\frac{5}{9} < \frac{7}{9}$

B Compare. $\frac{3}{8}$ ▧ $\frac{1}{3}$

Rewrite the fractions with a common denominator. Compare numerators.

$\frac{3}{8}$ ▧ $\frac{1}{3}$ $\frac{3 \times 3}{8 \times 3} = \frac{9}{24}$

$\frac{9}{24}$ ▧ $\frac{8}{24}$ $\frac{1 \times 8}{3 \times 8} = \frac{8}{24}$

$\frac{3}{8} > \frac{1}{3}$ $9 > 8$

C Compare. $2\frac{5}{8}$ ▧ $2\frac{4}{7}$

First compare the whole number parts. Since $2 = 2$, compare the fractions.

$\frac{5}{8}$ ▧ $\frac{4}{7}$ $\frac{5 \times 7}{8 \times 7} = \frac{35}{56}$

$\frac{35}{56}$ ▧ $\frac{32}{56}$ $\frac{4 \times 8}{8 \times 8} = \frac{32}{56}$

$\frac{5}{8} > \frac{4}{7}$ $35 > 32$

▶ Think and Discuss

1. Compare $\frac{5}{6}$ and $\frac{1}{6}$.

2. Compare $\frac{5}{7}$ and $\frac{7}{10}$.

3. Refer to the introduction to this lesson. Which of Korinne's recipes required less honey?

4. Name three jobs that require knowledge of fractions. Describe how fractions are used in these jobs.

Exercises

Compare. Use <, >, or =. (See Example A.)

5. $\frac{3}{8}$ ▨ $\frac{6}{8}$

6. $\frac{6}{7}$ ▨ $\frac{5}{7}$

7. $\frac{9}{10}$ ▨ $\frac{10}{10}$

8. $\frac{9}{16}$ ▨ $\frac{5}{16}$

9. $\frac{2}{12}$ ▨ $\frac{5}{12}$

10. $\frac{75}{78}$ ▨ $\frac{76}{78}$

Compare. Use <, >, or =. (See Example B.)

11. $\frac{3}{5}$ ▨ $\frac{6}{10}$

12. $\frac{7}{12}$ ▨ $\frac{7}{14}$

13. $\frac{8}{9}$ ▨ $\frac{9}{16}$

14. $\frac{7}{3}$ ▨ $\frac{9}{5}$

15. $\frac{3}{4}$ ▨ $\frac{3}{5}$

16. $\frac{5}{15}$ ▨ $\frac{1}{3}$

17. $\frac{2}{7}$ ▨ $\frac{6}{21}$

18. $\frac{2}{9}$ ▨ $\frac{3}{10}$

Compare. Use <, >, or =. (See Example C.)

19. $4\frac{1}{2}$ ▨ $4\frac{2}{4}$

20. $3\frac{3}{7}$ ▨ $4\frac{3}{7}$

21. $2\frac{7}{9}$ ▨ $2\frac{7}{8}$

22. $\frac{7}{3}$ ▨ $2\frac{1}{3}$

23. $1\frac{3}{10}$ ▨ $1\frac{4}{9}$

24. $3\frac{1}{2}$ ▨ $\frac{9}{2}$

25. $2\frac{1}{4}$ ▨ $\frac{5}{3}$

26. $8\frac{2}{5}$ ▨ $5\frac{2}{5}$

▶ **Mixed Practice** (For more practice, see page 416.)

Compare. Use <, >, or =.

27. $\frac{3}{5}$ ▨ $\frac{4}{5}$

28. $4\frac{1}{4}$ ▨ $\frac{12}{4}$

29. $3\frac{2}{3}$ ▨ $2\frac{2}{3}$

30. $2\frac{2}{7}$ ▨ $1\frac{9}{7}$

31. $7\frac{1}{3}$ ▨ $7\frac{3}{8}$

32. $\frac{9}{10}$ ▨ $\frac{7}{10}$

33. $2\frac{1}{2}$ ▨ $\frac{5}{2}$

34. $8\frac{2}{5}$ ▨ $5\frac{2}{5}$

35. $\frac{4}{4}$ ▨ $\frac{7}{7}$

36. $3\frac{1}{3}$ ▨ $3\frac{1}{2}$

37. $6\frac{5}{6}$ ▨ $\frac{42}{6}$

38. 6 ▨ $\frac{36}{6}$

▶ **Applications**

39. Tony has completed two-thirds of his homework, Sue has completed three-fourths of her homework, and Greg has completed five-sixths of his homework. If they all have the same amount of homework, who has completed the most?

40. Coralee has found these 3 pieces of fabric in the remnant bin. She wants to buy the longest piece. Which piece should she buy?

$2\frac{1}{2}$ yds. $2\frac{2}{3}$ yds. $2\frac{3}{8}$ yds.

▶ **Review** (Lessons 4.3, 4.5, 4.6)

Complete each sentence.

41. A chopstick has a length of about 24 ▨.   cm   km

42. A large thermos has a capacity of about 1 ▨.   L   mL

43. A baby has a mass of about 10 ▨.   g   kg

5.5 Factoring to Find the Greatest Common Factor

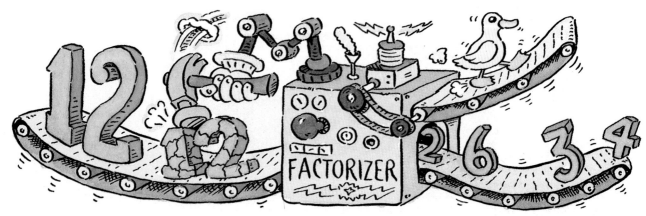

Breaking a number into its factors is the first step in writing fractions in lowest terms.

To find the greatest common factor (GCF) of 2 or more numbers, list the factors of each. The GCF is the greatest factor that is on both lists.

A List the factors of 20.

$1 \times \underline{20} = 20$	List the factors in order.
$2 \times \underline{10} = 20$	
$3 \times \underline{} = 20$	3 is not a factor of 20.
$4 \times \underline{5} = 20$	
$5 \times \underline{4} = 20$	When factors begin to repeat, you have found all factors.

The factors of 20 are **1, 2, 4, 5, 10, and 20.**

B Find the GCF of 12 and 16.

$12 = 1 \times 12$
$12 = 2 \times 6$
$12 = 3 \times 4$ The factors of 12 are 1, 2, 3, **4**, 6, and 12.

$16 = 1 \times 16$
$16 = 2 \times 8$
$16 = 4 \times 4$ The factors of 16 are 1, 2, **4**, 8, and 16.

The GCF of 12 and 16 is **4.**

▶ Think and Discuss

1. List the factors of 42.

2. Find the GCF of 24 and 36.

3. Describe the steps you would take to find the GCF of 3 whole numbers.

4. The factors of 2 are 1 and 2. Find the other whole numbers less than 20 that have exactly 2 factors.

◣ Exercises

List the factors. (See Example A.)

   5. 6   6. 10   7. 14   8. 15   9. 18   10. 36   11. 24

Find the GCF of each pair of numbers. (See Example B.)

| 12. 16   24 | 13. 12   8 | 14. 24   6 | 15. 27   21 |
| 16. 17   13 | 17. 20   100 | 18. 12   18 | 19. 36   27 |

▶ Mixed Practice (For more practice, see page 417.)

List the factors.

   20. 5   21. 8   22. 45   23. 13   24. 48   25. 33   26. 64

Find the GCF of each pair of numbers.

| 27. 42   14 | 28. 7   38 | 29. 3   35 | 30. 10   15 |
| 31. 12   40 | 32. 9   33 | 33. 6   30 | 34. 18   60 |

▶ Applications

35. Square tiles are packed in cartons like the one shown here. To fit the cartons exactly, the length of each side must be a factor of both 16 and 24. What four sizes of tile fit the cartons exactly?

36. Mark's eldest sister, Pam, is eighteen years old. The GCF of Mark's and Pam's ages is nine. How old is Mark?

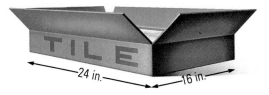

▶ Review (Lessons 1.1, 1.2)

Write the standard numeral.

37. 66 thousand, 8 hundred 3

38. 4 and 35 hundredths

39. two million, ninety

40. 23 and 8 thousandths

5.6 Writing Fractions in Lowest Terms

"Let's see," said Tanya. "It's 4:30 now. I'll meet you in thirty-sixtieths of an hour."

Tanya's friend may not know exactly when to meet her. Fortunately, most people express fractions of hours in lowest terms.

Examples

To write a fraction in lowest terms, find the GCF of the numerator and denominator. Divide both by their GCF.

A Write $\frac{8}{12}$ in lowest terms.

The factors of 8 are 1, 2, **4**, and 8.
The factors of 12 are 1, 2, 3, **4**, 6, and 12.
The GCF is 4.
$\frac{8}{12} = \frac{8 \div 4}{12 \div 4}$ Divide by the GCF.

$\frac{8}{12} = \frac{2}{3}$

B List the factors of 9 and 25. Write $\frac{9}{25}$ in lowest terms.

The factors of 9 are **1**, 3, 9.
The factors of 25 are **1**, 5, 25.
The GCF is 1.

$\frac{9}{25}$ is in lowest terms.

▶ Think and Discuss

1. In Example A, what was the GCF of 8 and 12?

2. Refer to the introduction to this lesson. What is a clearer way to express thirty-sixtieths of an hour?

3. A case of motor oil contains 24 cans. What part of a case is 6 cans of oil? Write your answer in lowest terms.

4. Which of the following fractions are in lowest terms?
$\frac{3}{6}$ $\frac{12}{8}$ $\frac{14}{24}$ $\frac{19}{38}$ $\frac{5}{2}$ $\frac{1}{8}$ $\frac{18}{33}$

5. Write the following improper fractions and mixed numbers in lowest terms.
$\frac{25}{9}$ $1\frac{3}{6}$ $\frac{14}{12}$ $6\frac{6}{8}$

6. Why is 25¢ called a quarter?

Exercises

Write in lowest terms. (See Example A.)

7. $\frac{4}{6}$　　8. $\frac{9}{18}$　　9. $\frac{6}{10}$　　10. $\frac{100}{300}$　　11. $\frac{8}{12}$　　12. $\frac{6}{9}$

13. $\frac{8}{14}$　　14. $\frac{6}{15}$　　15. $\frac{6}{20}$　　16. $\frac{18}{24}$　　17. $\frac{30}{90}$　　18. $\frac{5}{12}$

19. $\frac{15}{45}$　　20. $\frac{80}{160}$　　21. $\frac{18}{80}$　　22. $\frac{40}{42}$　　23. $\frac{15}{18}$　　24. $\frac{32}{48}$

25. $\frac{28}{21}$　　26. $\frac{72}{81}$　　27. $\frac{50}{125}$　　28. $\frac{30}{36}$　　29. $\frac{24}{15}$　　30. $\frac{72}{100}$

List the factors and write in lowest terms. (See Example B.)

31. $\frac{2}{5}$　　32. $\frac{1}{4}$　　33. $\frac{13}{14}$　　34. $\frac{8}{3}$　　35. $\frac{15}{16}$　　36. $\frac{24}{25}$

▶ Mixed Practice (For more practice, see page 417.)

Write in lowest terms.

37. $\frac{9}{12}$　　38. $\frac{14}{20}$　　39. $\frac{28}{35}$　　40. $\frac{18}{12}$　　41. $\frac{3}{8}$　　42. $\frac{25}{40}$

43. $\frac{5}{9}$　　44. $\frac{18}{10}$　　45. $\frac{55}{65}$　　46. $\frac{48}{72}$　　47. $\frac{12}{16}$　　48. $\frac{75}{100}$

49. $\frac{21}{24}$　　50. $\frac{18}{48}$　　51. $\frac{60}{48}$　　52. $1\frac{7}{21}$　　53. $4\frac{4}{8}$　　54. $\frac{35}{15}$

▶ Applications

55. Every Friday afternoon Marylu spends about 20 minutes cleaning her desk. What part of an hour is 20 minutes? Write your answer in lowest terms.

56. Fifteen bus lines run from the bus depot. Five lines go north, three lines go west, and the rest go south. What fraction of lines goes in each direction? Write your answers in lowest terms.

57. Manny bought two dozen eggs. On his way home, he tripped and fell. Eight eggs broke. What fractional part of the total was unbroken? Write your answer in lowest terms.

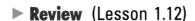

▶ Review (Lesson 1.12)

Estimate each sum or difference.

58. $\begin{array}{r} 2564 \\ + 8375 \end{array}$　　59. $\begin{array}{r} 9281 \\ - 6429 \end{array}$　　60. $\begin{array}{r} 7723 \\ + 8615 \end{array}$　　61. $\begin{array}{r} 10{,}942 \\ - 3{,}751 \end{array}$　　62. $\begin{array}{r} 15{,}180 \\ + 25{,}372 \end{array}$

5.7 Writing Fractions and Decimals

Kelly and Benjamin were shopping for party snacks. At the dairy case Kelly asked how much cheese they needed for nachos. "Well," said Benjamin, "the recipe calls for $\frac{3}{4}$ pound of cheese, but these packages show decimal weights. How much should I buy?"

To find the answer, Benjamin can convert $\frac{3}{4}$ to a decimal.

Examples

To write a fraction as a decimal, divide the numerator by the denominator.

To write a decimal as a fraction, use a denominator that is a power of 10. Write the fraction in lowest terms.

A Write $\frac{3}{4}$ as a decimal.

$$\begin{array}{r} 0.7\,5 \\ 4\overline{)3.0\,0} \\ 2\,8 \\ \hline 2\,0 \\ 2\,0 \\ \hline 0 \end{array}$$ Write zeros in the dividend.

$\frac{3}{4} = \mathbf{0.75}$

B Write $\frac{1}{3}$ as a decimal.

$$\begin{array}{r} 0.3\,3 \\ 3\overline{)1.0\,0} \\ 9 \\ \hline 1\,0 \\ 9 \\ \hline 1 \end{array}$$ If you continue, you will always get 3.

$\frac{1}{3} = \mathbf{0.\overline{3}}$ A bar over the 3 means the 3 repeats.

C Write 3.25 as a fraction.

3.25 is 3 and 25 **hundredths**.

$3.25 = 3 + \frac{25}{100} = 3\frac{25}{100} = 3\frac{1}{4}$

$3.25 = 3\frac{1}{4}$

D Write $2\frac{1}{2}$ as a decimal.

$2\frac{1}{2} = 2 + \frac{1}{2}$

$2\frac{1}{2} = 2 + 0.5$

$2\frac{1}{2} = \mathbf{2.5}$

$$\begin{array}{r} 0.5 \\ 2\overline{)1.0} \\ 1.0 \\ \hline 0 \end{array}$$

▶ Think and Discuss

1. Write $\frac{1}{8}$ as a decimal.

2. Write 0.08 as a fraction in lowest terms.

3. On a calculator, $\frac{7}{11}$ is 0.6363636. Write this decimal as shown in Example B.

4. Refer to the introduction to this lesson. How much cheese should Benjamin buy?

Exercises

Write as a decimal. (See Example A.)

5. $\frac{1}{2}$ 6. $\frac{1}{4}$ 7. $\frac{2}{5}$ 8. $\frac{5}{8}$ 9. $\frac{3}{5}$ 10. $\frac{7}{10}$ 11. $\frac{9}{16}$

Write as a decimal. (See Example B.)

12. $\frac{2}{3}$ 13. $\frac{1}{6}$ 14. $\frac{2}{9}$ 15. $\frac{5}{11}$ 16. $\frac{5}{6}$ 17. $\frac{9}{11}$ 18. $\frac{1}{15}$

Write as a fraction or mixed number in lowest terms.
(See Example C.)

19. 4.75   20. 0.9   21. 0.201   22. 0.036   23. 0.001   24. 3.08

Write as a decimal. (See Example D.)

25. $3\frac{1}{5}$ 26. $9\frac{8}{125}$ 27. $4\frac{7}{12}$ 28. $5\frac{3}{4}$ 29. $11\frac{1}{10}$ 30. $6\frac{1}{7}$ 31. $2\frac{1}{3}$

▶ Mixed Practice (For more practice, see page 418.)

Convert each fraction to a decimal and each decimal to a fraction.

32. $\frac{4}{5}$ 33. 0.019   34. $4\frac{8}{9}$ 35. 0.7   36. $\frac{11}{16}$ 37. 1.64

38. 8.15   39. $5\frac{14}{25}$ 40. $\frac{7}{30}$ 41. $\frac{4}{15}$ 42. 0.048   43. $\frac{29}{60}$

▶ Applications

44. Every morning Mrs. Sneedly walks her pet chihuahua, Smily. Their route is seven-eighths mile long. Write the distance as a decimal.

45. June needs $\frac{1}{4}$ pound of walnuts for a dip recipe. If she buys a package marked 0.45 pound, what fractional part of a pound will she have left over?

▶ Review (Lessons 3.10, 3.11, 3.12)

Divide. Round each quotient to the nearest hundredth.

46. $2\overline{)365.99}$ 47. $12\overline{)67.48}$ 48. $0.8\overline{)49.33}$ 49. $0.06\overline{)0.977}$

5.8 Finding Patterns with Fractions and Repeating Decimals

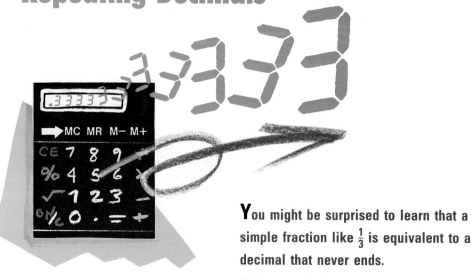

You might be surprised to learn that a simple fraction like $\frac{1}{3}$ is equivalent to a decimal that never ends.

Using a calculator, you can uncover surprises and patterns in decimals.

Here are some examples of repeating decimals.

$\frac{2}{11} = 0.18181818.\,.\,.$
pattern: 18 repeats

$\frac{1}{7} = 0.142857142857.\,.\,.$
pattern: 142857 repeats

$\frac{467}{990} = 0.4717171.\,.\,.$
pattern: 71 repeats

To indicate the pattern of repeating decimals, place a bar over the digits that repeat.

$\frac{4}{11} = 0.363636.\,.\,.$
$= 0.\overline{36}$

$\frac{3}{7} = 0.428571428571.\,.\,.$
$= 0.\overline{428571}$

$\frac{511}{990} = 0.51616.\,.\,.$
$= 0.5\overline{16}$

Use a calculator to help you answer the following questions.

1. What is the pattern for $\frac{2}{11}$, $\frac{1}{7}$, and $\frac{467}{990}$? Use a bar over the repeating digits.

► **The Harmonious Sevenths**

2. Find the decimals for $\frac{1}{7}$, $\frac{2}{7}$, $\frac{3}{7}$, $\frac{4}{7}$, $\frac{5}{7}$, and $\frac{6}{7}$. Use a bar over the repeating digits.

3. Diana, exploring the sevenths, was so impressed by the pattern in their decimals that she drew the diagram shown at the right. Explain how the diagram works.

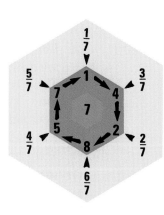

▶ Nines and Ones

4. Find the decimals for $\frac{1}{9}$, $\frac{1}{99}$, and $\frac{1}{999}$. Describe the pattern you see in these repeating decimals.

5. Based on the pattern you found, predict the decimal for $\frac{1}{9999}$. Can you check your prediction with a calculator?

6. Repeat this activity for $\frac{1}{11}$, $\frac{1}{111}$, and $\frac{1}{1111}$. Is the pattern similar to the one you found in Question 4? Predict the decimal for $\frac{1}{11,111}$.

▶ The Terminators

Terminating decimals are decimals that do not repeat forever. Examples of terminating decimals are:

$\frac{1}{2} = 0.5$ $\frac{1}{4} = 0.25$ $\frac{1}{5} = 0.2$ $\frac{1}{8} = 0.125$

Work with 2 or 3 classmates to solve the following problems.

7. Find the decimals for the fractions $\frac{1}{2}$, $\frac{1}{3}$, $\frac{1}{4}$, and so on down to $\frac{1}{25}$. Use a bar over repeating digits.

8. Find four decimals from Question 7 that you know for certain are terminating decimals. Explain how you know.

9. Is $\frac{1}{17}$ a terminating or repeating decimal? Does your calculator help you answer this question? Discuss.

▶ Review

Write which operation you would use to solve each problem. Then solve.

10. How many inches are in 17 feet?

11. Look at the class roster on the right. How many students are studying French?

12. There are nine hundred calories in six cups of milk. How many calories are there in one cup of milk?

Foreign Language Students

Spanish I	24
Spanish II	32
German I	20
German II	18
French I	27
French II	24

Complete each sentence. (Lesson 5.1)

1. The top part of a fraction is called the ___.

2. The part of a fraction that shows how many parts make up the whole is called the ___.

Write a fraction or a mixed number for the shaded part in each picture. (Lesson 5.1)

3.

4.

5.

Find 3 fractions equivalent to each fraction. (Lesson 5.2)

6. $\frac{5}{8}$

7. $\frac{7}{10}$

8. $\frac{3}{16}$

9. $\frac{4}{9}$

Use division to find a fraction equivalent to each fraction. (Lesson 5.2)

10. $\frac{5}{15}$

11. $\frac{6}{33}$

12. $\frac{27}{30}$

13. $\frac{25}{50}$

14. $\frac{4}{14}$

15. $\frac{18}{32}$

Write as an improper fraction. (Lesson 5.3)

16. $4\frac{1}{2}$

17. 6

18. $3\frac{1}{3}$

19. $7\frac{9}{10}$

Compare. Use <, >, or =. (Lesson 5.4)

20. $\frac{3}{8}$ ___ $\frac{3}{16}$

21. $2\frac{1}{2}$ ___ $\frac{5}{2}$

22. $4\frac{3}{4}$ ___ $5\frac{1}{8}$

23. $\frac{11}{12}$ ___ $\frac{13}{15}$

Find the GCF of each pair of numbers. (Lesson 5.5)

24. 6 5
25. 9 18
26. 10 15
27. 4 16
28. 14 21
29. 12 32

Write each fraction in lowest terms. (Lesson 5.6)

30. $\frac{12}{10}$
31. $\frac{14}{27}$
32. $\frac{80}{100}$
33. $\frac{48}{24}$
34. $\frac{30}{45}$
35. $\frac{10}{16}$
36. $1\frac{18}{32}$

Convert each fraction to a decimal and each decimal to a fraction. (Lesson 5.7)

37. $\frac{1}{4}$
38. $\frac{1}{3}$
39. 9.3
40. $\frac{4}{5}$
41. 0.125
42. $6\frac{5}{8}$

Multiplying and Dividing Fractions and Mixed Numbers

You multiply or divide fractions when you
- **determine the amount of lumber you should buy for a carpentry project.**
- **double or triple a recipe.**
- **have to figure out how much material to buy for a sewing project.**

Planning a party can also involve multiplying or dividing fractions and mixed numbers. Often you have to adjust recipes depending on the size of your party. For instance, if a recipe for chili serves 8, but you have invited 16 people, you would have to double the recipe. What would you do if you had invited only 4 people and did not want leftovers?

Party Planning

Choose 2 or 3 classmates to work with. Describe how to adjust the recipes below to serve the following numbers of people: 4 people, 16 people, and 32 people.

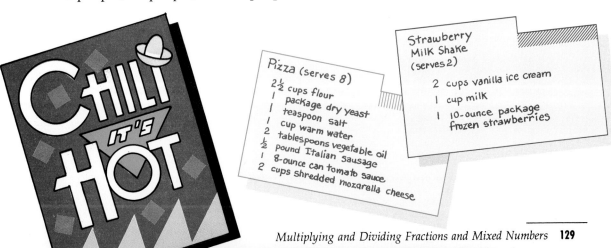

Pizza (serves 8)
2½ cups flour
1 package dry yeast
1 teaspoon salt
1 cup warm water
2 tablespoons vegetable oil
½ pound Italian sausage
1 8-ounce can tomato sauce
2 cups shredded mozarella cheese

Strawberry Milk Shake
(serves 2)
2 cups vanilla ice cream
1 cup milk
1 10-ounce package frozen strawberries

6.1 Multiplying Fractions

Smoked Salmon and Spinach Salad   Serves 4

3/4 cup smoked salmon
1 pound spinach
1 small onion, minced
1 tablespoon lemon juice
1/4 cup minced celery
1/2 cup salad dressing
salt and pepper

Sharon's Aunt Zenia was coming to visit, and Sharon needed a special dish to serve for lunch. "How about this Smoked Salmon and Spinach Salad recipe?" her roommate suggested. "Great," said Sharon.

Sharon can determine how much of each ingredient to buy for lunch with Aunt Zenia by multiplying the given amounts by $\frac{1}{2}$.

Examples

To multiply fractions, multiply the numerators. Then multiply the denominators. Write the answer in lowest terms.

A Multiply. $\frac{3}{4} \times \frac{1}{2}$

$\frac{3}{4} \times \frac{1}{2} = \frac{3}{8}$

B Multiply. $\frac{4}{9} \times \frac{3}{8}$

$\frac{4}{9} \times \frac{3}{8} = \frac{12}{72} = \frac{1}{6}$

C Multiply. $5 \times \frac{3}{4}$

$\frac{5}{1} \times \frac{3}{4} = \frac{15}{4}$, or $3\frac{3}{4}$

▶ Think and Discuss

1. Refer to the introduction to this lesson. How much minced celery does Sharon need?

2. Fill in the missing numbers. $\frac{5}{6} \times \frac{3}{\text{▦}} = \frac{15}{48} = \frac{5}{\text{▦}}$

3. Multiply 7 by $\frac{5}{8}$.

4. Find $\frac{1}{4}$ of 16.

5. Describe the steps you use when you multiply a fraction by a whole number.

Exercises

Multiply. Write the answers in lowest terms. (See Example A.)

6. $\frac{2}{3} \times \frac{4}{5}$ **7.** $\frac{1}{5} \times \frac{1}{3}$ **8.** $\frac{5}{8} \times \frac{1}{6}$ **9.** $\frac{3}{4} \times \frac{3}{5}$ **10.** $\frac{1}{9} \times \frac{1}{3}$

11. $\frac{4}{9} \times \frac{2}{5}$ **12.** $\frac{5}{8} \times \frac{3}{4}$ **13.** $\frac{3}{5} \times \frac{2}{5}$ **14.** $\frac{4}{5} \times \frac{3}{7}$ **15.** $\frac{2}{5} \times \frac{1}{9}$

Multiply. Write the answers in lowest terms. (See Example B.)

16. $\frac{3}{7} \times \frac{2}{9}$ **17.** $\frac{5}{8} \times \frac{2}{3}$ **18.** $\frac{2}{5} \times \frac{3}{4}$ **19.** $\frac{5}{6} \times \frac{3}{4}$ **20.** $\frac{7}{12} \times \frac{2}{3}$

21. $\frac{3}{8} \times \frac{4}{9}$ **22.** $\frac{7}{10} \times \frac{5}{6}$ **23.** $\frac{8}{9} \times \frac{5}{6}$ **24.** $\frac{5}{7} \times \frac{4}{15}$ **25.** $\frac{5}{16} \times \frac{4}{5}$

26. $\frac{1}{4} \times \frac{4}{7}$ **27.** $\frac{2}{5} \times \frac{1}{2}$ **28.** $\frac{5}{7} \times \frac{3}{5}$ **29.** $\frac{9}{15} \times \frac{2}{3}$ **30.** $\frac{1}{4} \times \frac{8}{11}$

Multiply. Write the answers in lowest terms. (See Example C.)

31. $4 \times \frac{3}{4}$ **32.** $7 \times \frac{9}{10}$ **33.** $3 \times \frac{2}{7}$ **34.** $8 \times \frac{2}{3}$ **35.** $5 \times \frac{7}{9}$

36. $10 \times \frac{1}{3}$ **37.** $2 \times \frac{11}{16}$ **38.** $9 \times \frac{1}{2}$ **39.** $4 \times \frac{2}{11}$ **40.** $8 \times \frac{2}{5}$

41. $3 \times \frac{3}{7}$ **42.** $11 \times \frac{1}{2}$ **43.** $4 \times \frac{2}{3}$ **44.** $15 \times \frac{3}{5}$ **45.** $2 \times \frac{7}{10}$

▶ Mixed Practice (For more practice, see page 418.)

Multiply. Write the answers in lowest terms.

46. $\frac{1}{6} \times \frac{7}{10}$ **47.** $7 \times \frac{8}{9}$ **48.** $\frac{3}{5} \times \frac{1}{8}$ **49.** $\frac{7}{8} \times \frac{6}{7}$ **50.** $5 \times \frac{2}{3}$

51. $3 \times \frac{1}{8}$ **52.** $\frac{6}{10} \times \frac{4}{9}$ **53.** $\frac{3}{8} \times \frac{1}{8}$ **54.** $9 \times \frac{3}{5}$ **55.** $\frac{2}{3} \times \frac{1}{12}$

56. $\frac{7}{8} \times \frac{3}{4}$ **57.** $\frac{9}{10} \times \frac{5}{6}$ **58.** $\frac{9}{16} \times \frac{2}{3}$ **59.** $4 \times \frac{5}{8}$ **60.** $\frac{8}{9} \times \frac{8}{9}$

▶ Applications

61. A printer's assistant earns $\frac{2}{5}$ of the total profit of every job he works on. How much does he earn on a profit of $45?

62. A pudding recipe calling for one-half cup of sugar serves four. How much sugar would you use to make three servings?

▶ Review (Lesson 5.5)

Find the GCF of each pair of numbers.

63. 12   18   **64.** 21   28   **65.** 20   11   **66.** 9   15   **67.** 25   5

68. 8   32   **69.** 26   10   **70.** 40   16   **71.** 34   17   **72.** 13   5

73. 12   16   **74.** 50   25   **75.** 21   3   **76.** 15   10   **77.** 16   18

6.2 Multiplying Fractions: A Shortcut

$$\frac{250}{9} \times \frac{67}{750} =$$

Sharif rushed into mathematics class late one day. The other students were working furiously to find a solution to the problem shown above. "I'm sorry I'm late," Sharif said. "Oh, by the way, the answer to the problem is $\frac{67}{27}$." How do you think Sharif solved the problem so quickly?

Examples

To multiply fractions using a shortcut method, first divide numerators and denominators by a common factor.

A **Multiply.** $\frac{4}{10} \times \frac{3}{7}$

Write fractions in lowest terms. Multiply.

$$4 \div 2$$

$$\frac{4}{10} \times \frac{3}{7} = \frac{\overset{2}{\cancel{4}}}{\underset{5}{\cancel{10}}} \times \frac{3}{7} = \frac{6}{35}$$

$$10 \div 2$$

$$\frac{4}{10} \times \frac{3}{7} = \frac{6}{35}$$

B **Multiply.** $\frac{250}{9} \times \frac{67}{750}$

Divide by the GCF of a numerator and a denominator.

$$250 \div 250$$

$$\frac{250}{9} \times \frac{67}{750} = \frac{\overset{1}{\cancel{250}}}{9} \times \frac{67}{\underset{3}{\cancel{750}}} = \frac{67}{27}, \text{ or } 2\frac{13}{27}$$

$$750 \div 250$$

$$\frac{250}{9} \times \frac{67}{750} = 2\frac{13}{27}$$

▶ Think and Discuss

1. Multiply $\frac{1}{4}$ by $\frac{4}{5}$. How did you simplify the problem?

2. Multiply $\frac{6}{9}$ by $\frac{3}{12}$. How did you simplify the problem?

3. Explain how to simplify in the problem $\frac{6}{8} \times \frac{24}{27}$.

4. Explain why it helps to simplify fractions before multiplying.

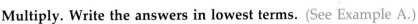

Exercises

Multiply. Write the answers in lowest terms. (See Example A.)

5. $\frac{4}{12} \times \frac{5}{7}$ **6.** $\frac{3}{9} \times \frac{4}{5}$ **7.** $\frac{5}{9} \times \frac{2}{8}$ **8.** $\frac{7}{14} \times \frac{3}{4}$ **9.** $\frac{4}{16} \times \frac{3}{7}$

10. $\frac{2}{3} \times \frac{10}{18}$ **11.** $\frac{1}{2} \times \frac{5}{15}$ **12.** $\frac{4}{10} \times \frac{4}{5}$ **13.** $\frac{6}{9} \times \frac{1}{5}$ **14.** $\frac{3}{21} \times \frac{1}{7}$

15. $\frac{3}{11} \times \frac{6}{8}$ **16.** $\frac{1}{7} \times \frac{10}{15}$ **17.** $\frac{7}{21} \times \frac{8}{24}$ **18.** $\frac{9}{63} \times \frac{64}{80}$ **19.** $\frac{6}{15} \times \frac{13}{15}$

Multiply. Write the answers in lowest terms. (See Example B.)

20. $\frac{8}{9} \times \frac{3}{10}$ **21.** $\frac{1}{3} \times \frac{6}{7}$ **22.** $\frac{7}{9} \times \frac{9}{13}$ **23.** $\frac{2}{15} \times \frac{10}{11}$ **24.** $\frac{6}{7} \times \frac{11}{15}$

25. $\frac{3}{4} \times \frac{12}{15}$ **26.** $\frac{4}{15} \times \frac{9}{11}$ **27.** $\frac{1}{15} \times \frac{72}{73}$ **28.** $\frac{8}{45} \times \frac{15}{7}$ **29.** $\frac{9}{17} \times \frac{13}{18}$

30. $\frac{7}{8} \times \frac{16}{17}$ **31.** $\frac{3}{4} \times \frac{2}{3}$ **32.** $\frac{8}{13} \times \frac{39}{41}$ **33.** $\frac{2}{7} \times \frac{7}{16}$ **34.** $\frac{8}{35} \times \frac{14}{25}$

▶ Mixed Practice (For more practice, see page 419.)

Multiply. Write the answers in lowest terms.

35. $\frac{4}{6} \times \frac{7}{11}$ **36.** $9 \times \frac{2}{3}$ **37.** $\frac{5}{14} \times \frac{35}{17}$ **38.** $\frac{1}{9} \times \frac{63}{65}$ **39.** $\frac{36}{54} \times \frac{5}{8}$

40. $\frac{15}{60} \times \frac{7}{8}$ **41.** $\frac{9}{10} \times \frac{7}{14}$ **42.** $\frac{1}{27} \times \frac{39}{41}$ **43.** $\frac{3}{4} \times \frac{11}{55}$ **44.** $\frac{18}{19} \times \frac{4}{15}$

45. $\frac{11}{16} \times \frac{8}{9}$ **46.** $\frac{1}{2} \times \frac{62}{71}$ **47.** $\frac{1}{4} \times 16$ **48.** $\frac{2}{3} \times \frac{5}{15}$ **49.** $\frac{25}{27} \times \frac{21}{40}$

▶ Applications

50. John normally swims $\frac{4}{5}$ mile every day. Yesterday he didn't have much time, so he swam half his usual distance. How far did John swim?

51. Ruby has a punch recipe that serves forty people. She is having a dinner party for ten and wants to serve punch. The recipe calls for four quarts of grape juice. How much juice should Ruby use?

▶ Review (Lesson 5.2)

Write each whole or mixed number as an improper fraction.

52. $3\frac{3}{4}$ **53.** $7\frac{3}{8}$ **54.** $5\frac{7}{10}$ **55.** 9   **56.** $6\frac{2}{3}$

57. $7\frac{4}{5}$ **58.** $2\frac{11}{12}$ **59.** 4   **60.** $11\frac{1}{2}$ **61.** $8\frac{5}{6}$

62. $4\frac{1}{4}$ **63.** 20   **64.** $5\frac{1}{3}$ **65.** $8\frac{3}{5}$ **66.** $1\frac{15}{16}$

6.3 Using Guess and Check to Solve Problems

Bernice dozed off while watching a movie on TV and woke up just as the movie was over. As it turned out, she was asleep 25 minutes more than she was awake. If the movie ran for 109 minutes, how long was Bernice awake?

- **Read**   The movie was 109 minutes long. Bernice was asleep 25 minutes more than she was awake. You are trying to find the number of minutes she was awake.

- **Plan**   You can use "Guess and Check" to solve this problem. First you guess and then you figure out whether your guess is too high or too low. This helps you make a better second guess. You guess and check until you find the correct answer.

- **Do**   Since Bernice was awake for less than half of the movie, choose as your first guess a number less than half of 109.

First guess: 50 minutes
Awake: 50 minutes
Asleep: 50 + 25 = 75 minutes
Total: 50 + 75 = 125 minutes
125 > 109
That number is too high, since the movie ran for 109 minutes.
The first guess is too high. Try a guess smaller than 50 minutes.

Second guess: 40 minutes
Awake: 40 minutes
Asleep: 40 + 25 = 65 minutes
Total: 40 + 65 = 105 minutes
105 < 109
The second guess is too low, but it is very close. Try another guess close to but greater than 40 minutes.

Third guess: 42 minutes.
Awake: 42 minutes
Asleep: 42 + 25 = 67 minutes
Total: 42 + 67 = 109 minutes
109 = 109
That works! Bernice was awake for 42 minutes of the movie before she fell asleep.

- **Check** Put your calculation into words. Bernice was awake for 42 minutes and asleep for 67 minutes, which adds up to 109 minutes. And 67 is 25 more than 42.

Refer to the problem about Bernice to answer the following questions.

1. Describe in your own words why your first guess should be a number less than half of 109.

2. When should your second guess be lower than your first guess? When should your second guess be higher than your first guess?

3. Explain why you know the guess of 40 is closer to the correct answer than the guess of 50.

Use Guess and Check to solve each problem.

4. Miss Quintana's Spanish class is next door to Mr. Alouette's French class. The two classes have a total of 72 students. There are 16 more students in Spanish than in French. Is a guess of 20 students in French class too high or too low? How many students are in each class?

5. José and Cecile went out for dinner. Together they spent $28. José spent $5 more than Cecile. Is a guess of $18 for José too high or too low? How much money did each spend?

▶ **Review** (Lesson 5.7)

Change each fraction to a decimal.

6. $\frac{5}{18}$ 7. $\frac{11}{18}$ 8. $\frac{13}{44}$ 9. $\frac{3}{22}$

6.4 Multiplying Mixed Numbers

"Sorry, I've got to run," Etta said. "I don't want to be late. I make $6 an hour today!" "Wait a second," exclaimed Annie. "You told me you made $4 an hour." "I do on most days, but today is Sunday and I get time and a half."

"Time and a half" means $1\frac{1}{2}$ times the usual pay. To find out whether Etta is correct, multiply $1\frac{1}{2}$ by 4.

Examples

To multiply mixed numbers, first write each mixed number as an improper fraction.

A Multiply. $2\frac{3}{4} \times 2\frac{1}{2}$

$2\frac{3}{4} = \frac{11}{4}$ $2\frac{1}{2} = \frac{5}{2}$

$\frac{11}{4} \times \frac{5}{2} = \frac{55}{8}$, or $6\frac{7}{8}$

$2\frac{3}{4} \times 2\frac{1}{2} = 6\frac{7}{8}$

B Multiply. $\frac{3}{4} \times 2\frac{2}{5}$

$\frac{3}{4} \times 2\frac{2}{5} = \frac{3}{4} \times \frac{12}{5}$

$\frac{3}{4} \times \frac{\overset{3}{\cancel{12}}}{5} = \frac{9}{5}$, or $1\frac{4}{5}$

$\frac{3}{4} \times 2\frac{2}{5} = 1\frac{4}{5}$

C Multiply. $1\frac{1}{2} \times 4$

$1\frac{1}{2} \times 4 = \frac{3}{2} \times \frac{4}{1}$

$\frac{3}{\cancel{2}} \times \frac{\overset{2}{\cancel{4}}}{1} = \frac{6}{1} = 6$

$1\frac{1}{2} \times 4 = 6$

▶ Think and Discuss

1. Write $2\frac{2}{5}$ as an improper fraction.

2. Multiply $1\frac{2}{3}$ by $3\frac{3}{4}$.

3. What shortcut is used in Example B?

4. Refer to the introduction to this lesson. Was Etta's calculation correct? Show the problem and solution.

5. Select two mixed numbers between 3 and 4. Explain why their product will be greater than 9 and less than 16.

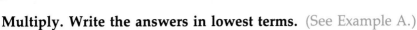

Exercises

Multiply. Write the answers in lowest terms. (See Example A.)

6. $1\frac{1}{3} \times 2\frac{1}{2}$ **7.** $2\frac{2}{3} \times 5\frac{1}{4}$ **8.** $6\frac{1}{2} \times 1\frac{2}{3}$ **9.** $3\frac{5}{7} \times 9\frac{1}{3}$

10. $1\frac{1}{10} \times 4\frac{1}{3}$ **11.** $4\frac{3}{4} \times 5\frac{1}{5}$ **12.** $8\frac{1}{5} \times 4\frac{2}{7}$ **13.** $4\frac{8}{9} \times 2\frac{4}{5}$

Multiply. Write the answers in lowest terms. (See Example B.)

14. $\frac{1}{3} \times 2\frac{1}{2}$ **15.** $\frac{6}{15} \times 1\frac{1}{3}$ **16.** $\frac{2}{5} \times 8\frac{1}{7}$ **17.** $\frac{1}{2} \times 4\frac{5}{8}$ **18.** $\frac{3}{7} \times 1\frac{2}{5}$

19. $6\frac{1}{3} \times \frac{1}{2}$ **20.** $5\frac{3}{4} \times \frac{2}{9}$ **21.** $\frac{4}{7} \times 5\frac{1}{5}$ **22.** $10\frac{2}{3} \times \frac{1}{6}$ **23.** $4\frac{4}{9} \times \frac{2}{11}$

Multiply. Write the answers in lowest terms. (See Example C.)

24. $4 \times 2\frac{1}{3}$ **25.** $15\frac{4}{5} \times 5$ **26.** $3\frac{14}{15} \times 4$ **27.** $9\frac{1}{2} \times 5$

28. $18\frac{1}{10} \times 20$ **29.** $10 \times 5\frac{7}{10}$ **30.** $7 \times 4\frac{5}{6}$ **31.** $1\frac{1}{4} \times 3$

▶ Mixed Practice (For more practice, see page 419.)

Multiply. Write the answers in lowest terms.

32. $2\frac{1}{3} \times 6$ **33.** $\frac{2}{3} \times 5\frac{1}{2}$ **34.** $1\frac{1}{5} \times 2\frac{2}{3}$ **35.** $11 \times 6\frac{2}{3}$

36. $7\frac{1}{24} \times 7$ **37.** $\frac{2}{5} \times 2\frac{3}{4}$ **38.** $1\frac{4}{5} \times 6\frac{7}{8}$ **39.** $2 \times 4\frac{3}{4}$

▶ Applications

40. Bill normally works seven-hour days. This week he worked four full days and one half-day. How many hours did he work?

41. How much oats and water do you need to make $2\frac{1}{2}$ cups of hot oatmeal?

> **Oatmeal Cooking Instructions**
> Add 2/3 cup rolled oats and 3/4 tsp salt to 2 cups boiling water.
> Boil 5-8 minutes.
> Makes two 1/2 cup servings

▶ Review (Lessons 2.2, 2.4, 2.8, 2.10)

Multiply.

42. $\begin{array}{r} 36 \\ \times\ 4.5 \\ \hline \end{array}$ **43.** $\begin{array}{r} 0.09 \\ \times\ 0.004 \\ \hline \end{array}$ **44.** $\begin{array}{r} 8925 \\ \times\ \ \ \ 6 \\ \hline \end{array}$ **45.** $\begin{array}{r} 1.24 \\ \times\ 0.53 \\ \hline \end{array}$ **46.** $\begin{array}{r} 307 \\ \times\ 207 \\ \hline \end{array}$

47. $\begin{array}{r} 2.7 \\ \times\ 6.3 \\ \hline \end{array}$ **48.** $\begin{array}{r} 9.5 \\ \times\ 0.5 \\ \hline \end{array}$ **49.** $\begin{array}{r} 66.6 \\ \times\ 403 \\ \hline \end{array}$ **50.** $\begin{array}{r} 0.001 \\ \times\ 0.001 \\ \hline \end{array}$ **51.** $\begin{array}{r} 78 \\ \times\ 81 \\ \hline \end{array}$

6.5 Estimating Products of Fractions and Mixed Numbers

"**W**hen you use math, you don't always need an exact answer," Mrs. Oxtoby told her class. "Let's look at the problem $\frac{7}{8} \times \frac{13}{25}$. Can you estimate that product mentally?"

Melanie raised her hand. "7 times 13 is about 100. 8 times 25 is 200. So the product is about $\frac{1}{2}$."

$$\frac{7}{8} \times \frac{13}{25} \approx \frac{100}{200} = \frac{1}{2}$$

Thomas said, "I got $\frac{1}{2}$ too, but I worked with each fraction first. Since $\frac{7}{8}$ is about 1 and $\frac{13}{25}$ is about $\frac{1}{2}$, the product is about $\frac{1}{2}$.

$$\frac{7}{8} \times \frac{13}{25} \approx 1 \times \frac{1}{2} = \frac{1}{2}$$

1. Is $\frac{7}{8}$ really about 1? Write $\frac{7}{8}$ as a decimal.

2. Is $\frac{13}{25}$ really about $\frac{1}{2}$? Write $\frac{13}{25}$ as a decimal.

3. Which of the fractions in the following list are close to 1? Which are close to $\frac{1}{2}$? Which are close to neither 1 nor $\frac{1}{2}$?

   $\frac{11}{10} \quad \frac{7}{29} \quad \frac{5}{9} \quad \frac{21}{20} \quad \frac{15}{16} \quad \frac{1}{20} \quad \frac{9}{20} \quad \frac{100}{205}$

4. How can you tell if a fraction is close to 1? To $\frac{1}{2}$? Discuss.

"Estimation becomes even more important when the numbers get bigger. How would you estimate $6\frac{1}{5} \times 3\frac{5}{7}$ mentally?" Mrs. Oxtoby asked.

Rafael raised his hand. "I'd just use whole numbers. $6\frac{1}{5}$ is close to 6. $3\frac{5}{7}$ is close to 4. The product is about 24."

$$6\frac{1}{5} \times 3\frac{5}{7} \approx 6 \times 4 = 24$$

5. One student said her estimate for $8\frac{3}{8} \times 9\frac{11}{16}$ is somewhere between 72 and 90. How do you think she arrived at her estimate? What would your estimate be?

6. Using estimation, determine if the following statement is true: $3\frac{1}{8} \times 8\frac{1}{7} > 24$. Explain your reasoning.

Exercises

Estimate each product mentally.

7. $\frac{3}{8} \times \frac{5}{12}$ **8.** $4\frac{9}{10} \times 2\frac{1}{4}$ **9.** $\frac{7}{8} \times 4\frac{7}{16}$ **10.** $\frac{8}{11} \times \frac{3}{7}$ **11.** $\frac{5}{8} \times 10\frac{2}{9}$

12. $1\frac{1}{8} \times 1\frac{7}{8}$ **13.** $\frac{10}{19} \times 5\frac{3}{4}$ **14.** $\frac{9}{20} \times \frac{19}{20}$ **15.** $8\frac{12}{21} \times 2\frac{1}{8}$ **16.** $\frac{21}{50} \times \frac{17}{30}$

17. $5\frac{4}{5} \times \frac{19}{21}$ **18.** $25\frac{1}{3} \times 2\frac{1}{4}$ **19.** $\frac{7}{9} \times \frac{16}{33}$ **20.** $13\frac{2}{15} \times 2\frac{8}{9}$ **21.** $\frac{11}{20} \times 8\frac{1}{2}$

▶ Applications

22. Mr. Ras works $37\frac{1}{2}$ hours a week and earns \$11 per hour. Estimate his gross pay.

23. Rodney is going backpacking. His route is $23\frac{5}{8}$ miles long. He plans to walk $\frac{1}{5}$ of the total distance each day. About how many miles will he walk each day?

$23\frac{5}{8}$

▶ Review (Lessons 3.2, 3.5)

Divide.

24. $4\overline{)17}$ **25.** $11\overline{)123}$ **26.** $8\overline{)69}$ **27.** $7\overline{)49}$ **28.** $9\overline{)852}$

29. $15\overline{)316}$ **30.** $22\overline{)78}$ **31.** $25\overline{)555}$ **32.** $16\overline{)271}$ **33.** $32\overline{)467}$

6.6 Dividing Fractions

When a community play was proposed, Greg suggested that half of the play's profits be divided equally among the three organizations involved. What part of the total profit should each get?

Divide to determine each group's part of the total profit.

To divide a fraction, multiply by the reciprocal of the divisor.
Reciprocals are two numbers whose product is 1.
For example, $\frac{3}{8} \times \frac{8}{3} = \frac{24}{24} = 1$.   The reciprocal of $\frac{3}{8}$ is $\frac{8}{3}$.

A Find the reciprocal of $\frac{5}{6}$.	**B** Divide. $\frac{2}{3} \div \frac{1}{2}$	**C** Divide. $\frac{1}{2} \div 3$

A Find the reciprocal of $\frac{5}{6}$.

$\frac{5}{6} \times \frac{6}{5}$ Invert the numerator and the denominator.

Since $\frac{5}{6} \times \frac{6}{5} = 1$, the reciprocal of $\frac{5}{6}$ is $\frac{6}{5}$.

B Divide. $\frac{2}{3} \div \frac{1}{2}$

$\frac{2}{3} \div \frac{1}{2} = \frac{2}{3} \times \frac{2}{1}$

$\frac{2}{3} \div \frac{1}{2} = \frac{4}{3}$, or $1\frac{1}{3}$

$\frac{2}{3} \div \frac{1}{2} = 1\frac{1}{3}$

C Divide. $\frac{1}{2} \div 3$

$\frac{1}{2} \div 3 = \frac{1}{2} \times \frac{1}{3}$

$\frac{1}{2} \div 3 = \frac{1}{6}$

▶ Think and Discuss

1. Write 3 as a fraction. Find the reciprocal of 3.

2. Divide $\frac{3}{8}$ by $\frac{5}{6}$.

3. Refer to the introduction to this lesson. What part of the profits did each group get?

4. What is the reciprocal of 1?

SKILLS

5. What is the product of 6 and its reciprocal?

6. What is the product of any number multiplied by its reciprocal?

Exercises

Find the reciprocal. (See Example A.)

7. $\frac{2}{3}$ **8.** $\frac{5}{7}$ **9.** 4 **10.** $\frac{1}{2}$ **11.** $\frac{5}{16}$ **12.** 10 **13.** $\frac{1}{5}$ **14.** $\frac{7}{300}$

Divide. Write the answers in lowest terms. (See Example B.)

15. $\frac{8}{9} \div \frac{4}{9}$ **16.** $\frac{3}{4} \div \frac{5}{8}$ **17.** $\frac{4}{5} \div \frac{8}{15}$ **18.** $\frac{8}{9} \div \frac{5}{6}$ **19.** $\frac{7}{11} \div \frac{11}{12}$

20. $\frac{2}{3} \div \frac{5}{8}$ **21.** $\frac{8}{11} \div \frac{1}{2}$ **22.** $\frac{9}{16} \div \frac{3}{8}$ **23.** $\frac{5}{6} \div \frac{1}{4}$ **24.** $\frac{4}{7} \div \frac{4}{9}$

Divide. Write the answers in lowest terms. (See Example C.)

25. $\frac{3}{4} \div 3$ **26.** $\frac{5}{6} \div 7$ **27.** $\frac{9}{10} \div 9$ **28.** $\frac{2}{3} \div 6$ **29.** $\frac{1}{6} \div 2$

30. $\frac{11}{12} \div 7$ **31.** $\frac{6}{7} \div 3$ **32.** $\frac{8}{15} \div 4$ **33.** $\frac{12}{25} \div 5$ **34.** $\frac{4}{5} \div 10$

▶ **Mixed Practice** (For more practice, see page 420.)

Divide. Write the answers in lowest terms.

35. $\frac{4}{9} \div 2$ **36.** $\frac{5}{6} \div \frac{3}{4}$ **37.** $\frac{15}{16} \div 10$ **38.** $\frac{7}{12} \div \frac{7}{8}$ **39.** $\frac{15}{16} \div \frac{2}{3}$

40. $\frac{3}{8} \div \frac{1}{4}$ **41.** $\frac{7}{8} \div 14$ **42.** $\frac{4}{5} \div \frac{1}{2}$ **43.** $\frac{4}{5} \div 2$ **44.** $\frac{4}{11} \div 7$

▶ **Applications**

45. Cris is building a record holder that is three-quarters of a yard long. He wants to divide it into three equal sections. How wide will each section be?

46. About $\frac{1}{18}$ of the world's population lives in the United States. About $\frac{1}{216}$ of the world's population lives in New York State. About what part of the U.S. population lives in New York State?

▶ **Review** (Lessons 1.12, 2.6)

Estimate each sum or product.

47. $\begin{array}{r} 428 \\ \times\ \ 75 \end{array}$ **48.** $\begin{array}{r} 669 \\ +\ 741 \end{array}$ **49.** $\begin{array}{r} 189 \\ \times\ 537 \end{array}$ **50.** $\begin{array}{r} 919 \\ +\ 850 \end{array}$ **51.** $\begin{array}{r} 647 \\ \times\ 636 \end{array}$

6.7 Dividing Mixed Numbers

ROADRUNNER CLUB MARATHON

Magda is training for a race. One day she ran $4\frac{1}{2}$ miles in $\frac{1}{2}$ hour. To find how many miles per hour she ran, Magda can divide $4\frac{1}{2}$ by $\frac{1}{2}$.

Examples

To divide mixed numbers, first write each mixed number as an improper fraction. Then proceed as with fractions.

A Divide. $3\frac{3}{4} \div 4\frac{2}{5}$

$$3\frac{3}{4} \div 4\frac{2}{5} = \frac{15}{4} \div \frac{22}{5}$$

$$3\frac{3}{4} \div 4\frac{2}{5} = \frac{15}{4} \times \frac{5}{22}$$

$$3\frac{3}{4} \div 4\frac{2}{5} = \frac{75}{88}$$

B Divide. $\frac{4}{5} \div 1\frac{1}{2}$

$$\frac{4}{5} \div 1\frac{1}{2} = \frac{4}{5} \div \frac{3}{2}$$

$$\frac{4}{5} \div 1\frac{1}{2} = \frac{4}{5} \times \frac{2}{3}$$

$$\frac{4}{5} \div 1\frac{1}{2} = \frac{8}{15}$$

C Divide. $4\frac{1}{2} \div \frac{1}{2}$

$$4\frac{1}{2} \div \frac{1}{2} = \frac{9}{2} \div \frac{1}{2}$$

$$4\frac{1}{2} \div \frac{1}{2} = \frac{9}{2} \times \frac{\overset{1}{2}}{\underset{1}{1}} = \frac{9}{1}$$

$$4\frac{1}{2} \div \frac{1}{2} = 9$$

▶ Think and Discuss

1. Divide $\frac{1}{4}$ by $1\frac{1}{2}$.

2. Refer to the introduction to this lesson. How many miles per hour did Magda run?

3. Suppose you were asked to tutor a younger student in mixed-number division. Explain how to solve this problem: $4\frac{1}{5} \div \frac{1}{4}$.

4. If you estimate $3\frac{1}{8} \div 4\frac{3}{7}$, will the quotient be less than 1 or greater than 1? Explain.

5. When is 1 the quotient of two mixed numbers? Explain.

Exercises

Divide. Write the answers in lowest terms. (See Example A.)

6. $1\frac{1}{5} \div 3\frac{1}{2}$ **7.** $3\frac{1}{7} \div 1\frac{1}{4}$ **8.** $1\frac{3}{5} \div 3\frac{2}{3}$ **9.** $2\frac{2}{3} \div 4\frac{1}{2}$ **10.** $1\frac{9}{10} \div 1\frac{2}{5}$

11. $3\frac{1}{3} \div 2\frac{1}{2}$ **12.** $5\frac{1}{3} \div 1\frac{3}{5}$ **13.** $5\frac{2}{5} \div 2\frac{2}{9}$ **14.** $8\frac{1}{3} \div 3\frac{1}{3}$ **15.** $1\frac{1}{5} \div 3\frac{3}{5}$

Divide. Write the answers in lowest terms. (See Example B.)

16. $\frac{6}{9} \div 2\frac{1}{2}$ **17.** $\frac{1}{3} \div 4\frac{2}{7}$ **18.** $\frac{2}{5} \div 10\frac{1}{12}$ **19.** $\frac{1}{4} \div 1\frac{1}{4}$ **20.** $\frac{7}{9} \div 2\frac{1}{3}$

21. $\frac{4}{8} \div 5\frac{1}{5}$ **22.** $\frac{1}{2} \div 1\frac{1}{9}$ **23.** $\frac{12}{18} \div 3\frac{3}{5}$ **24.** $\frac{2}{3} \div 4\frac{3}{8}$ **25.** $\frac{4}{5} \div 9\frac{1}{10}$

Divide. Write the answers in lowest terms. (See Example C.)

26. $4\frac{3}{5} \div \frac{2}{3}$ **27.** $5\frac{3}{4} \div \frac{1}{2}$ **28.** $2\frac{1}{2} \div \frac{4}{5}$ **29.** $3\frac{5}{8} \div \frac{2}{3}$ **30.** $6\frac{9}{10} \div \frac{4}{9}$

31. $1\frac{3}{4} \div \frac{10}{11}$ **32.** $7\frac{1}{7} \div \frac{5}{7}$ **33.** $2\frac{7}{8} \div \frac{1}{3}$ **34.** $1\frac{7}{16} \div \frac{1}{4}$ **35.** $8\frac{4}{5} \div \frac{5}{6}$

▶ **Mixed Practice** (For more practice, see page 420.)

Divide. Write the answers in lowest terms.

36. $\frac{1}{5} \div 1\frac{2}{3}$ **37.** $2\frac{5}{8} \div \frac{7}{8}$ **38.** $\frac{5}{8} \div 2\frac{1}{12}$ **39.** $4\frac{1}{2} \div 2\frac{2}{3}$ **40.** $10\frac{1}{10} \div \frac{2}{7}$

41. $7\frac{1}{2} \div 1\frac{1}{2}$ **42.** $11\frac{1}{4} \div \frac{1}{8}$ **43.** $\frac{2}{5} \div 4\frac{4}{5}$ **44.** $3\frac{3}{7} \div 2\frac{1}{6}$ **45.** $\frac{4}{9} \div 9\frac{4}{9}$

46. $8\frac{1}{4} \div \frac{1}{2}$ **47.** $1\frac{2}{3} \div 1\frac{1}{4}$ **48.** $\frac{3}{4} \div 1\frac{2}{3}$ **49.** $5\frac{2}{5} \div 1\frac{2}{5}$ **50.** $\frac{6}{11} \div 6\frac{4}{5}$

▶ **Applications**

51. A carpenter is constructing a $21\frac{1}{3}$-foot wall from panels that are $1\frac{1}{3}$ feet wide. How many panels are needed?

52. Sara is typing a photo caption for the school paper. If each typed character is $\frac{1}{12}$ inch wide, how many characters will fit on a line that measures $7\frac{1}{2}$ inches long?

▶ **Review** (Lessons 2.3, 2.7)

Multiply mentally.

53. 60×800 **54.** 300×900 **55.** 8.57×10 **56.** 100×0.9

57. 50×80 **58.** 7000×80 **59.** 5.4×1000 **60.** 10×0.065

61. 1000×0.6 **62.** 70×40 **63.** 55.4×10 **64.** 400×600

6.8 Fractions: A Business Application

Rae makes life-size dolls of celebrities, politicians, and famous people from history. Recently, Rae got an order from Desert Community College. Their school mascot is Laurence the Lion, and they want 250 Laurence dolls to sell in the student center bookstore.

Rae's first task is to plan what supplies she will need for each lion. The table below shows her figures:

Materials for 1 Lion	Amount	Unit Cost
gold velour (body)	$\frac{5}{8}$ yard	$3.69/yd.
white fleece (chest)	$\frac{1}{3}$ yard	$4.25/yd.
brown fleece (mane)	$\frac{3}{8}$ yard	$4.25/yd.
blue satin (school letter and cap)	$\frac{1}{4}$ yard	$8.99/yd.
yellow cord (trim)	$\frac{2}{3}$ yard	$0.45/yd.
buttons (eyes)	2	$2\frac{1}{2}$¢
stuffing	$\frac{1}{2}$ bag	$0.89/bag

• ▶ Think and Discuss

1. What does "unit cost" in the table above mean?

2. How much gold velour does Rae need for each lion? What is the unit cost of the velour? How much must Rae spend on the velour?

3. Describe how Rae can find what one lion will cost to make. How many calculations does she need to make?

4. Find the cost of materials for one lion.

5. Rae charges $12.50 per hour for labor (the time she spends sewing). When she made a sample doll for the community college, it took her 3 hours. Once she begins production, though, she estimates that she can cut that time in half. How much will each lion cost in labor alone?

6. What does one lion cost in supplies and labor together?

Rae must buy large quantities of expensive fabric, so she wants to waste as little fabric as possible.

7. Make a table like the one Rae made. Add two new columns: "Total Amount" and "Total Cost." Use Rae's table to figure the amounts of materials that are needed to make 250 lions. Write the answers under "Total Amount." Calculate the cost of each of the materials for 250 lions. Write the answers under "Total Cost."

8. Find the total cost of materials for 250 lions.

9. Find the total time it will take Rae to produce the entire order. How much will she charge for her labor?

Rae always adds on a profit of $\frac{1}{5}$ the total cost of materials when estimating her total charge.

10. What will Rae's profit be?

11. How will Rae calculate her total charge for 250 Laurence the Lion dolls? Find the total charge.

12. Rae's next project will be to make 7 large tropical fish for a pet store. Use the following table to figure out what each fish will cost to make and what the total of 7 will cost.

Materials for 1 Fish	Amount	Unit Cost
multicolored satin	$1\frac{1}{2}$ yards	$8.99/yd.
gold cord for trim	$3\frac{1}{4}$ yards	$0.45/yd.
gold buttons for eyes	2	$8\frac{1}{2}$¢
gold thread for detailing	$\frac{1}{4}$ spool	$0.99/spool
stuffing	$1\frac{1}{2}$ bags	$0.89/bag

13. If it takes $\frac{3}{4}$ hour to make 1 fish and Rae charges $12.50 an hour, what will be her total labor charge for 7 fish?

14. If Rae adds on a profit of $\frac{1}{5}$ the total cost of materials, what will her profit be for 7 fish?

▶ **Review** (Lessons 2.8, 2.10)

Multiply.

15. 42×0.46

16. 2.9×4.5

17. 9.125×45

18. 603×9.55

19. 0.75×0.99

20. 475×0.5

21. 91×0.1

22. 7654×2.2

Chapter 6 Review

Multiply. Write the answers in lowest terms. (Lessons 6.1, 6.2)

1. $\frac{1}{2} \times \frac{3}{4}$ 2. $\frac{5}{6} \times \frac{1}{8}$ 3. $\frac{2}{3} \times \frac{4}{5}$ 4. $\frac{1}{12} \times \frac{1}{3}$ 5. $\frac{3}{8} \times \frac{4}{9}$

6. $\frac{2}{3} \times \frac{3}{10}$ 7. $9 \times \frac{7}{8}$ 8. $\frac{5}{12} \times 6$ 9. $\frac{3}{5} \times \frac{5}{12}$ 10. $\frac{5}{8} \times \frac{5}{6}$ 11. $\frac{14}{15} \times \frac{5}{7}$

Use Guess and Check to solve. (Lesson 6.3)

12. Melissa went out to lunch twice last week. She spent $7.65 in all. The first lunch cost $0.45 less than the second lunch cost. How much money did she spend on each lunch?

13. Video tapes cost $3.99. Audio tapes cost $2.49. Today Hernán spent $39.87 on tapes. The number of audio tapes he bought was 3 more than the number of video tapes he bought. How many of each kind did he buy?

Multiply. Write the answers in lowest terms. (Lesson 6.4)

14. $\frac{3}{4} \times 1\frac{2}{3}$ 15. $5\frac{4}{5} \times 10$ 16. $1\frac{7}{8} \times 1\frac{1}{6}$ 17. $7\frac{1}{2} \times 2\frac{1}{5}$ 18. $4\frac{1}{3} \times 3\frac{1}{4}$

Estimate each product. (Lesson 6.5)

19. $\frac{11}{12} \times \frac{5}{6}$ 20. $4\frac{1}{2} \times \frac{9}{10}$ 21. $7\frac{1}{4} \times 5\frac{1}{3}$ 22. $6\frac{5}{8} \times 3\frac{2}{5}$ 23. $2\frac{7}{8} \times 4\frac{1}{9}$

Find the reciprocal. (Lesson 6.6)

24. $\frac{4}{5}$ 25. $\frac{13}{16}$ 26. $\frac{7}{15}$ 27. $\frac{9}{22}$ 28. $\frac{13}{17}$ 29. $\frac{16}{27}$ 30. $\frac{23}{32}$

Divide. Write the answers in lowest terms. (Lessons 6.6, 6.7)

31. $\frac{7}{8} \div \frac{7}{12}$ 32. $\frac{3}{4} \div 8$ 33. $3\frac{1}{5} \div \frac{2}{5}$ 34. $\frac{7}{10} \div 5\frac{1}{2}$ 35. $\frac{9}{10} \div \frac{1}{3}$

36. $6 \div 5\frac{3}{4}$ 37. $5\frac{5}{6} \div 2\frac{1}{2}$ 38. $9\frac{1}{3} \div 4\frac{2}{3}$ 39. $\frac{7}{8} \div 8\frac{1}{6}$ 40. $\frac{4}{5} \div 1\frac{1}{4}$

Multiply or divide. Write each answer in lowest terms. (Lessons 6.4, 6.7)

41. Kevin's room is $8\frac{3}{4}$ feet wide. How many posters $1\frac{3}{4}$ feet wide can he fit along the wall?

42. Mary is doubling a recipe that calls for $2\frac{2}{3}$ cups of flour. How much flour should she use?

Chapter 6 Test

Estimate each product.

1. $3\frac{7}{10} \times 2\frac{1}{6}$ 2. $\frac{11}{12} \times \frac{7}{8}$ 3. $5\frac{1}{3} \times 5\frac{3}{4}$ 4. $1\frac{5}{6} \times 2\frac{3}{8}$

Find the reciprocal.

5. $\frac{10}{36}$ 6. $\frac{33}{56}$ 7. $\frac{9}{17}$

Multiply. Write the answers in lowest terms.

8. $\frac{3}{8} \times \frac{1}{3}$ 9. $5\frac{1}{2} \times 4$ 10. $\frac{7}{12} \times 6$ 11. $4\frac{4}{9} \times 1\frac{4}{5}$ 12. $\frac{1}{3} \times \frac{6}{7}$

13. $9 \times 1\frac{5}{6}$ 14. $3\frac{1}{3} \times 6\frac{1}{4}$ 15. $\frac{7}{10} \times \frac{4}{5}$ 16. $2\frac{5}{8} \times 1\frac{2}{3}$ 17. $\frac{4}{5} \times \frac{4}{7}$

18. Marcy is making 3 skirts. Each skirt requires $1\frac{3}{8}$ yards of fabric. How much fabric does she need altogether?

19. To can 1 quart of apples you need $2\frac{1}{2}$ pounds of fresh apples. How many pounds of apples are needed to can 8 quarts?

Use Guess and Check to solve.

20. Yahna has 15 coins in her hand worth $3.00. Some are dimes and the rest are quarters. How many quarters and dimes does she have?

21. Tickets to the championship game are $2 for students and $5 for adults. Ten times as many students as adults bought tickets. Ticket sales were $250. How many of each kind of ticket were sold?

Divide. Write the answers in lowest terms.

22. $\frac{15}{16} \div \frac{5}{8}$ 23. $\frac{3}{7} \div \frac{7}{10}$ 24. $6\frac{1}{6} \div \frac{2}{3}$ 25. $1\frac{3}{4} \div 2\frac{4}{5}$ 26. $\frac{2}{3} \div 1\frac{1}{2}$

27. $\frac{13}{16} \div 6\frac{1}{2}$ 28. $\frac{8}{9} \div \frac{7}{12}$ 29. $3\frac{1}{5} \div 5\frac{1}{3}$ 30. $\frac{4}{5} \div \frac{5}{6}$ 31. $\frac{7}{8} \div 9$

"I never thought I'd use fractions until I got a job at a fabric store. My first customer bought the same material in 3 different colors. I had to add $1\frac{5}{8}$ yards, $3\frac{2}{3}$ yards, and $2\frac{3}{4}$ yards to find the total amount of fabric."

Adding and Subtracting Fractions and Mixed Numbers

Knowing how to add and subtract fractions and mixed numbers can help you
- determine how many miles you run in a week.
- determine the correct size of matting for a photo.
- figure out the number of letters that will fit on a poster you want to make.

Fraction Magic

Materials: a centimeter ruler

Do you know the trick in which a magician takes two or three pieces of rope and mysteriously combines them into one piece? Let's try this trick with line segments. Work with 2 or 3 classmates. Each person should draw a line segment. Measure each line segment to the nearest $\frac{1}{10}$ centimeter. Then draw the line segments next to each other to form one long line segment. Then measure it. Write an addition statement. See the example below.

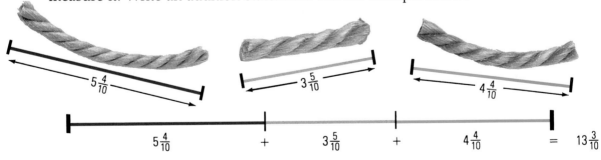

$$5\frac{4}{10} \quad + \quad 3\frac{5}{10} \quad + \quad 4\frac{4}{10} \quad = \quad 13\frac{3}{10}$$

7.1 Adding and Subtracting Fractions with Like Denominators

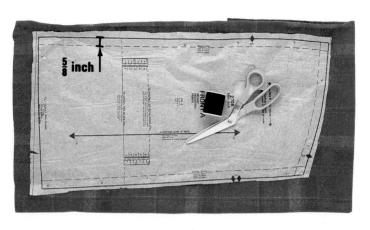

Lila is altering the skirt pattern shown here. She needs to make the waist $\frac{3}{8}$ inch larger on each side of the skirt. She can do this by decreasing the $\frac{5}{8}$-inch seam allowance that is indicated on the pattern. What will each seam allowance measure?

To answer this question, you can subtract $\frac{3}{8}$ from $\frac{5}{8}$.

Examples

To add or subtract fractions with like denominators, add or subtract the numerators. The denominator remains the same.

A Add. $\frac{5}{12} + \frac{11}{12}$

$\frac{5}{12} + \frac{11}{12} = \frac{16}{12} = 1\frac{4}{12} = 1\frac{1}{3}$ Write the answer in lowest terms.

B Subtract. $\frac{5}{8} - \frac{3}{8}$

$\frac{5}{8} - \frac{3}{8} = \frac{2}{8} = \frac{1}{4}$ Write the answer in lowest terms.

▶ Think and Discuss

1. Add. $\frac{7}{9} + \frac{5}{9}$

2. Subtract. $\frac{11}{12} - \frac{1}{12}$

3. Refer to the introduction to this lesson. How wide should Lila make the seam allowance on the pattern?

4. What is wrong with the statement "$\frac{2}{3} + \frac{1}{3} = \frac{3}{6}$"? What is the correct answer?

Exercises

Add. Write the answers in lowest terms. (See Example A.)

5. $\frac{6}{13}$ $+ \frac{4}{13}$

6. $\frac{3}{11}$ $+ \frac{4}{11}$

7. $\frac{2}{7}$ $+ \frac{5}{7}$

8. $\frac{7}{12}$ $+ \frac{11}{12}$

9. $\frac{9}{30}$ $+ \frac{23}{30}$

10. $\frac{5}{18}$ $+ \frac{17}{18}$

11. $\frac{5}{6} + \frac{5}{6}$

12. $\frac{9}{14} + \frac{5}{14}$

13. $\frac{11}{16} + \frac{9}{16} + \frac{7}{16}$

14. $\frac{17}{20} + \frac{13}{20} + \frac{9}{20}$

Subtract. Write the answers in lowest terms. (See Example B.)

15. $\frac{3}{5}$ $- \frac{1}{5}$

16. $\frac{9}{20}$ $- \frac{2}{20}$

17. $\frac{7}{9}$ $- \frac{1}{9}$

18. $\frac{9}{10}$ $- \frac{3}{10}$

19. $\frac{9}{16}$ $- \frac{5}{16}$

20. $\frac{17}{24}$ $- \frac{9}{24}$

21. $\frac{5}{9} - \frac{2}{9}$

22. $\frac{11}{18} - \frac{7}{18}$

23. $\frac{19}{25} - \frac{4}{25}$

24. $\frac{34}{35} - \frac{7}{35}$

25. $\frac{47}{50} - \frac{11}{50}$

▶ Mixed Practice (For more practice, see page 421.)

Add or subtract. Write the answers in lowest terms.

26. $\frac{15}{17}$ $- \frac{9}{17}$

27. $\frac{13}{20}$ $+ \frac{17}{20}$

28. $\frac{35}{36}$ $- \frac{32}{36}$

29. $\frac{21}{25}$ $- \frac{16}{25}$

30. $\frac{19}{50}$ $+ \frac{11}{50}$

31. $\frac{43}{48}$ $+ \frac{41}{48}$

32. $\frac{13}{18} + \frac{5}{18}$

33. $\frac{33}{36} - \frac{13}{36}$

34. $\frac{23}{27} - \frac{14}{27}$

35. $\frac{89}{100} - \frac{57}{100}$

36. $\frac{19}{35} + \frac{23}{35}$

37. $\frac{3}{4} + \frac{1}{4} + \frac{3}{4}$

38. $\frac{89}{90} - \frac{19}{90}$

39. $\frac{7}{9} + \frac{5}{9} + \frac{1}{9}$

▶ Applications

40. Look at the charms from Cori's bracelet. How much longer is the first charm than the second?

$\frac{9}{16}$ inch $\frac{5}{16}$ inch

41. Roosevelt spends six hours in school, two hours at basketball practice, and eight hours sleeping. What fraction of his day is left for other activities?

▶ Review (Lesson 1.8)

Round to the nearest hundredth.

42. 8.164   43. 3.095   44. 6.847   45. 0.9325   46. 2.6891   47. 1.842

Round to the nearest tenth.

48. 16.79   49. 7.515   50. 9.053   51. 4.36   52. 0.829   53. 4.654

7.2 Finding the Lowest Common Denominator

A school track is shown here. Ken runs each lap in 6 minutes. Roger runs each lap in 8 minutes. Suppose that both boys begin at the starting line at the same time. After how many minutes will they both be at the starting line at the same time?

The answer is a multiple of 6 and 8.

Examples

To find the lowest common denominator (LCD), list multiples of each denominator. The LCD is the smallest number that appears on all lists.

A **List the first five multiples of 6.**

$6 \times 1 = \mathbf{6}$ $6 \times 2 = \mathbf{12}$ $6 \times 3 = \mathbf{18}$ $6 \times 4 = \mathbf{24}$ $6 \times 5 = \mathbf{30}$
The first five multiples of 6 are **6, 12, 18, 24,** and **30.**

B **Find the LCD of $\frac{2}{6}$ and $\frac{5}{8}$.**

Multiples of 6:   6   12   18   **24**
Multiples of 8:   8   16   **24**   32   40
The LCD of $\frac{2}{6}$ and $\frac{5}{8}$ is **24.**

C **Find the LCD of $\frac{1}{2}$, $\frac{1}{4}$, and $\frac{1}{5}$.**

Multiples of 2:   2   4   6   8   10   12   14   16   18   **20**
Multiples of 4:   4   8   12   16   **20**   24   28
Multiples of 5:   5   10   15   **20**   25   30
The LCD of $\frac{1}{2}$, $\frac{1}{4}$, and $\frac{1}{5}$ is **20.**

▶ Think and Discuss

1. Find the LCD of $\frac{2}{3}$, $\frac{5}{6}$, and $\frac{7}{8}$.

2. Find the LCD of $\frac{2}{5}$ and $\frac{3}{10}$.

3. What are the first five multiples of 7? Of 9?

4. Refer to the introduction to this lesson. After how many minutes will both boys be at the starting line at the same time?

5. When would the LCD of two fractions be one of the denominators?

Exercises

List the first five multiples. (See Example A.)

6. 10   7. 15   8. 12   9. 22   10. 13   11. 18

Find the LCD. (See Example B.)

12. $\frac{1}{2}$ $\frac{3}{4}$ 13. $\frac{1}{3}$ $\frac{5}{9}$ 14. $\frac{7}{8}$ $\frac{5}{6}$ 15. $\frac{9}{10}$ $\frac{3}{4}$ 16. $\frac{5}{7}$ $\frac{2}{3}$

Find the LCD. (See Example C.)

17. $\frac{1}{3}$ $\frac{1}{2}$ $\frac{1}{4}$ 18. $\frac{1}{2}$ $\frac{1}{5}$ $\frac{1}{8}$ 19. $\frac{1}{4}$ $\frac{3}{5}$ $\frac{5}{6}$ 20. $\frac{3}{7}$ $\frac{1}{4}$ $\frac{1}{6}$ 21. $\frac{2}{7}$ $\frac{4}{5}$ $\frac{21}{35}$

▶ **Mixed Practice** (For more practice, see page 421.)

List the first five multiples.

22. 20   23. 11   24. 25   25. 30   26. 17   27. 32

Find the LCD.

28. $\frac{2}{3}$ $\frac{7}{10}$ 29. $\frac{1}{6}$ $\frac{2}{9}$ 30. $\frac{1}{3}$ $\frac{3}{4}$ $\frac{5}{6}$ 31. $\frac{2}{9}$ $\frac{13}{27}$ 32. $\frac{4}{5}$ $\frac{7}{8}$

33. $\frac{1}{10}$ $\frac{2}{3}$ $\frac{2}{15}$ 34. $\frac{1}{12}$ $\frac{3}{8}$ 35. $\frac{1}{2}$ $\frac{2}{5}$ $\frac{3}{7}$ 36. $\frac{1}{6}$ $\frac{1}{7}$ $\frac{1}{2}$ 37. $\frac{7}{11}$ $\frac{5}{7}$

▶ **Applications**

38. John bought stock at $\frac{3}{4}$ (of a dollar). It went up $\frac{1}{2}$ the first day. If John wants to figure out the stock's new value, he first needs to find the LCD of the two fractions. What is it?

39. In football, you get six points for a touchdown, seven points for a touchdown plus an extra point, three points for a field goal, and two points for a safety. List the eight different ways you can score fourteen points.

▶ **Review** (Lesson 1.9)

Add.

40. 6.7
    $+\ 9.88$

41. 6006
    $+\ 1984$

42. 11
    $+\ 7.45$

43. $86{,}471$
    $+\ 89{,}799$

44. 4.638
    $+\ 0.975$

7.3 Adding Fractions with Unlike Denominators

Rainfall

	Daily	Total to Date
Monday 10/3	1/4 inch	1/4 inch
Tuesday 10/4	1/4 inch	1/2 inch
Wednesday 10/5	0 inch	1/2 inch
Thursday 10/6	7/8 inch	_____

Seth records rainfall for a science project in a table like the one shown here. What should he write in the second column for Thursday, October 6th?

To find the *Total to Date*, Seth needs to add $\frac{1}{2}$ inch and $\frac{7}{8}$ inch.

Examples

To add fractions with unlike denominators, rewrite them using their LCD. Then add the numerators.

A Add. $\frac{1}{5} + \frac{2}{3}$

$$\frac{1}{5} = \frac{1}{5} \times \frac{3}{3} = \frac{3}{15}$$
$$+ \frac{2}{3} = \frac{2}{3} \times \frac{5}{5} = \frac{10}{15}$$
$$\overline{\qquad\qquad\qquad \frac{13}{15}}$$

Rewrite using the LCD.

$$\frac{1}{5} + \frac{2}{3} = \frac{13}{15}$$

B Add. $\frac{1}{2} + \frac{7}{8}$

$$\frac{1}{2} = \frac{1}{2} \times \frac{4}{4} = \frac{4}{8}$$
$$+ \frac{7}{8} = \quad\frac{7}{8}\quad = \frac{7}{8}$$
$$\overline{\qquad\qquad\qquad \frac{11}{8}, \text{ or } 1\frac{3}{8}}$$

Rewrite using the LCD.

Rewrite as a mixed number.

$$\frac{1}{2} + \frac{7}{8} = 1\frac{3}{8}$$

▶ Think and Discuss

1. Add. $\frac{9}{10} + \frac{3}{4}$

2. Add. $\frac{7}{12} + \frac{2}{3}$

3. Refer to the introduction to this lesson. What is the *Total to Date* for October 6th?

4. To add fractions with unlike denominators, why must you first find equivalent fractions?

5. How is adding fractions with unlike denominators different from adding fractions with like denominators?

6. Why did $\frac{7}{8}$ remain the same in Example B?

Exercises

Add. Write the answers in lowest terms. (See Example A.)

7. $\frac{1}{5}$
    $+\frac{1}{2}$

8. $\frac{7}{12}$
    $+\frac{1}{6}$

9. $\frac{3}{8}$
    $+\frac{5}{16}$

10. $\frac{7}{30}$
    $+\frac{3}{5}$

11. $\frac{1}{2}$
    $+\frac{2}{7}$

12. $\frac{2}{5}$
    $+\frac{1}{4}$

13. $\frac{4}{9} + \frac{1}{3}$

14. $\frac{7}{10} + \frac{1}{5} + \frac{1}{2}$

15. $\frac{5}{12} + \frac{3}{8}$

16. $\frac{5}{11} + \frac{1}{4}$

Add. Write the answers in lowest terms. (See Example B.)

17. $\frac{1}{2}$
    $+\frac{3}{4}$

18. $\frac{5}{6}$
    $+\frac{2}{3}$

19. $\frac{2}{5}$
    $+\frac{3}{4}$

20. $\frac{9}{10}$
    $+\frac{3}{8}$

21. $\frac{7}{9}$
    $+\frac{5}{12}$

22. $\frac{7}{11}$
    $+\frac{4}{7}$

23. $\frac{4}{5} + \frac{5}{6}$

24. $\frac{5}{8} + \frac{13}{24} + \frac{1}{3}$

25. $\frac{9}{32} + \frac{3}{4}$

26. $\frac{8}{9} + \frac{3}{5}$

▶ **Mixed Practice** (For more practice, see page 422.)

Add. Write the answers in lowest terms.

27. $\frac{1}{4}$
    $+\frac{3}{8}$

28. $\frac{6}{7}$
    $+\frac{3}{14}$

29. $\frac{9}{10}$
    $+\frac{3}{4}$

30. $\frac{4}{9}$
    $+\frac{1}{2}$

31. $\frac{7}{16}$
    $+\frac{2}{3}$

32. $\frac{9}{10}$
    $+\frac{1}{12}$

33. $\frac{3}{4} + \frac{2}{11}$

34. $\frac{7}{15} + \frac{1}{2}$

35. $\frac{7}{8} + \frac{1}{2} + \frac{1}{4}$

36. $\frac{3}{10} + \frac{1}{5} + \frac{1}{6}$

▶ **Applications**

37. Two partial sticks of butter are marked $\frac{2}{3}$ cup and $\frac{3}{8}$ cup. Susan needs 1 cup of butter for baking. Does she have enough?

38. Calvin took a study break. He watched a half-hour show and then talked for twenty minutes. What part of an hour was his break?

▶ **Review** (Lessons 4.2, 4.5, 4.6)

Convert each measure.

39. 72.6 cm to mm

40. 8 mL to L

41. 28.1 kg to g

7.4 Adding Mixed Numbers

Willy's best score for the high jump is $71\frac{1}{2}$ inches. He must improve his best jump by $2\frac{3}{4}$ inches to match the school record. What is the school record?

Examples

To add mixed numbers, add the fractions and the whole numbers separately.

A Add. $7\frac{3}{5} + 4\frac{4}{5}$

$$
\begin{aligned}
7\frac{3}{5} &= \quad 7 + \frac{3}{5} \\
+\ 4\frac{4}{5} &= \quad 4 + \frac{4}{5} \\
\hline
& 11 + \frac{7}{5} = 11 + 1\frac{2}{5} \\
& \qquad\qquad = 12\frac{2}{5}
\end{aligned}
$$

$7\frac{3}{5} + 4\frac{4}{5} = \mathbf{12\frac{2}{5}}$

B Add. $71\frac{1}{2} + 2\frac{3}{4}$

$$
\begin{aligned}
71\frac{1}{2} &= 71 + \frac{1}{2} = 71 + \frac{2}{4} \\
+\ 2\frac{3}{4} &= \ \ 2 + \frac{3}{4} = \ \ 2 + \frac{3}{4} \\
\hline
& \qquad\qquad\qquad 73 + \frac{5}{4} \\
& \qquad\qquad\quad = 73 + 1\frac{1}{4} \\
& \qquad\qquad\quad = 74\frac{1}{4}
\end{aligned}
$$

$71\frac{1}{2} + 2\frac{3}{4} = \mathbf{74\frac{1}{4}}$

▶ Think and Discuss

1. Add. $8\frac{5}{8} + 3\frac{3}{8}$

2. Write $10\frac{11}{7}$ so that it does not contain an improper fraction.

3. Refer to the introduction to this lesson. What is the record?

4. When is the sum of two mixed numbers a whole number?

Exercises

Add. Write the answers in lowest terms. (See Example A.)

5. $2\frac{2}{5}$
$+\ 3\frac{1}{5}$

6. $4\frac{1}{4}$
$+\ \ \frac{1}{4}$

7. $3\frac{5}{8}$
$+\ 6\frac{1}{8}$

8. $5\frac{1}{6}$
$+\ 7\frac{2}{6}$

9. $2\frac{4}{7}$
$+\ 3\frac{5}{7}$

10. $4\frac{1}{2} + 8\frac{1}{2}$

11. $3\frac{3}{8} + 4\frac{1}{8}$

12. $5\frac{1}{5} + \frac{4}{5}$

13. $6\frac{7}{8} + 9\frac{5}{8}$

Add. Write the answers in lowest terms. (See Example B.)

14. $3\frac{1}{2}$
$+\ 2\frac{1}{4}$

15. $9\frac{1}{2}$
$+\ 3\frac{1}{5}$

16. $3\frac{1}{6}$
$+\ 5\frac{1}{4}$

17. $7\frac{2}{5}$
$+\ 9\frac{1}{4}$

18. $5\frac{1}{2}$
$+\ \ \frac{1}{3}$

19. $4\frac{5}{6}$
$+\ 6\frac{5}{8}$

20. $7\frac{1}{2} + 4\frac{5}{8}$

21. $6\frac{2}{3} + 4\frac{5}{12}$

22. $8\frac{3}{5} + 2\frac{5}{6}$

23. $3\frac{5}{7} + \frac{1}{2}$

▶ Mixed Practice (For more practice, see page 422.)

Add. Write the answers in lowest terms.

24. $7\frac{7}{8} + 8\frac{5}{6}$

25. $8\frac{1}{4} + 6\frac{1}{6}$

26. $6\frac{3}{7} + \frac{4}{9}$

27. $6\frac{7}{15} + 3\frac{13}{15}$

28. $3\frac{4}{7}$
$+\ 6\frac{2}{7}$

29. $9\frac{2}{3}$
$+\ 3\frac{1}{6}$

30. $5\frac{3}{4}$
$+\ 2\frac{1}{8}$

31. $4\frac{3}{10}$
$+\ \ \frac{1}{2}$

32. $5\frac{3}{8}$
$+\ 9\frac{4}{5}$

33. $7\frac{7}{9}$
$+\ 2\frac{5}{9}$

34. $2\frac{4}{5}$
$+\ \ \frac{7}{8}$

35. $8\frac{3}{4}$
$+\ 6\frac{1}{4}$

36. $5\frac{5}{9}$
$+\ 7\frac{1}{2}$

37. $2\frac{7}{10}$
$4\frac{1}{10}$
$+\ 5\frac{3}{10}$

38. $3\frac{3}{8}$
$9\frac{1}{4}$
$+\ 1\frac{2}{3}$

39. $7\frac{3}{4}$
$5\frac{3}{7}$
$+\ 4\frac{1}{2}$

▶ Applications

40. A baby grew $1\frac{1}{4}$ inches. If it was $22\frac{5}{8}$ inches long before, how long is the baby now?

41. An antenna was added to the Eiffel Tower, shown here. What is the new height of the Eiffel Tower?

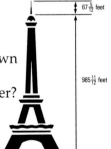

$67\frac{5}{12}$ feet

$985\frac{11}{12}$ feet

▶ Review (Lesson 5.7)

Convert each fraction to a decimal.

42. $\frac{1}{5}$

43. $\frac{7}{10}$

44. $\frac{3}{4}$

45. $\frac{7}{8}$

46. $\frac{2}{3}$

47. $\frac{5}{6}$

7.5 Subtracting Fractions with Unlike Denominators

Deluxe Paneling

8 ft.
8 ft.
4 ft.
4 ft.
$\frac{1}{4}$ inch thick
$\frac{7}{16}$ inch thick
$7.88
$9.95

Look at the measurements for the two types of paneling shown above. How much thicker is the more expensive paneling?

Examples

To subtract fractions with unlike denominators, first rewrite them using their LCD. Then subtract the numerators.

A Subtract. $\frac{7}{16} - \frac{1}{4}$

$$\frac{7}{16} = \frac{7}{16} = \frac{7}{16}$$
$$-\ \frac{1}{4} = \frac{1}{4} \times \frac{4}{4} = \frac{4}{16}$$
$$\frac{3}{16}$$

Rewrite using the LCD.

$$\frac{7}{16} - \frac{1}{4} = \frac{3}{16}$$

B Subtract. $\frac{3}{4} - \frac{2}{3}$

$$\frac{3}{4} = \frac{3}{4} \times \frac{3}{3} = \frac{9}{12}$$
$$-\ \frac{2}{3} = \frac{2}{3} \times \frac{4}{4} = \frac{8}{12}$$
$$\frac{1}{12}$$

Rewrite using the LCD.

$$\frac{3}{4} - \frac{2}{3} = \frac{1}{12}$$

▶ Think and Discuss

1. Subtract. $\frac{17}{18} - \frac{5}{6}$

2. Refer to the introduction to this lesson. How much thicker is the more expensive paneling?

3. What is the LCD of $\frac{3}{4}$ and $\frac{1}{6}$?

4. Explain the following mistake: $\frac{2}{5} - \frac{1}{3} = \frac{1}{2}$.

Exercises

Subtract. Write the answers in lowest terms. (See Example A.)

5. $\frac{3}{8}$
$-\frac{1}{4}$

6. $\frac{2}{5}$
$-\frac{1}{10}$

7. $\frac{1}{2}$
$-\frac{1}{4}$

8. $\frac{2}{3}$
$-\frac{7}{12}$

9. $\frac{7}{9}$
$-\frac{2}{3}$

10. $\frac{7}{10}$
$-\frac{1}{5}$

11. $\frac{8}{15} - \frac{1}{3}$

12. $\frac{5}{6} - \frac{1}{2}$

13. $\frac{1}{2} - \frac{3}{8}$

14. $\frac{3}{4} - \frac{5}{12}$

15. $\frac{11}{18} - \frac{1}{2}$

Subtract. Write the answers in lowest terms. (See Example B.)

16. $\frac{1}{2}$
$-\frac{1}{3}$

17. $\frac{1}{2}$
$-\frac{1}{5}$

18. $\frac{2}{3}$
$-\frac{1}{4}$

19. $\frac{1}{4}$
$-\frac{1}{5}$

20. $\frac{1}{5}$
$-\frac{1}{6}$

21. $\frac{2}{3}$
$-\frac{2}{5}$

22. $\frac{3}{5} - \frac{1}{2}$

23. $\frac{5}{6} - \frac{3}{4}$

24. $\frac{5}{8} - \frac{2}{5}$

25. $\frac{5}{9} - \frac{1}{4}$

26. $\frac{6}{7} - \frac{2}{3}$

▶ **Mixed Practice** (For more practice, see page 423.)

Subtract. Write the answers in lowest terms.

27. $\frac{1}{2}$
$-\frac{3}{8}$

28. $\frac{1}{2}$
$-\frac{2}{9}$

29. $\frac{5}{8}$
$-\frac{1}{4}$

30. $\frac{1}{5}$
$-\frac{1}{10}$

31. $\frac{3}{4}$
$-\frac{2}{7}$

32. $\frac{2}{3}$
$-\frac{3}{7}$

33. $\frac{17}{21}$
$-\frac{1}{3}$

34. $\frac{9}{10}$
$-\frac{5}{6}$

35. $\frac{8}{9}$
$-\frac{3}{4}$

36. $\frac{4}{5}$
$-\frac{7}{25}$

37. $\frac{6}{7}$
$-\frac{3}{5}$

38. $\frac{10}{11}$
$-\frac{3}{5}$

39. $\frac{7}{10} - \frac{1}{5}$

40. $\frac{5}{7} - \frac{1}{2}$

41. $\frac{11}{20} - \frac{1}{4}$

42. $\frac{4}{9} - \frac{1}{6}$

43. $\frac{7}{8} - \frac{4}{5}$

▶ **Applications**

44. Rachel spent $\frac{3}{4}$ hour on her homework. Monty spent $\frac{1}{6}$ hour less than Rachel. What part of an hour did Monty spend on his homework?

45. The bags of chips shown here both cost the same. Which bag weighs more? How much more?

▶ **Review** (Lessons 2.3, 3.3)

Multiply or divide mentally.

46. 70×10

47. $600 \div 100$

48. 54×100

49. $4900 \div 70$

50. $8100 \div 900$

51. 33×1000

52. $48{,}000 \div 100$

53. $67 \times 10{,}000$

7.6 Working with Fractions: An Alternative Strategy

Mr. Lloyd began math class with the problem shown here. The students solved the problem by finding the LCD and then adding.

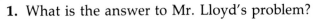

$$\frac{1}{2} + \frac{2}{3} + \frac{3}{4} =$$

1. What is the answer to Mr. Lloyd's problem?

Pleased with the students' response, Mr. Lloyd was ready to move on to another topic. "But wait!" Carla exclaimed. "I found another way to do these problems, and it's easier. Let me show you."

Carla went to the board. "You don't always have to find the LCD," she explained. Carla then taught the class her approach to adding and subtracting fractions.

STEP 1: Instead of finding the least common denominator, simply multiply all of the denominators. The product is a common denominator.

$$2 \times 3 \times 4 = 24$$

STEP 2: Rewrite each fraction using the new common denominator.

$$\frac{1}{2} = \frac{12}{24} \qquad \frac{2}{3} = \frac{16}{24} \qquad \frac{3}{4} = \frac{18}{24}$$

STEP 3: Add

STEP 4: Write the answer in lowest terms.

2. Complete Steps 3 and 4 of Carla's problem. Does her approach give the correct answer?

3. How is Carla's approach different from using the LCD? How is it easier? Discuss.

The class began to question Carla about her approach. "Does it work for subtraction, too?" one student asked. "Of course," she replied. "Is it easier?" "Not always, but try both approaches on some problems," she answered. "Then you can decide when each approach is better."

Exercises

Solve each problem using the LCD approach and Carla's approach.

4. $\frac{1}{2} + \frac{1}{4} + \frac{1}{8}$ **5.** $\frac{3}{8} - \frac{1}{5}$ **6.** $\frac{2}{9} + \frac{1}{6}$ **7.** $\frac{1}{7} + \frac{3}{5} + \frac{1}{2}$

8. $\frac{2}{3} + \frac{8}{9} + \frac{26}{27}$ **9.** $\frac{7}{12} - \frac{1}{7}$ **10.** $\frac{7}{12} - \frac{1}{8}$ **11.** $\frac{1}{3} + \frac{2}{3}$

12. $\frac{1}{4} - \frac{1}{8}$ **13.** $\frac{2}{7} + \frac{3}{5}$ **14.** $\frac{1}{2} + \frac{4}{11} + \frac{1}{3}$ **15.** $\frac{3}{4} + \frac{5}{7}$

16. $\frac{3}{4} + \frac{3}{4}$ **17.** $\frac{1}{2} - \frac{4}{9}$ **18.** $\frac{1}{4} + \frac{3}{8} + \frac{5}{16}$ **19.** $\frac{3}{13} + \frac{3}{4}$

20. $\frac{1}{5} + \frac{1}{25}$ **21.** $\frac{3}{7} - \frac{2}{14}$ **22.** $\frac{1}{2} + \frac{2}{3} + \frac{4}{5}$ **23.** $\frac{5}{8} - \frac{1}{8}$

24. $\frac{5}{9} - \frac{4}{11}$ **25.** $\frac{2}{3} + \frac{1}{5} + \frac{4}{7}$ **26.** $\frac{3}{4} + \frac{7}{8} + \frac{15}{16}$ **27.** $\frac{1}{3} + \frac{5}{6} - \frac{3}{7}$

28. Sometimes Carla's approach gives the LCD. Examine the completed exercises. Write a rule that states when Carla's approach gives the LCD.

▶ Applications

29. A share of stock rose $\frac{1}{8}$ on Monday, $\frac{3}{4}$ on Tuesday, and $\frac{1}{2}$ on Wednesday. By how much did it rise over the three days?

30. On January 6, Gina cut $\frac{1}{4}$ inch off her hair. By March 6, her hair had grown $\frac{3}{8}$ inch. Did the length of her hair increase or decrease over the two months?

▶ Review (Lessons 6.1, 6.2)

Multiply. Write the answers in lowest terms.

31. $\frac{1}{8} \times 7$ **32.** $\frac{5}{12} \times \frac{2}{3}$ **33.** $6 \times \frac{8}{9}$ **34.** $\frac{1}{3} \times \frac{9}{10}$ **35.** $10 \times \frac{3}{16}$

36. $9 \times \frac{3}{4}$ **37.** $\frac{1}{2} \times \frac{15}{16}$ **38.** $\frac{4}{5} \times \frac{11}{12}$ **39.** $\frac{8}{15} \times 5$ **40.** $\frac{1}{4} \times \frac{7}{10}$

7.7 Subtracting Mixed Numbers from Whole Numbers

Compare the times shown above. By how much did Timely Writer beat the record?

Examples

To subtract a mixed number from a whole number, first rename the whole number. Then subtract the fractions and whole numbers separately.

A Rename. $12 = 11\frac{\boxed{}}{3}$

$12 = 11 + 1 = 11 + \frac{3}{3} = 11\frac{3}{3}$
$12 = 11\frac{3}{3}$

B Subtract. $58 - 56\frac{4}{5}$

$$\begin{array}{r} 58 \;=\; 57\frac{5}{5} \\ -\; 56\frac{4}{5} \;=\; 56\frac{4}{5} \\ \hline 1\frac{1}{5} \end{array}$$

Rename 58 as a mixed number with a denominator of 5.

▶ Think and Discuss

1. Rename 17 as a mixed number with a denominator of 10.

2. Subtract. $9 - 4\frac{5}{8}$

3. Refer to the introduction to this lesson. By how much did Timely Writer beat the record?

4. If you are subtracting a mixed number from a whole number, how do you determine what denominator to use?

Exercises

Rename each whole number. (See Example A.)

5. $3 = 2\frac{\boxed{}}{4}$　　**6.** $9 = 8\frac{\boxed{}}{6}$　　**7.** $10 = 9\frac{\boxed{}}{8}$　　**8.** $7 = 6\frac{\boxed{}}{12}$　　**9.** $6 = 5\frac{\boxed{}}{7}$

Subtract. Write the answers in lowest terms. (See Example B.)

10. $\begin{array}{r} 5 \\ -\ 1\frac{2}{3} \\ \hline \end{array}$　**11.** $\begin{array}{r} 8 \\ -\ 6\frac{1}{5} \\ \hline \end{array}$　**12.** $\begin{array}{r} 9 \\ -\ 4\frac{5}{7} \\ \hline \end{array}$　**13.** $\begin{array}{r} 12 \\ -\ 3\frac{7}{10} \\ \hline \end{array}$　**14.** $\begin{array}{r} 15 \\ -\ 10\frac{5}{6} \\ \hline \end{array}$　**15.** $\begin{array}{r} 27 \\ -\ 14\frac{5}{8} \\ \hline \end{array}$

16. $18 - 11\frac{3}{4}$　　**17.** $20 - 16\frac{5}{11}$　　**18.** $33 - 26\frac{4}{9}$　　**19.** $45 - 19\frac{17}{20}$

▶ Mixed Practice (For more practice, see page 423.)

Subtract. Write the answers in lowest terms.

20. $\begin{array}{r} 7 \\ -\ 2\frac{1}{4} \\ \hline \end{array}$　**21.** $\begin{array}{r} 11 \\ -\ 7\frac{3}{8} \\ \hline \end{array}$　**22.** $\begin{array}{r} 15 \\ -\ 2\frac{9}{15} \\ \hline \end{array}$　**23.** $\begin{array}{r} 21 \\ -\ 14\frac{3}{10} \\ \hline \end{array}$　**24.** $\begin{array}{r} 37 \\ -\ 20\frac{13}{16} \\ \hline \end{array}$

25. $\begin{array}{r} 53 \\ -\ 27\frac{6}{25} \\ \hline \end{array}$　**26.** $\begin{array}{r} 42 \\ -39\frac{5}{13} \\ \hline \end{array}$　**27.** $\begin{array}{r} 62 \\ -47\frac{13}{30} \\ \hline \end{array}$　**28.** $\begin{array}{r} 75 \\ -\ 32\frac{11}{50} \\ \hline \end{array}$　**29.** $\begin{array}{r} 87 \\ -\ 68\frac{9}{48} \\ \hline \end{array}$

30. $8 - 3\frac{6}{7}$　　**31.** $10 - 5\frac{4}{5}$　　**32.** $17 - 13\frac{12}{13}$　　**33.** $35 - 21\frac{4}{21}$

34. $42 - 28\frac{7}{12}$　　**35.** $50 - 19\frac{9}{14}$　　**36.** $61 - 13\frac{5}{18}$　　**37.** $72 - 47\frac{37}{100}$

▶ Applications

38. The limbo record is $6\frac{1}{8}$ inches from the floor. Georgette's best try is 12 inches from the floor. How far away from the record is she?

39. Earl is not supposed to eat more than four ounces of cheese a week. On Monday, he ate three snack packs like the one shown. How much more cheese can he eat this week?

▶ Review (Lessons 1.1, 1.2)

Write each number in words.

40. 7.5　　**41.** 8013.06　　**42.** 469,570　　**43.** 0.395　　**44.** 200,000.43

Find the place value of the underlined 5 in each number.

45. <u>5</u>324　　**46.** 12.0<u>5</u>9　　**47.** <u>5</u>97,307　　**48.** 0.385<u>5</u>　　**49.** <u>5</u>,205,769

7.8 Simplifying the Problem

Alicia learned that every minute she roller skates, she burns $4\frac{2}{5}$ calories. She timed herself one day and skated $19\frac{1}{2}$ minutes before resting.

Would you add, subtract, multiply, or divide to find how many calories are burned?

Most people find it easier to work with whole numbers than fractions. Sometimes whole numbers are easier to work with and make it easier to see which operations to use. For example, suppose Alicia burned 4 calories a minute for 20 minutes. Solving this simpler problem can help you solve the original problem. The operation used for both the simpler problem and the original problem should be the same.

1. What operation would you use to solve the simpler problem?

2. What is the answer to the simpler problem?

3. Use the same operation used in the simpler problem to solve the original problem.

4. Is the answer to the original problem equal to the answer to the simpler problem? Why or why not?

5. If Alicia skated for $35\frac{1}{2}$ minutes, what whole numbers would you substitute in this problem to make a simpler one?

6. How many calories would she burn in $35\frac{1}{2}$ minutes?

One day Ramzi spent $6\frac{1}{2}$ hours working, $\frac{5}{6}$ hour for a lunch break, and $1\frac{1}{4}$ hours riding the bus. How much of his day was spent from the time he left for work to the time he arrived home?

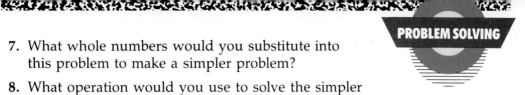
7. What whole numbers would you substitute into this problem to make a simpler problem?

8. What operation would you use to solve the simpler problem?

9. Solve the simpler problem.

10. Solve the original problem.

Each problem below involves fractions. A simpler problem using only whole numbers may help you see how to solve it. Write the operation used for each problem below, and then solve.

11. Edna's office is being remodeled to include a closet across the entire width of the room. If each closet door is $1\frac{1}{2}$ feet wide, how many doors are needed?

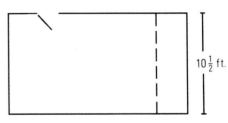

$10\frac{1}{2}$ ft.

12. Connie watched the neighbor's children for three hours and fifteen minutes at two-and-a-half dollars an hour. How much did Connie earn?

13. When Mickey's family moved in to a new house, the bathroom entrance was widened $6\frac{5}{8}$ inches to allow for his wheelchair. The door was $24\frac{7}{8}$ inches wide. How wide is it now?

14. Ian claims that he can lift $2\frac{1}{2}$ times his weight. If he weighs 142 pounds, how much can he lift?

15. The odometer in Louise's car was at $12,657\frac{9}{10}$ miles when she started her trip. When she arived home, it was at $13,243\frac{1}{10}$ miles. How far did she travel?

16. Sue wanted to cover her bulletin board with album covers like the one shown. Her bulletin board is $49\frac{1}{2}$ inches wide. If no space is left between the album covers, how many albums will fit across the bulletin board?

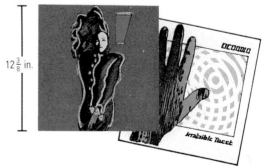

$12\frac{3}{8}$ in.

▶ **Review** (Lesson 3.9)

Simplify.

17. $12 - 5 \div 5 \times 2$ **18.** $16 + (15 - 3)$ **19.** $9 \times 7 - (6 + 2)$

7.9 Subtracting Mixed Numbers

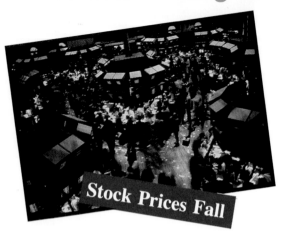

Stock Prices Fall

In 1929 they called it a **crash**. In 1987 they called it a **meltdown**. How much did the price of IBM stock drop in 1987?

1987 Prices per Share

	High	Low
IBM	$175\frac{7}{8}$	$102\frac{1}{4}$
General Motors	$94\frac{1}{8}$	50
General Electric	$66\frac{3}{8}$	$38\frac{3}{4}$

Examples

To subtract mixed numbers, rewrite the fractions using their LCD. If necessary, rename the larger mixed number.

A Rename. $8\frac{3}{5} = 7\frac{▨}{5}$

$$8\frac{3}{5} = 8 + \frac{3}{5}$$
$$8\frac{3}{5} = 7 + 1 + \frac{3}{5}$$
$$8\frac{3}{5} = 7 + \frac{5}{5} + \frac{3}{5} = 7\frac{8}{5}$$

B Subtract. $175\frac{7}{8} - 102\frac{1}{4}$

$$175\frac{7}{8} = 175 + \frac{7}{8} = 175 + \frac{7}{8}$$
$$- \ 102\frac{1}{4} = 102 + \frac{1}{4} \times \frac{2}{2} = 102 + \frac{2}{8}$$
$$\overline{\qquad\qquad\qquad\qquad 73 + \frac{5}{8}}$$

$$175\frac{7}{8} - 102\frac{1}{4} = 73\frac{5}{8}$$

C Subtract. $6\frac{1}{5} - 2\frac{3}{4}$

$$6\frac{1}{5} = 6 + \frac{1}{5} = 6 + \frac{1}{5} \times \frac{4}{4} = 6 + \frac{4}{20} = 5 + \frac{24}{20}$$
$$- \ 2\frac{3}{4} = 2 + \frac{3}{4} = 2 + \frac{3}{4} \times \frac{5}{5} = 2 + \frac{15}{20} = 2 + \frac{15}{20}$$
$$\overline{\qquad\qquad\qquad\qquad\qquad\qquad\qquad\quad 3 + \frac{9}{20}}$$

Rewrite using the LCD.

Rename $6\frac{4}{20}$ as $5\frac{24}{20}$.

$$6\frac{1}{5} - 2\frac{3}{4} = 3\frac{9}{20}$$

▶ Think and Discuss

1. Rename $94\frac{1}{8}$.

2. Subtract. $66\frac{3}{8} - 38\frac{3}{4}$

3. Refer to the introduction to this lesson. By how much did the price of IBM stock drop in 1987?

4. When is it necessary to rename a mixed number?

Exercises

Rename each mixed number. (See Example A.)

5. $7\frac{1}{3} = 6\frac{\blacksquare}{3}$ 6. $3\frac{2}{5} = 2\frac{\blacksquare}{5}$ 7. $6\frac{8}{9} = 5\frac{\blacksquare}{9}$ 8. $10\frac{7}{12} = 9\frac{\blacksquare}{12}$

Subtract. Write the answers in lowest terms. (See Example B.)

9. $6\frac{7}{8}$
$-\ 2\frac{2}{3}$

10. $5\frac{4}{7}$
$-\ 1\frac{1}{2}$

11. $9\frac{3}{5}$
$-\ 6\frac{1}{4}$

12. $5\frac{7}{8}$
$-\ 2\frac{2}{5}$

13. $3\frac{3}{4}$
$-\ 1\frac{1}{9}$

14. $6\frac{5}{6}$
$-\ 3\frac{7}{10}$

Subtract. Write the answers in lowest terms. (See Example C.)

15. $9\frac{1}{3}$
$-\ 6\frac{2}{3}$

16. $5\frac{2}{5}$
$-\ 2\frac{7}{10}$

17. $8\frac{3}{8}$
$-\ 3\frac{3}{4}$

18. $7\frac{1}{4}$
$-\ 5\frac{5}{6}$

19. $4\frac{4}{5}$
$-\ 1\frac{7}{8}$

20. $6\frac{1}{6}$
$-\ 3\frac{9}{10}$

▶ Mixed Practice (For more practice, see page 424.)

Subtract. Write the answers in lowest terms.

21. $5\frac{1}{2}$
$-\ 1\frac{3}{4}$

22. $6\frac{3}{8}$
$-\ 2\frac{7}{8}$

23. $9\frac{1}{4}$
$-\ 3\frac{2}{3}$

24. $6\frac{7}{8}$
$-\ 4\frac{1}{3}$

25. $5\frac{1}{4}$
$-\ 3\frac{4}{5}$

26. $3\frac{1}{8}$
$-\ 2\frac{1}{3}$

27. $8\frac{3}{5} - 7\frac{4}{5}$ 28. $5\frac{3}{4} - 2\frac{1}{7}$ 29. $3\frac{2}{3} - \frac{5}{7}$ 30. $7\frac{1}{8} - 5\frac{2}{5}$

▶ Applications

31. The men's world speed-skating record for 500 meters is $45\frac{2}{25}$ seconds. The women's record is $48\frac{89}{100}$ seconds. How much faster is the men's record than the women's?

32. Jenny threw the discus 101 feet $4\frac{1}{8}$ inches. Jami threw it 101 feet $1\frac{3}{4}$ inches. Who threw the discus farther? How much farther?

▶ Review (Lessons 6.6, 6.7)

Divide.

33. $\frac{9}{10} \div \frac{5}{6}$ 34. $\frac{11}{12} \div 6$ 35. $3\frac{1}{2} \div \frac{3}{4}$ 36. $1\frac{1}{5} \div 2\frac{1}{4}$

7.10 Estimating Sums and Differences of Fractions and Mixed Numbers

Martin set a goal to run about 5 miles a day during summer vacation. Which combinations of paths shown here can Martin run in order to accomplish his daily goal?

Martin can estimate sums to figure out which paths are at least 5 miles long.

Examples

To estimate sums and differences of fractions and mixed **numbers**, determine the nearest $\frac{1}{2}$ or the nearest whole number for each fraction or mixed number. Then add or subtract.

A Estimate to the nearest whole number. $7\frac{8}{19} - 4\frac{21}{25}$	**B** Estimate to the nearest $\frac{1}{2}$. $1\frac{3}{5} + 1\frac{9}{10}$
$7\frac{8}{19}$ → closer to **7** than 8 → 7 $4\frac{21}{25}$ → closer to **5** than 4 → $\dfrac{-5}{2}$ $7\frac{8}{19} - 4\frac{21}{25}$ is **about 2.**	$1\frac{3}{5}$ → closer to **$1\frac{1}{2}$** than 2 → $1\frac{1}{2}$ $1\frac{9}{10}$ → closer to **2** than $1\frac{1}{2}$ → $\dfrac{+\,2}{3\frac{1}{2}}$ $1\frac{3}{5} + 1\frac{9}{10}$ is **about $3\frac{1}{2}$.**

▶ Think and Discuss

1. Is $\frac{7}{8}$ closer to $\frac{1}{2}$ or 1? How can you tell?

2. Estimate to the nearest whole number. $1\frac{19}{20} + 4\frac{3}{7}$

3. Estimate to the nearest $\frac{1}{2}$. $5\frac{1}{9} - \frac{8}{13}$

4. Refer to the introduction to this lesson. What are three combinations of paths Martin can run to accomplish his goal?

Exercises

Estimate to the nearest whole number. (See Example A.)

5. $\dfrac{9}{16}$
$+ \dfrac{7}{10}$

6. $1\dfrac{7}{10}$
$+ 2\dfrac{4}{9}$

7. $6\dfrac{12}{13}$
$+ 5\dfrac{3}{16}$

8. $4\dfrac{5}{8}$
$- 2\dfrac{3}{13}$

9. $6\dfrac{8}{9}$
$- 3\dfrac{7}{15}$

10. $9\dfrac{1}{12}$
$- 5\dfrac{4}{17}$

11. $\dfrac{9}{10} + \dfrac{8}{9}$

12. $4\dfrac{3}{7} + 3\dfrac{5}{9}$

13. $2\dfrac{1}{3} - \dfrac{5}{8}$

14. $4\dfrac{4}{17} - 2\dfrac{1}{9}$

Estimate to the nearest $\dfrac{1}{2}$. (See Example B.)

15. $7\dfrac{9}{11}$
$+ 1\dfrac{7}{16}$

16. $2\dfrac{4}{9}$
$+ 7\dfrac{8}{11}$

17. $9\dfrac{5}{37}$
$+ 8\dfrac{21}{23}$

18. $2\dfrac{3}{16}$
$- 1\dfrac{5}{12}$

19. $4\dfrac{7}{9}$
$- 2\dfrac{13}{15}$

20. $9\dfrac{3}{10}$
$- 6\dfrac{1}{24}$

21. $1\dfrac{1}{5} + 2\dfrac{1}{10}$

22. $3\dfrac{4}{9} + 5\dfrac{10}{21}$

23. $3\dfrac{1}{6} - 1\dfrac{1}{7}$

24. $5\dfrac{4}{9} - 2\dfrac{1}{8}$

▶ **Mixed Practice** (For more practice, see page 424.)

Estimate to the nearest whole number and to the nearest $\dfrac{1}{2}$.

25. $2\dfrac{9}{11}$
$- 1\dfrac{10}{21}$

26. $3\dfrac{20}{21}$
$+ 4\dfrac{5}{19}$

27. $6\dfrac{5}{48}$
$- 3\dfrac{13}{25}$

28. $9\dfrac{5}{16}$
$+ 8\dfrac{1}{7}$

29. $4\dfrac{5}{11}$
$+ \dfrac{9}{16}$

30. $7\dfrac{9}{20}$
$+ 4\dfrac{3}{20}$

31. $1\dfrac{1}{7} + 2\dfrac{5}{9}$

32. $3\dfrac{3}{5} - 1\dfrac{3}{9}$

33. $1\dfrac{1}{8} + \dfrac{7}{9}$

34. $4\dfrac{5}{8} - 1\dfrac{3}{7}$

▶ **Applications**

35. It rained $\dfrac{7}{8}$ inch Monday, $1\dfrac{1}{4}$ inches Tuesday, and $\dfrac{3}{8}$ inch Wednesday. Estimate the total rainfall for the 3 days to the nearest $\dfrac{1}{2}$ inch.

36. Look at the picture. Estimate the distance each player's hand is from the rim.

$1\dfrac{1}{2}$ feet

▶ **Review** (Lessons 4.2, 4.5)

Complete each statement.

37. A marathon is about 45 ____ long.   m   km

38. The mass of a feather is about 2 ____.   g   kg

39. The child grew about 6 ____ in a year.   cm   m

Add or subtract. Write the answers in lowest terms.
(Lesson 7.1)

1. $\frac{5}{6} + \frac{5}{6}$ **2.** $\frac{7}{8} + \frac{3}{8} + \frac{5}{8}$ **3.** $\frac{11}{12} - \frac{5}{12}$ **4.** $\frac{9}{10} - \frac{1}{10}$

List the first five multiples. (Lesson 7.2)

5. 10   **6.** 4   **7.** 9   **8.** 12   **9.** 15   **10.** 30

Find the LCD. (Lesson 7.2)

11. $\frac{1}{2}$ $\frac{3}{8}$ **12.** $\frac{5}{6}$ $\frac{3}{5}$ **13.** $\frac{3}{4}$ $\frac{4}{5}$ $\frac{7}{10}$ **14.** $\frac{6}{7}$ $\frac{7}{9}$ **15.** $\frac{1}{3}$ $\frac{5}{8}$ $\frac{1}{6}$

Add. Write the answers in lowest terms. (Lessons 7.3, 7.4)

16. $\frac{1}{4}$ $+ \frac{1}{3}$ **17.** $\frac{5}{8}$ $+ \frac{1}{3}$ **18.** $\frac{5}{6}$ $+ \frac{3}{4}$ **19.** $9\frac{11}{12}$ $+ 6\frac{7}{8}$ **20.** $2\frac{1}{2}$ $+ 3\frac{7}{8}$ **21.** $4\frac{3}{5}$ $+ 5\frac{2}{3}$

22. Carl lost $\frac{3}{4}$ pound one week and $1\frac{1}{2}$ pounds the next week. How many pounds did he lose in 2 weeks?

23. Pam swam $\frac{11}{12}$ mile, Maria swam $\frac{5}{6}$ mile, and Joan swam $\frac{3}{4}$ mile. How many miles did they swim altogether?

Rename each whole number. (Lesson 7.7)

24. $9 = 8\frac{⬚}{12}$ **25.** $5 = 4\frac{⬚}{6}$ **26.** $7 = 6\frac{⬚}{10}$ **27.** $2 = 1\frac{⬚}{7}$ **28.** $8 = 7\frac{⬚}{25}$

Subtract. Write the answers in lowest terms. (Lessons 7.5, 7.7, 7.9)

29. $\frac{7}{10}$ $- \frac{1}{5}$ **30.** $\frac{11}{12}$ $- \frac{1}{8}$ **31.** 10 $- 4\frac{1}{4}$ **32.** 8 $- 2\frac{3}{10}$ **33.** 4 $- 1\frac{3}{5}$ **34.** 7 $- 5\frac{5}{9}$

35. $5\frac{1}{5}$ $- 1\frac{1}{2}$ **36.** $6\frac{7}{8}$ $- 3\frac{3}{4}$ **37.** $9\frac{1}{4}$ $- 2\frac{3}{5}$ **38.** $7\frac{1}{12}$ $- 4\frac{2}{3}$ **39.** $2\frac{2}{3}$ $- 1\frac{4}{5}$ **40.** $8\frac{2}{7}$ $- 6\frac{3}{4}$

Estimate to the nearest whole number. (Lesson 7.10)

41. $5\frac{1}{4} + 8\frac{1}{10}$ **42.** $9\frac{3}{4} - 3\frac{1}{6}$ **43.** $2\frac{7}{12} + 4\frac{3}{5}$ **44.** $12\frac{5}{8} - 9\frac{7}{8}$

Chapter 7 Test

Subtract. Write the answers in lowest terms.

1. $\dfrac{4}{5}$ $-\dfrac{1}{3}$

2. 6 $-3\dfrac{2}{3}$

3. $9\dfrac{1}{8}$ $-5\dfrac{3}{4}$

4. $\dfrac{11}{12}$ $-\dfrac{5}{12}$

5. $3\dfrac{1}{6}$ $-\dfrac{7}{8}$

Estimate to the nearest whole number.

6. $2\dfrac{5}{6} - 1\dfrac{1}{16}$

7. $6\dfrac{7}{16} + 3\dfrac{3}{8}$

8. $5\dfrac{9}{10} - 2\dfrac{3}{4}$

9. $8\dfrac{1}{3} + 7\dfrac{4}{5}$

Rename each whole number.

10. $12 = 11\dfrac{\boxed{}}{16}$

11. $6 = 5\dfrac{\boxed{}}{9}$

12. $9 = 8\dfrac{\boxed{}}{10}$

13. $15 = 14\dfrac{\boxed{}}{20}$

Find the LCD.

14. $\dfrac{3}{4}$ $\dfrac{1}{5}$

15. $\dfrac{1}{2}$ $\dfrac{2}{3}$

16. $\dfrac{7}{10}$ $\dfrac{2}{15}$ $\dfrac{1}{6}$

17. $\dfrac{11}{12}$ $\dfrac{5}{8}$

18. $\dfrac{7}{20}$ $\dfrac{3}{5}$

Add. Write the answers in lowest terms.

19. $1\dfrac{1}{3}$ $+ 2\dfrac{1}{4}$

20. $\dfrac{3}{16}$ $+ \dfrac{5}{16}$

21. $5\dfrac{1}{2}$ $+ 6\dfrac{2}{5}$

22. $2\dfrac{5}{8}$ $+ 6\dfrac{3}{8}$

23. Stock ABC went up $3\dfrac{7}{8}$. If it was at $13\dfrac{1}{4}$ before, what is the new price?

24. Brad worked $2\dfrac{1}{4}$ hours on Monday and $1\dfrac{1}{5}$ hours on Tuesday. How many hours did he work?

List the first five multiples.

25. 11

26. 17

27. 21

28. 25

29. 50

Add or subtract. Write the answers in lowest terms.

30. $\dfrac{1}{12} + \dfrac{1}{2}$

31. $5\dfrac{9}{10} - 2\dfrac{3}{10}$

32. $\dfrac{9}{10} + \dfrac{2}{3}$

33. $\dfrac{5}{6} - \dfrac{1}{4}$

"...$\frac{2}{3}$ cup shortening, $3\frac{1}{2}$ cups flour, $1\frac{1}{2}$ teaspoons cinnamon. To bake this cake, I need to know a lot about measurement. Would you believe that I have to measure the length and width of the baking pan to be sure that it's the right size?"

Chapter 8

Customary Measurement

Did you know that you use customary measurements when you find:
- the amount of juice in a cup?
- the distance a football is thrown on a playing field?
- the size of a piece of property?
- how tall you have grown?

What are some other ways you use customary measurement?

Length in Objects

Materials: classroom objects

Work with 2 or 3 classmates. Use your hand span as shown here as a unit of length. Estimate the number of hand spans between two desks in your classroom. Then measure the distance in hand spans. Copy and complete the table below.

Estimate	Measurement

Repeat the procedure for several different objects. How close were your estimates?

8.1 Introduction to the Customary System

Miles *Feet*

Months

Degrees

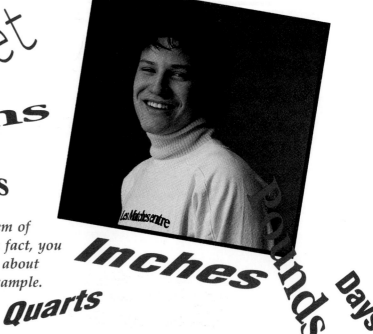

You use the customary system of measurement all the time. In fact, you use it often to describe facts about yourself. Take Brett as an example.

Inches *Pounds* *Days*

Quarts

Brett is 5 feet, $6\frac{1}{2}$ inches tall and weighs 145 pounds. He is 14 years, 3 months, and 12 days old. His normal body temperature is 98.6° Fahrenheit, and his body contains about 5 quarts of blood. He lives $5\frac{1}{2}$ miles from downtown Chicago. Brett works 20 hours a week and earns $4 an hour.

1. List the units of measurement that you recognize in the description of Brett.

2. List some units of measurement that were not included in the description of Brett.

3. Which of the following types of measurement are included in the description of Brett?

   capacity   height   temperature   time   weight

4. Write a description of yourself using several units of measurement.

5. What types of measurement would you use to describe a baseball game?

6. What types of measurement do you use each day?

▶ Express Yourself

You are familiar with many customary measurement terms
already. Use the glossary and a dictionary to write
definitions for the following terms.

7. capacity   **8.** length   **9.** temperature   **10.** weight

List all the units you can think of that describe the following
types of measurement.

11. distance   **12.** time   **13.** weight   **14.** height   **15.** capacity

For each description in Column A, find the appropriate unit of
measurement listed in Column B.

Column A

16. The distance from your home to
Yellowstone National Park

17. The amount of milk used for a cake
recipe

18. The height of a box of cereal

19. The time it takes to run a 100-yard dash

20. The amount you weigh

21. The temperature outside

Column B

a. degrees Fahrenheit
b. inches
c. miles
d. cups
e. seconds
f. pounds

▶ Practice What You Know

You may remember that the metric system is based on powers
of 10. The division of measurements in the customary system is
not so regular. The most common numbers you must work with
in converting from one customary unit to another are multiples of
4, such as 8, 12, 16, 24, 36, and 60. Other numbers you might
have to work with are 3, 1760, and 5280.

Solve.

22. $3\frac{1}{2} \times 12$ **23.** 4×60 **24.** $78 \div 36$ **25.** $500 \div 1760$

26. 4×8 **27.** $25,000 \div 5280$ **28.** 17×3 **29.** $92 \div 12$

8.2 Converting Customary Units of Length

Which roller coaster drops farther?

To convert customary units of length, first determine if you are converting to a larger or smaller unit.

Equivalents
12 inches   = 1 foot
3 feet   = 1 yard
1760 yards = 1 mile

Abbreviations
inch **in.**   yard **yd.**
foot **ft.**   mile **mi.**

A Convert 50 yards to feet.

Multiply since you are changing to a smaller unit.

1 yd. = 3 ft., so
50 yd. = 50 × 3 = 150 ft.
50 yd. = **150 ft.**

B Convert 456 inches to yards.

Divide since you are changing to a larger unit.

12 in. = 1 ft., so
456 in. = 456 ÷ 12 = 38 ft.
3 ft. = 1 yd., so
38 ft. = 38 ÷ 3 = $12\frac{2}{3}$ yd.
456 in. = **$12\frac{2}{3}$ yd.**

▶ Think and Discuss

1. Complete the table below.

    1 ft. = ____ in.   1 yd. = ____ ft.   1 mi. = ____ yd.
                               1 yd. = ____ in.   1 mi. = ____ ft.
                                                         1 mi. = ____ in.

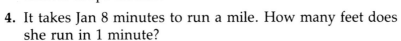

2. Convert 5 miles to feet.

3. Refer to the introduction to this lesson. Which roller coaster drops farther?

4. It takes Jan 8 minutes to run a mile. How many feet does she run in 1 minute?

5. Convert 5.1 feet to inches in two ways. First, round to the nearest foot and then convert. Second, convert and then round to the nearest inch. Which answer is a better estimate? Why?

6. Explain how you would order 35 yards, 1392 inches, and 114 feet from shortest to longest.

Exercises

Convert each measure. (See Example A.)

7. 4 ft. to in.	**8.** 25 yd. to ft.	**9.** 2 mi. to ft.	**10.** 7 yd. to in.
11. 20 ft. to in.	**12.** 3 yd. to ft.	**13.** 10 yd. to in.	**14.** 9 mi. to ft.

Convert each measure. (See Example B.)

15. 561 ft. to yd.	**16.** 540 in. to yd.	**17.** 15,840 ft. to mi.
18. 228 in. to ft.	**19.** 7040 yd. to mi.	**20.** 36,960 ft. to mi.

▶ Mixed Practice (For more practice, see page 425.)

Convert each measure.

21. 4 ft. to in.	**22.** 108 in. to yd.	**23.** 297 ft. to yd.
24. 10 ft. to in.	**25.** 1800 in. to ft.	**26.** 21,600 in. to yd.

▶ Applications

27. A marathon can be 26 miles 385 yards long. How many yards is that?

28. Bill told his mother he was going to race in the 440 this weekend. "Don't you usually run the $\frac{1}{4}$-mile race?" she asked. Was Bill running in a different race than usual? Explain.

▶ Review (Lessons 7.1, 7.3, 7.4)

Add.

29. $\begin{array}{r} \frac{1}{6} \\ + \frac{5}{6} \\ \hline \end{array}$

30. $\begin{array}{r} 2\frac{1}{2} \\ + 1\frac{3}{4} \\ \hline \end{array}$

31. $\begin{array}{r} \frac{3}{10} \\ + \frac{1}{4} \\ \hline \end{array}$

32. $\frac{4}{5} + \frac{1}{2}$

33. $2\frac{2}{3} + 6\frac{1}{3}$

8.3 Measuring with Customary Units of Length

When Maury was at the doctor's office, the nurse measured his height and said, "5 feet $7\frac{3}{4}$ inches." Then she wrote 5 feet 8 inches on Maury's chart. What happened to the $\frac{1}{4}$ inch?

Examples

To measure length, line up the zero mark on a ruler with one end of an object. Depending on the precision you need, you can round to the nearest $\frac{1}{4}$ inch, $\frac{1}{2}$ inch, inch, foot, or yard.

A Measure the length of the pencil to the nearest inch, $\frac{1}{2}$ inch, and $\frac{1}{4}$ inch.

Nearest inch: $4\frac{9}{16}$ inches is closer to **5 inches** than 4 inches.

Nearest $\frac{1}{2}$ inch: $4\frac{9}{16}$ inches is closer to $4\frac{1}{2}$ **inches** than 5 inches.

Nearest $\frac{1}{4}$ inch: $4\frac{9}{16}$ inches is closer to $4\frac{1}{2}$ **inches** than $4\frac{3}{4}$ inches.

B Choose the more reasonable measure.

The height of a classroom is about ___.   40 ft. 4 yd.

The height of a 4-story building is about 40 feet.
The height of a basketball rim is a little over 3 yards.

The more reasonable answer is **4 yards**.

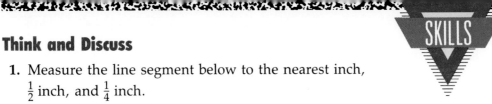

▶ Think and Discuss

1. Measure the line segment below to the nearest inch, $\frac{1}{2}$ inch, and $\frac{1}{4}$ inch.

   _____

2. Refer to the introduction to this lesson. If the nurse rounds to the nearest inch, someone listed as 5 feet 8 inches could be as short as ___※___ feet ___※___ inches tall.

3. What unit is used for distances between two cities? Why?

Exercises

Measure each item to the nearest inch, $\frac{1}{2}$ inch, and $\frac{1}{4}$ inch.
(See Example A.)

4. the length of your shoe

5. the line segment at the right _____

Complete each statement. Choose the more reasonable measure.
(See Example B.)

6. The width of a classroom is about ___※___.   1500 in.   15 ft.

7. The length of a golf club is about ___※___.   10 ft.   1 yd.

▶ Mixed Practice (For more practice, see page 425.)

Measure each item to the nearest inch, $\frac{1}{2}$ inch, and $\frac{1}{4}$ inch.

8. the width of your desk   9. the length of your thumbnail

Complete each statement. Choose the more reasonable measure.

10. The length of a football field is ___※___.   30 yd.   360 ft.

11. The height of a floor lamp is about ___※___.   60 in.   6 yd.

▶ Applications

12. Measure the length and width of the comb shown to the nearest $\frac{1}{4}$ inch.

13. Measure the length and width of this book's cover to the nearest $\frac{1}{4}$ inch.

▶ Review (Lessons 7.1, 7.5, 7.7, 7.9)

Subtract.

14. $\frac{9}{10} - \frac{1}{2}$ 15. $7 - 4\frac{1}{3}$ 16. $\frac{15}{16} - \frac{7}{16}$ 17. $6\frac{3}{4} - 1\frac{1}{6}$

8.4 Customary Units of Weight

Abdul wanted to buy potato chips for a party. He preferred Stacko Chips, but the bag of Crunchy Chips looked larger. He decided to compare the net weights written on the packages. Is one package heavier?

Examples

To convert customary units of weight, first determine if you are converting to a larger or smaller unit.

Equivalents
16 ounces   = 1 pound
2000 pounds = 1 short ton

Abbreviations
ounce **oz.**   pound **lb.**
short ton **T.**

A Convert 3 pounds 2 ounces to ounces.

Multiply since you are changing to a smaller unit.

16 oz. = 1 lb., so
3 lb. = 3 × 16 = 48 oz.
3 lb. 2 oz. = 48 + 2 = **50 oz.**

B Convert 4500 pounds to tons.

Divide since you are changing to a larger unit.

1 T. = 2000 lb., so
4500 lb. = 4500 ÷ 2000 = $2\frac{500}{2000}$ T.

The numerator is the number of pounds.

4500 lb. = **2 T. 500 lb.**

C Choose the more reasonable measure.

Chris's hamburger weighs 3 ___. lb. oz.

A 3-pound hamburger could exist, but would be awfully big (and hard to handle).
A 3-ounce hamburger is a little less than a quarter-pounder.

The more reasonable measure is **ounces.**

▶ **Think and Discuss**

1. Convert 3 tons 248 pounds to pounds.

2. How much does a $\frac{1}{4}$-pound hamburger weigh in ounces?

3. Refer to the introduction to this lesson. Which container of potato chips weighs more?

4. How many pounds equal 0.25 ton? Explain.

Exercises

Convert each measure. (See Example A.)

5. 19 lb. to oz.   **6.** 33 lb. to oz.   **7.** 11 T. to oz.

8. 9 lb. 13 oz. to oz.   **9.** 21 lb. 8 oz. to oz.   **10.** 2 T. 850 lb. to lb.

Convert each measure. (See Example B.)

11. 48 oz. to lb.   **12.** 176 oz. to lb.   **13.** 98 oz. to lb.

14. 119 oz. to lb.   **15.** 8000 lb. to T.   **16.** 18,000 lb. to T.

Complete each statement. Choose the more reasonable measure.
(See Example C.)

17. A newborn baby weighs about 8 ___.   lb.   oz.

18. A whale weighs about 10 ___.   T.   lb.

▶ Mixed Practice (For more practice, see page 426.)

Convert each measure.

19. 2 lb. 14 oz. to oz.   **20.** 13 T. to lb.   **21.** 47 oz. to lb.

22. 6321 lb. to T.   **23.** 17,860 lb. to T.   **24.** 64,000 oz. to T.

Complete each statement. Choose the more reasonable measure.

25. When Rafael was 14, he weighed about 100 ___.   lb.   oz.

26. A pair of blue jeans weighs about 14 ___.   lb.   oz.

▶ Applications

27. A farm supply company has 3 tons 756 pounds of seed on hand. How many 1-pound bags can they fill with the seed?

28. "On my diet, I've gone from weighing $\frac{1}{8}$ ton to $\frac{1}{10}$ ton!" Mrs. Lati exclaimed. How many pounds has Mrs. Lati lost?

▶ Review (Lessons 1.5, 1.6, 5.4)

Compare. Use >, <, or =.

29. $\frac{7}{8}$ ___ $\frac{5}{8}$ **30.** 6853 ___ 6935   **31.** 0.073 ___ 0.0076

32. 4.32 ___ 4.3200   **33.** 8.745 ___ 8.8   **34.** 343 ___ 334

8.5 Using Tables to Solve Problems

Lucy weighs 100 pounds. She claims she can lose 15 pounds overnight. "It's easy," she remarked. "All I have to do is move to Venus!"

Lucy knows that weight depends on gravity.

The weight of an object varies from planet to planet because the force of gravity is different on different planets. The table below shows how the weights of Lucy and her dog vary on the planets and on the Moon.

Location	Lucy's Weight	Her Dog's Weight
Earth	100 lb.	45.0 lb.
Moon	16 lb.	7.2 lb.
Mercury	28 lb.	12.6 lb.
Venus	85 lb.	38.3 lb.
Mars	38 lb.	17.1 lb.
Jupiter	260 lb.	117.0 lb.
Saturn	110 lb.	49.5 lb.
Uranus	80 lb.	36.0 lb.
Neptune	120 lb.	54.0 lb.
Pluto	1 lb.	0.5 lb.

How much would a 200-pound man on Earth weigh on the Moon?

First find 100 pounds on Earth from the table. Then look down the column to see that this would equal 16 pounds on the Moon. Therefore, 1 pound on Earth is 0.16 pound on the Moon.

200 lb. on Earth × 0.16 = 32 lb. on the Moon.

Use the table on page 182 to answer the following questions.

1. Gina weighs 28 pounds on Mercury. How much does she weigh on Earth?

2. How much more does Lucy's dog weigh on Jupiter than on Venus?

3. If an object weighs 32 pounds on the Moon, how much will it weigh on Earth?

4. If a spacecraft weighs four tons on Earth, how much will it weigh on the Moon in pounds?

5. How much will 1 pound of raisins on Earth weigh on Mars?

6. How much will 5 pounds of potatoes on Earth weigh on Jupiter?

Now suppose you move to Mars and want a weight conversion table for a 100-pound piece of lidanium crystal that you found. To do this, you can divide all entries from Lucy's Earth table by 0.38 on a calculator. Round your answers to the nearest pound. Fill in the table to the right.

Location	Rock's Weight
Mars	100 lb.
Moon	42 lb.
7. Mercury	▩
8. Venus	▩
9. Earth	▩
10. Jupiter	▩

Use the completed table to answer the following questions.

11. If a person weighs 150 pounds on Mars, how much will that person weigh on Earth?

12. If a boulder weighs 1 ton on Mars, how many pounds will it weigh on Jupiter?

13. Next you move to Jupiter and adopt an interterrestrial cat that weighs 100 pounds. Complete a weight conversion table for your pet cat.

▶ **Review** (Lesson 3.8)

Write which operation you would use to solve each problem. Then solve.

14. Vito bought 5 large packs of gum for $0.69 each. How much was the gum (not including tax)?

15. Marty weighs 175 pounds. He gained 8 pounds by lifting weights. How much did he weigh before?

8.6 Customary Units of Capacity

Georgette helps prepare meals at a nursing home. She is constantly amazed at the amount of food consumed. She would like to figure out how many gallons of soup the cooking pot contains. The cook makes enough to give 60 patients each an 8-ounce serving.

Georgette needs to convert fluid ounces to gallons.

Examples

To convert customary units of capacity, first determine if you are converting to a larger or smaller unit.

Equivalents	Abbreviations
8 fluid ounces = 1 cup	fluid ounce **fl. oz.**
16 fluid ounces = 2 cups = 1 pint	cup **c.** pint **pt.**
32 fluid ounces = 2 pints = 1 quart	quart **qt.** gallon **gal.**
128 fluid ounces = 4 quarts = 1 gallon	

A Convert 480 fl. oz. to gal.

Divide since you are changing to a larger unit.

1 gal. = 128 fl. oz., so

480 fl oz. = 480 ÷ 128 = $3\frac{3}{4}$ gal.

480 fl. oz. = $\mathbf{3\frac{3}{4}}$ **gal.**

B Convert 3 c. to fl. oz..

Multiply since you are changing to a smaller unit.

1 c. = 8 fl. oz., so

3 c. = 3 × 8 = **24 fl. oz.**

C Choose the more reasonable measure.

A small bottle of perfume holds 4 ___ . fl. oz.   c.

The more reasonable measure is **fluid ounces**.

A pitcher would hold about 4 cups.

▶ Think and Discuss

1. Refer to the introduction to this lesson. How many gallons of soup does the cook make?

2. Convert 20 fluid ounces to cups.

3. Complete the table below.

1 cup = ___ fl. oz.   1 qt. = ___ pt.   1 gal. = ___ qt.
1 pt. = ___ c.   1 qt. = ___ c.   1 gal. = ___ pt.
1 pt. = ___ fl. oz.   1 qt. = ___ fl. oz.   1 gal. = ___ c.
1 gal. = ___ fl. oz.

Exercises

Convert each measure. (See Example A.)

4. 32 fl. oz. to c.
5. 28 c. to pt.
6. 42 pt. to qt.

7. 48 qt. to gal.
8. 70 fl. oz. to c.
9. 19 c. to pt.

Convert each measure. (See Example B.)

10. 7 c. to fl. oz.
11. 13 pt. to c.
12. 28 qt. to pt.

13. 19 gal. to qt.
14. 19 c. to fl. oz.
15. 9 gal. 2 qt. to qt.

Complete each statement. Choose the more reasonable measure.
(See Example C.)

16. A water cooler holds 10 ___ of water.   fl. oz.   gal.

17. A serving of orange juice is 6 ___.   fl. oz.   pt.

▶ **Mixed Practice** (For more practice, see page 426.)

Convert each measure.

18. 27 c. to pt.
19. 5 pt. to fl. oz.
20. 9 qt. to c.

21. 8 pt. 1 c. to c.
22. 7 gal. to c.
23. 4 gal. 2 qt. to pt.

Complete each statement. Choose the more reasonable measure.

24. An automobile tank will hold 12 ___ of gasoline.   qt.   gal.

25. A large lemonade pitcher holds 1 ___ of liquid.   c.   gal.

▶ **Applications**

26. Kim needs five quarts of oil for her car. If the store sells only gallon containers of oil, how many gallons must she buy?

27. A 6-pack of 12-ounce cans and a $\frac{1}{2}$-gallon bottle each cost $1.19. Which is the better buy?

▶ **Review** (Lessons 2.3, 2.7, 3.3)

Multiply or divide mentally.

28. $89.4 \div 10$
29. 500×80
30. $9013 \div 100$
31. 5.75×1000

8.7 Computing with Customary Units

Elliot tapes comedy routines off the radio. Can he fit the three routines described here on the 15 minutes left on a tape?

Examples

To compute with customary units, add or subtract like units. Regroup if necessary.

A Add.   4 min. 12 sec. + 3 min. 55 sec.

 4 min. 12 sec.
+ 3 min. 55 sec.
 7 min. 67 sec.   Rename 67 sec. as 1 min. 7 sec. Add to 7 min.

4 min. 12 sec. + 3 min. 55 sec. = **8 min. 7 sec.**

B Subtract.   12 yd. 1 ft. − 9 yd. 2 ft.

 12 yd. 1 ft. →   11 yd. 4 ft.   Regroup 12 yd.
− 9 yd. 2 ft. →   9 yd. 2 ft.
                      2 yd. 2 ft.

12 yd. 1 ft. − 9 yd. 2 ft. = **2 yd. 2 ft.**

▶ Think and Discuss

1. Refer to the introduction to this lesson. Will the three routines fit on Elliot's tape?

2. Add.   9 hr. 25 min. 55 sec. + 2 hr. 45 min. 10 sec.

3. Regroup 25 yards 2 feet so you could subtract 7 feet from it.

4. Explain why it is sometimes necessary to regroup when subtracting with customary units. Give an example.

Exercises

Add. (See Example A.)

5.	1 hr. 19 min. + 2 hr. 32 min.	**6.**	16 lb. 7 oz. + 9 lb. 8 oz.	**7.**	19 gal. 1 qt. + 13 gal. 2 qt.
8.	18 ft. 5 in. + 6 ft. 9 in.	**9.**	13 hr. 32 min. + 6 hr. 39 min.	**10.**	27 lb. 13 oz. + 13 lb. 12 oz.

Subtract. (See Example B.)

11.	3 hr. 28 min. − 2 hr. 19 min.	**12.**	6 gal. 2 qt. − 5 gal. 3 qt.	**13.**	29 lb. 12 oz. − 13 lb. 15 oz.
14.	9 yd. 10 in. − 5 yd. 5 in.	**15.**	19 hr. 23 min. − 11 hr. 29 min.	**16.**	31 lb. 6 oz. − 15 lb. 11 oz.

► Mixed Practice (For more practice, see page 427.)

Add or subtract.

17.	13 min. 15 sec. + 26 min. 29 sec.	**18.**	9 yd. 2 ft. 3 in. − 7 yd. 2 ft. 9 in.	**19.**	15 gal. 1 qt. − 9 gal. 3 qt.
20.	3 T. 843 lb. − 1 T. 1650 lb.	**21.**	9 yd. 2 ft. + 3 yd. 1 ft. 7 in.	**22.**	19 min. − 6 min. 8 sec.

► Applications

23. The winning time for the marathon was 2 hours 26 minutes 18 seconds. This time was 2 hours 36 minutes 53 seconds faster than that of the last finisher. What was the slowest time?

24. An empty delivery truck weighs 6 tons 650 pounds. The truck weighs 8 tons when completely full. What is the weight of the truck's full load in pounds?

25. A program manager wrote the list at the right for a 30-minute TV show. How much time is planned for commercials?

TV₃

First Segment	2:30	Third Segment	9:00
Theme Song	1:30	Commercials	
Commercials		Closing Credits	0:30
Second Segment	12:15	Commercials	
Commercials			

► Review (Lesson 3.9)

Simplify.

26. $63 \div 7 + 8 \times 6$

27. $9 \times (8 - 6) \times 7$

28. $12 + 60 \div 10 \times 5$

29. $56 \div 8 + 10 - 4$

30. $28 \div 4 \times (95 + 4)$

31. $45 \div (9 + 6) \times 7$

8.8 Measuring Temperature

Before leaving on a trip to France, Mr. Kuhn telephoned to find out about the weather there. The overseas operator told him the temperature was 30 degrees. Mr. Kuhn was surprised, but to be safe he wore his winter coat, boots, and gloves. When he got off the plane, Mr. Kuhn realized that what he really needed was his swimsuit. What caused the misunderstanding?

Mr. Kuhn assumed that the temperature scale was Fahrenheit, but it actually was Celsius.

Examples

To determine temperatures on the Fahrenheit or Celsius scale, use the drawing of a thermometer on page 189.

A Determine the room temperature in degrees Fahrenheit (°F).	**B** Determine the temperature of a hot day in Los Angeles in degrees Celsius (°C).
68°F	32°C

▶ Think and Discuss

1. If you were going ice-skating outside, would you want the temperature to be 30 degrees Celsius? Explain your answer.

2. Would you want to swim outdoors if the temperature were 10 degrees Celsius? Explain your answer.

3. When a temperature is given, why is it important to know which scale is being used?

4. Explain why it is incorrect to say that 20°F is "twice as warm" as 10°F.

Exercises

Use the thermometer to determine each temperature in degrees Fahrenheit. (See Example A.)

5. a cold winter day in Chicago

6. freezing point of water

7. normal body temperature

8. boiling point of water

Use the thermometer to determine each temperature in degrees Celsius. (See Example B.)

9. room temperature

10. a cold winter day in Denver

11. normal body temperature

12. freezing point of water

▶ **Mixed Practice** (For more practice, see page 427.)

Use the thermometer to determine each temperature in degrees Fahrenheit and Celsius.

13. a hot day in Miami

14. warm dinner rolls

15. a warm shower

16. hot soup

▶ **Applications**

17. The high temperature for the day in Tucson was ninety-eight degrees Fahrenheit. What would the temperature be if it dropped nineteen degrees?

18. There is a 12-degree difference between 20 degrees Celsius and 32 degrees Celsius. What is the difference between the equivalent temperatures on the Fahrenheit scale?

▶ **Review** (Lessons 6.1, 6.2, 6.4, 6.6, 6.7)

Multiply or divide.

19. $\frac{7}{10} \times \frac{4}{5}$

20. $3\frac{1}{3} \div 6\frac{1}{4}$

21. $2\frac{2}{3} \times 9$

22. $\frac{7}{8} \div 3$

23. $8 \div 1\frac{3}{4}$

24. $5 \times 5\frac{3}{5}$

25. $\frac{5}{8} \div \frac{5}{12}$

26. $1\frac{1}{10} \times 3\frac{1}{5}$

8.9 Investigating Elapsed Time

School begins at 8:10 in the morning
and ends at 2:25 in the afternoon.
How long is the school day?

1. Find the length of the school day. Discuss how
you computed your answer.

You may find that people solve problems
involving time in different ways. Did anyone in
your class solve the problem using one of the
methods shown below?

Method 1

8:10 to 12:10 → 4 hours
12:10 to 2:10 → 2 hours
2:10 to 2:25 → _15 minutes_   Add the times.

6 hours, 15 minutes

Method 2

8:10 to noon → 3 hours, 50 minutes
Noon to 2:25 → _2 hours, 25 minutes_   Add the times.

5 hours, 75 minutes = 6 hours, 15 minutes

Method 3

Convert 2:25 to 14:25.   Add 12 hours and 2 hours.
14:25 − 8:10 = 6:15   Subtract.

In each case,
the total time is **6 hours,
15 minutes**.

No one of the three methods shown is always the best one to
use. Other methods are also possible. Was any method used by
your class different from those shown?

Solve each problem below using any method.

2. A baseball game began at 7:35 p.m. and ended at 10:42 p.m. How long was the game?

3. A woman began work at 9:15 a.m. and ended at 5:30 p.m. How long did she work?

4. Mr. Button came over to borrow some sugar at 10:45 a.m. He didn't leave until 4:20 p.m. How long did he stay?

5. The marathon race began at 6:30 a.m. The last runner crossed the finish line at 6:36 p.m. How long did this runner take?

6. A man began repeatedly singing "You Light Up My Life" at 8:12 a.m. on May 7. He did not stop singing until 3:21 p.m. on May 10. How long did he sing?

Knowing a variety of different ways to attack elapsed time problems helps you solve other problems involving time. Discuss and solve each problem below.

7. Can you find the elapsed time from 8:40 a.m. to 1:15 p.m. using Method 3? Discuss how this method is similar to subtracting decimals. How is it different?

8. Find the elapsed time from 1:10 a.m. to 12:55 a.m.

9. A concert begins at 8:00 p.m. It takes an hour and 15 minutes to get there. When should you leave if you want to be there an hour early?

10. You are taking an all-day exam. Your watch reads 10:48. The section you are on now ends at 11:06. How much time is left?

11. An opera begins at 7:00 p.m. and lasts $4\frac{1}{2}$ hours. You get home 20 minutes later. Are you home by midnight?

Copy and complete the table below.

	Starting Time	Finishing Time	Elapsed Time
12.	11:40 a.m.	6:42 p.m.	_____
13.	_____	3:20 a.m.	6 hours, 7 minutes
14.	6:30 p.m.	_____	17 hours, 5 minutes
15.	_____	4:11 p.m.	8 hours, 52 minutes

▶ **Review** (Lesson 5.8)

Write as a repeating decimal.

16. $\frac{1}{6}$ 17. $\frac{5}{6}$ 18. $\frac{1}{3}$ 19. $\frac{2}{3}$ 20. $\frac{2}{9}$ 21. $\frac{3}{11}$

8.10 Calculating Earnings

Maria and Jessie recently began weekend jobs. Maria earns $5.50 an hour and Jessie earns $4.75 an hour. On their first payday, Maria was surprised that her paycheck was smaller than Jessie's. Jessie remarked, "Remember, you worked a short day last Sunday."

Maria and Jessie could compare their gross earnings, or earnings before taxes, by reading their weekly timecards.

Name: Maria Cabrera				Total Hours
Day	In	Out	In	Out	
Sat.	10:00	12:00	1:00	5:00	
Sun.	12:00	3:00			

Name: Jessie Martin				Total Hours
Day	In	Out	In	Out	
Sat.	10:00	12:00	1:00	5:00	
Sun.	12:00	6:00			

Examples

To determine gross earnings, first figure the number of hours worked. Then multiply by the hourly rate.

A Determine the number of hours Maria worked.

Maria worked **9 hours**.

Saturday:	10–12	2 hours
	1–5	4
Sunday:	12–3	+ 3
		9 hours

B Determine Maria's gross earnings.

Maria earned **$49.50**.

Multiply Maria's hourly rate by the number of hours she worked.

$$\begin{array}{r} \$5.50 \leftarrow \text{rate per hour} \\ \times \quad 9 \leftarrow \text{hours worked} \\ \hline \$49.50 \leftarrow \text{gross earnings} \end{array}$$

▶ Think and Discuss

1. Determine the total weekly earnings for 20 hours at $7 an hour.

2. Refer to the introduction to this lesson. How much greater was Jessie's gross pay than Maria's?

3. Jean works 35 hours a week for $6.25 an hour. How much does she earn in 2 weeks?

4. Fran earns $7 an hour and works 20 hours a week. Her brother Rogelio earns $3.50 an hour. How many hours must Rogelio work to earn as much as Fran in a week? Solve mentally.

Exercises

Use the timecards below.
Determine the total hours worked each week. (See Example A.)

5.

Day	Mon.	Tue.	Wed.	Thu.	Fri.
Hours Worked	$5\frac{1}{2}$	$7\frac{3}{4}$	$6\frac{3}{4}$	9	$8\frac{1}{2}$

6.

Day	Mon.	Tue.	Wed.	Thu.	Fri.
Hours Worked	$3\frac{1}{2}$	2	$3\frac{1}{4}$	$6\frac{1}{2}$	$7\frac{1}{4}$

Determine the gross earnings for both timecards.
(See Example B.)

7. Rate: $4.35 an hour

8. Rate: $5.10 an hour

▶ Mixed Practice (For more practice, see page 428.)

Copy and complete the chart below. Then find the gross earnings.

Mergenthal Pharmacy, Inc. Employee Timecard			Name: Eric Obenza Hourly Wage: $ 5.25		
Date	In	Out	In	Out	Hours
9. 1-3	8:30	12:00	12:30	5:00	
10. 1-4	8:45	12:30	1:15	4:45	
11. 1-5	8:30	12:15	1:00	5:00	
12. 1-6	8:30	12:30	1:00	5:15	
13. 1-7	9:00	1:30			
14. 1-8	12:00	5:45			
15. Total Hours					
16. Gross Earnings					

▶ Review (Lesson 2.3)

Multiply mentally.

17. 8000 × 50

18. 30 × 700

19. 300 × 100

20. 600 × 9000

21. 10 × 60,000

22. 400 × 500

23. 6000 × 70

24. 25 × 300

Chapter 8 Review

Complete each statement. (Lesson 8.1)

1. The four basic units of length in the customary system are the
   ___ , the ___ , the ___ , and the ___ .

2. The Fahrenheit and Celsius scales measure ___ .

Convert each measure. (Lessons 8.2, 8.4, 8.6)

3. 7 ft. to in.   4. 75 yd. to ft.   5. 99 in. to ft.

6. 87 in. to yd.   7. 35 oz. to lb.   8. 10 qt. to gal.

Complete each statement. Choose the more reasonable measure.
(Lessons 8.3, 8.4, 8.6)

9. During practice, the athlete swam ___ .   30 yd.   3 mi.

10. The cake recipe called for 1 ___ of corn oil.   c.   gal.

11. Laura's new bracelet weighed 4 ___ .   oz.   lb.

Measure each line segment below to the nearest inch, $\frac{1}{2}$ inch, $\frac{1}{4}$ inch.
(Lesson 8.3)

12. _____   13. _____

14. _____

Use the table to solve the following. (Lesson 8.5)

15. If you buy 6 roses, how much
    does each rose cost?

16. If you buy 6 roses one day and
    2 roses the next, how much did
    you spend on the flowers?

Number	Cost
1	$3
2	$6
6	$15
12	$25

Add or subtract. (Lesson 8.7)

17.   7 yd. 2 ft. 8 in.   18.   1 gal. 2 qt.   19.   20 lb. 5 oz.
    + 9 yd. 2 ft. 9 in.   −   3 qt.   − 5 lb. 15 oz.

Determine the temperature of the following. (Lesson 8.8)

20. The normal body temperature is about 98.6°Fahrenheit,
    which is about ___ .   37°C   100°C   0°C

Chapter 8 Test

Add or subtract.

1. 2 gal. 3 qt. 1 pt.
    + 4 gal. 2 qt. 1 pt.

2. 8 yd. 1 ft. 8 in.
    − 6 yd. 2 ft. 9 in.

3. 36 lb. 14 oz.
    + 25 lb. 13 oz.

Complete each statement.

4. To convert pounds to ounces, you would multiply by ▦ .

5. There are ▦ inches in a foot and ▦ feet in a yard.

Convert each measure.

6. 67 oz. to lb.

7. 13 gal. to c.

8. 28 yd. to ft.

Use the table to answer the following.

9. Which is the highest priced call?

10. Which is the lowest priced call?

11. What is the difference in cost between these two calls?

Number of calls	Length of calls	Cost per call
2	3 min.	$1.80
1	6 min.	3.30
5	4 min.	2.30
1	18 min.	9.30

Complete each statement. Choose the more reasonable unit of measurement.

12. A newborn kitten weighs about 6 ▦ .   oz.   lb.

13. A caterpillar is about 3 ▦ long.   in.   ft.

14. Room temperature is about 20 degrees ▦ .
    Celsius   Fahrenheit

Measure each line segment to the nearest inch, $\frac{1}{2}$ inch, $\frac{1}{4}$ inch.

15. _____

16. _____

TEST

▶ **Choose the letter that shows the correct answer.**

1. $3\frac{1}{8} \times 2\frac{2}{5}$

   a. $2\frac{29}{48}$
   b. 15
   c. $7\frac{1}{2}$
   d. not given

2. $\frac{3}{4} + \frac{7}{8}$

   a. $1\frac{1}{4}$
   b. $1\frac{5}{8}$
   c. $\frac{10}{32}$
   d. not given

3. $13 - 4\frac{5}{12}$

   a. $8\frac{7}{12}$
   b. $9\frac{7}{12}$
   c. $8\frac{5}{12}$
   d. not given

4. $4\frac{1}{2} \div \frac{2}{9}$

   a. 1
   b. $4\frac{1}{2}$
   c. $\frac{18}{4}$
   d. not given

5. $6\frac{3}{5} \div 3\frac{2}{3}$

   a. $1\frac{4}{5}$
   b. $24\frac{3}{5}$
   c. $1\frac{1}{5}$
   d. not given

6. $6\frac{5}{6} + 3\frac{3}{4}$

   a. $10\frac{1}{2}$
   b. $9\frac{7}{12}$
   c. $10\frac{7}{12}$
   d. not given

7. $5\frac{1}{2} - 2\frac{9}{10}$

   a. $2\frac{2}{5}$
   b. $3\frac{3}{5}$
   c. $3\frac{3}{4}$
   d. not given

8. $\frac{8}{9} \times \frac{4}{5}$

   a. $\frac{6}{7}$
   b. $\frac{32}{45}$
   c. $1\frac{1}{7}$
   d. not given

9.
```
  9 yd. 1 ft.  4 in.
- 3 yd. 2 ft. 10 in.
```
   a. 5 yd. 1 ft. 6 in.
   b. 6 yd. 1 ft. 4 in.
   c. 6 yd. 2 ft. 6 in.
   d. not given

10.
```
  7 gal. 1 qt.
- 4 gal. 3 qt.
```
   a. 3 gal. 2 qt.
   b. 4 gal. 1 qt.
   c. 3 gal. 1 qt.
   d. not given

11.
```
  28 lb. 14 oz.
+ 35 lb. 11 oz.
```
   a. 63 lb. 9 oz.
   b. 53 lb. 25 oz.
   c. 64 lb. 9 oz.
   d. not given

▶ **Convert each measure.**

12. 127 in. to ft.
   a. $10\frac{7}{12}$ ft.
   b. $42\frac{1}{12}$ ft.
   c. $12\frac{7}{12}$ ft.
   d. not given

13. 35 c. to qt.
   a. $17\frac{1}{4}$ qt.
   b. $4\frac{3}{4}$ qt.
   c. $8\frac{3}{4}$ qt.
   d. not given

14. 8 lb. 9 oz. to oz.
   a. 105 oz.
   b. 137 oz.
   c. 17 oz.
   d. not given

15. 3 hr. 39 min. to min.
   a. 159 min.
   b. 42 min.
   c. 219 min.
   d. not given

▶ **Complete each statement.**

16. Water boils at 100 degrees ▩ .
    a. Fahrenheit   **b.** Celsius
    c. Fahrenheit and Celsius   **d.** not given

17. The metric system is organized by powers of ▩ .
    a. 5   **b.** 100   **c.** 12   **d.** not given

18. A(n) ▩ is one of a pair of numbers whose product is 1.
    a. difference   **b.** sum   **c.** reciprocal   **d.** not given

19. A fraction with a numerator that is greater than or equal to
the denominator is called a(n) ▩ fraction.
    a. improper   **b.** LCD   **c.** GCF   **d.** not given

20. 0.3333333 . . . is called a(n) ▩ decimal.
    a. terminating   **b.** repeating   **c.** nonrepeating   **d.** not given

▶ **Use Guess and Check to solve.**

21. Alfonso has $10 and wants to
buy an equal number of pounds
of oranges and bananas.
Oranges cost $1.59 for 3 pounds.
Bananas cost $0.39 for 3 pounds.
What is the largest number of
pounds he can buy?
    a. 5 pounds of each
    b. 10 pounds of each
    c. 15 pounds of each
    d. not given

22. Some tape storage boxes hold
10 tapes and some hold 16. What
kinds of boxes should Anne get
if she wants all of her 46 tapes to
be stored?
    a. 2 large
    b. 1 large and 3 small
    c. 4 small
    d. not given

▶ **Round each number to the nearest**

23. tenth.	**24.** hundred.	**25.** cent.	**26.** million.
9.638	25,251	$13.7973	11,499,105
a. 10	**a.** 25,200	**a.** $13.79	**a.** 11,000,000
b. 9.65	**b.** 25,600	**b.** $13.80	**b.** 12,000,000
c. 9.7	**c.** 25,300	**c.** $14.00	**c.** 11,500,000
d. not given	**d.** not given	**d.** not given	**d.** not given

▶ Choose the letter that shows the fraction or mixed
number that is NOT equivalent to the given fraction.

27. $\frac{3}{4}$

   a. $\frac{8}{9}$

   b. $\frac{12}{16}$

   c. $\frac{75}{100}$

   d. not given

28. $\frac{17}{4}$

   a. $3\frac{5}{4}$

   b. $\frac{34}{8}$

   c. $4\frac{1}{4}$

   d. not given

29. $\frac{2}{16}$

   a. $\frac{1}{8}$

   b. $\frac{4}{36}$

   c. $\frac{8}{64}$

   d. not given

30. $\frac{10}{50}$

   a. $\frac{20}{100}$

   b. $\frac{2}{5}$

   c. $\frac{1}{5}$

   d. not given

▶ Compare. Use $<$, $>$, or $=$.

31. $\frac{5}{6}$ ✳ $\frac{11}{18}$

   a. $<$

   b. $>$

   c. $=$

   d. not given

32. $2\frac{1}{4}$ ✳ $\frac{15}{4}$

   a. $<$

   b. $>$

   c. $=$

   d. not given

33. $\frac{8}{16}$ ✳ $\frac{10}{20}$

   a. $<$

   b. $>$

   c. $=$

   d. not given

34. $3\frac{5}{8}$ ✳ $3\frac{7}{12}$

   a. $<$

   b. $>$

   c. $=$

   d. not given

▶ Solve.

35. How many $1\frac{1}{2}$-foot pieces of
wood can Sheila cut from a
$9\frac{3}{4}$-foot board?

   a. 14 pieces

   b. 7 pieces

   c. 6 pieces

   d. not given

36. Cans of juice are packed 24 to a
box. How many full boxes can
be packed from 836 cans?

   a. 34 cases

   b. 30 cases

   c. 340 cases

   d. not given

37. In a Chicago snowstorm the
north side of town received
$8\frac{1}{4}$ inches of snow. The south
side received $2\frac{5}{8}$ inches. How
much more snow fell on the
north side?

   a. $10\frac{7}{8}$ in.

   b. $6\frac{3}{8}$ in.

   c. $5\frac{5}{8}$ in.

   d. not given

38. Mark wants to triple his chili
recipe. The recipe calls for
$2\frac{3}{4}$ teaspoons of chili powder.
How much chili powder should
he use for the larger recipe?

   a. 6 teaspoons

   b. $8\frac{1}{4}$ teaspoons

   c. $\frac{1}{12}$ of a teaspoon

   d. not given

TAKE 5

1 ▷ Squaring Off

What are the next two square numbers? Draw the squares.

2 ▷ Pyramid Puzzle

What are the next two triangular numbers? Draw the triangles.

3 ▷ Lucky 13

Copy the figure below and place 1, 2, 3, 4, 5, 6, 7, or 8 in each circle. The sum on each side must be 13.

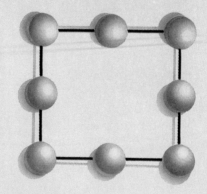

4 ▷ Half or Half Not

Pliny told his friend Virgil that half of 12 is 7. Can you show how this is possible?

5 ▷ Pole Pieces

If it takes 1 minute to make each cut, how long will it take to cut an 8-foot pole into 8 equal pieces?

What Is It?

- A three-point landing
- Ten-four
- Numero uno
- Four-leaf clover
- A two-by-four

"We just finished a report on the Civil War. We collected so much information—population figures, battle statistics, economic facts—that we were swimming in numbers! We finally decided to use several graphs."

Graphs

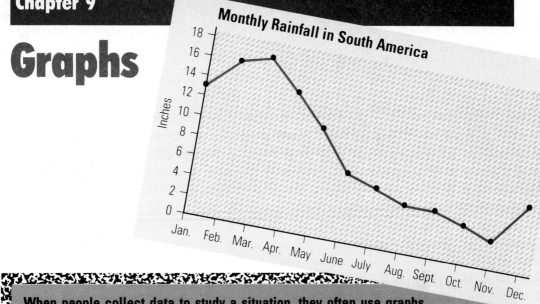

Monthly Rainfall in South America

When people collect data to study a situation, they often use graphs to present their data. Look in almost any magazine or newspaper, and you'll see a graph that describes
- the economy.
- the weather.
- population.
- the stock market.

Graphs can help you see the different ways numbers fit together.

Activity

Work with 2 or 3 students. In other textbooks or in magazines, find 5 graphs. Describe in a few short sentences what each graph is about. Then select the graph that you think is the most interesting. Show this graph to the class, explaining what it shows and why it interests you.

Monthly Rainfall in South America

9.1 Introduction to Graphs

This chapter is about graphs and graphing. A *graph* is a drawing or diagram that displays information. Although graphs cannot always show information as accurately as tables can, they can give a "snapshot view" of complex situations.

Sometimes graphs show information broken into groups. For example, the **bar graph** below shows how test scores were distributed among As, Bs, Cs, and Ds. The same information can also be shown in a **circle graph** or a **pictograph**.

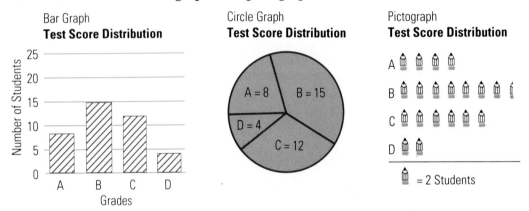

Line graphs are used to show trends, or changes over time. The line graph below shows how the average price of farmland in the United States changed between 1978 and 1987.

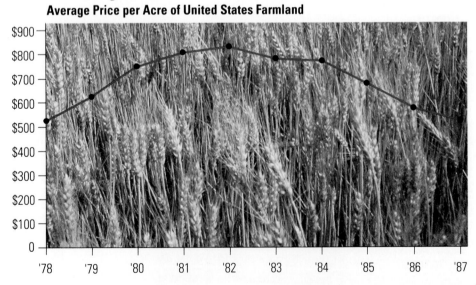

Average Price per Acre of United States Farmland

▶ Express Yourself

You read a line graph or bar graph by first examining the title. Then you examine the **vertical scale** and the **horizontal scale** of the graph. The scales of a graph often show numerical information such as years, dollar amounts, or numbers of people.

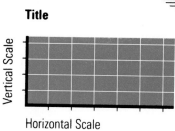

Title

Horizontal Scale

1. What does the vertical scale of the line graph on the opposite page show?

2. What does the horizontal scale of the bar graph on the opposite page show?

3. Examine the title and vertical and horizontal scales of the line graph on the opposite page. Describe what is shown.

When a scale shows numerical information, it is usually divided into **equal intervals**. For example, the interval on the vertical scale of the bar graph on the opposite page is 5—the difference between marks on the scale is always 5.

4. What is the interval on the vertical scale of the line graph on the opposite page?

5. What is the interval on the horizontal scale of the line graph?

▶ Practice What You Know

Often, when you read a graph, you must estimate values. For example, in the line graph on the opposite page the highest point is about one-quarter of the way between $800 and $900.

6. What is the difference between $800 and $900?

7. What is one-quarter of this difference?

8. Add the answer you found in Question 7 to $800 to find the approximate value of the highest point in the line graph.

Estimate each of the following.

9. a value one-half of the way between $500 and $1000

10. a value three-quarters of the way between 400 and 500

11. a value four-fifths of the way between $1000 and $2000

12. a value one-quarter of the way between 20 and 40

9.2 Reading Bar Graphs

If there's one thing Matthew loves, it's cars. He reads car magazines every chance he gets. To compare all the statistics, Matthew especially likes to read the graphs.

Examples

To read a bar graph, first read the labels to see what is being studied.

Acceleration, 0 – 60 MPH

Capella 2000	/////////////
Pollux DS	////////
Crucis	///////////
Sirius M2	/////

0 1 2 3 4 5 6 7 8 9 10 11
Seconds

A Find how long it takes the Pollux DS to accelerate from 0 to 60 miles per hour.

Find the bar for the Pollux DS. Go to the end of the bar and, with your finger, drop straight down to the horizontal scale. The number on the scale is a 7.

It takes **7 seconds** for the Pollux DS to accelerate from 0 to 60 miles per hour.

B Find the cost of the Crucis.

Find the bar for the Crucis. Go to the top of the bar and trace a line straight across to the vertical scale. The number on the scale is halfway between $10,000 and $20,000, or $15,000.

The Crucis costs **about $15,000.**

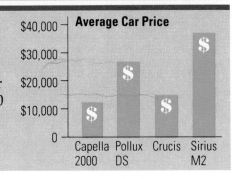

▶ Think and Discuss

1. Refer to Example A. How long does it take the Sirius M2 to accelerate from 0 to 60 miles per hour?

2. Refer to Example B. Estimate the cost of the Capella 2000.

3. Refer to Example B. About how much less is the cost of the Crucis than the Sirius M2?

Exercises

Use the graph below to answer each question.
(See Example A.)

Winter Olympic Gold Medals, 1932 – 1988

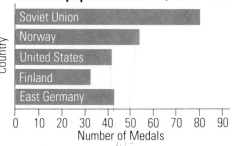

Country

Soviet Union
Norway
United States
Finland
East Germany

0 10 20 30 40 50 60 70 80 90
Number of Medals

4. Which country has won the second most gold medals in the Winter Olympic Games?

5. Which country has won about 45 gold medals?

6. About how many more gold medals has the Soviet Union won than the United States?

Use the graph to the right to answer each question. (See Example B.)

7. Which country has won the most gold medals in hockey?

8. How many more gold medals in hockey has Canada won than the U.S.?

9. How many gold medals have been awarded in hockey since 1932?

Olympic Gold Medals in Hockey, 1932– 1984

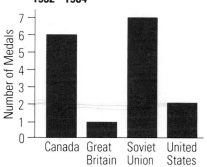

Number of Medals

7
6
5
4
3
2
1
0

Canada Great Britain Soviet Union United States

▶ **Mixed Practice** (For more practice, see page 428.)

Use the graphs above to answer each question.

10. Which country has won the fewest gold medals in hockey?

11. Which country has won the fewest gold medals in all sports?

12. Which country has won about 6 gold medals in hockey?

▶ **Applications**

13. Find a bar graph in a newspaper or magazine. Write a paragraph describing the contents of the graph.

▶ **Review** (Lessons 2.6, 3.4)

Estimate.

14. $436 \div 3$ 15. 8911×5 16. $922 \div 20$ 17. 348×19 18. 1987×8

19. $4321 \div 7$ 20. $791 \div 82$ 21. 22×4723 22. $8471 \div 9$ 23. 13×1111

9.3 Reading Double Bar Graphs

Mr. Jay's sociology class surveyed the sophomores to find out about after-school activities. The students then constructed graphs. Marty made a double bar graph to show how the boys and the girls differed.

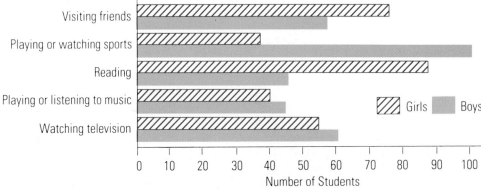

After School Activities

Girls   Boys

Number of Students

Examples

To read a double bar graph, read each bar separately.

Find the difference in the number of boys and girls who enjoy reading after school.

First, find the 2 bars for the reading category. According to the key, the ▬ bar corresponds to boys and the ▨ bar corresponds to girls. Determine the value for each bar.

Boys: 45   Girls: 86   86 − 45 = 41

41 more girls than boys said that they enjoy reading after school.

▶ Think and Discuss

1. Which activity is most popular among boys? Which is most popular among girls?

2. Which activity is least popular overall? How did you find it?

3. Describe how the information shown in Marty's graph could be represented on two single bar graphs. Could it be shown in one single bar graph?

4. When is a double bar graph useful?

Exercises   (For more practice, see page 429.)

Use the graph to the right to answer the following questions.

5. In which city was the median house price highest in 1977?

6. Approximately how much more did a house in Chicago cost in 1987 than in 1977? *about 55¢*

7. Estimate the difference in house prices between Los Angeles and Chicago in 1987.

8. In which city did house prices nearly triple between 1977 and 1987?

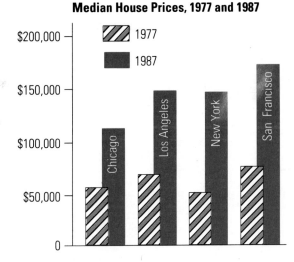

Median House Prices, 1977 and 1987

▶ **Applications**

9. You are doing a report on public land in the United States and plan to summarize the bar graph shown below in a paragraph. What will you write?

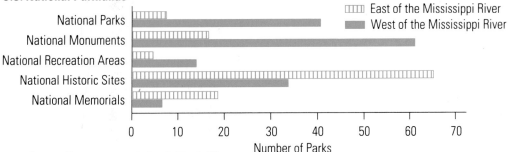

U.S. National Parklands

▶ **Review** (Lessons 2.3, 2.7, 3.3)

Divide or multiply.

10. $3797 \div 10$ 11. 0.9×3000 12. 72.3×100 13. $88 \div 400$

14. 2.5×8000 15. $0.6211 \div 100$ 16. 109×1000 17. 44.36×2000

9.4 Constructing Bar Graphs

Erica works at the Get Fit Gym. People often ask her how many calories they burn doing various activities. Erica has decided to draw a bar graph that shows the information.

Erica gathers and arranges the information in a chart as shown below.

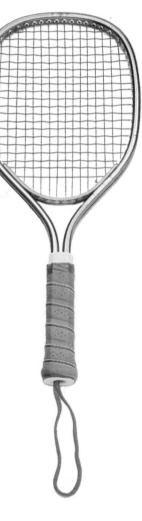

Activity	Calories Burned Per Hour
Aerobic Dance	778
Basketball	564
Jumping Rope	922
Racquetball	732
Running	786
Swimming	320
Weight Training	294

Then Erica must decide how to set up the scales of her graph.

1. Look at the headings in Erica's chart. Could these headings be made into the labels for the vertical and horizontal scales? Explain how.

2. Erica decides to make a vertical bar graph. What will the horizontal scale show? What will the vertical scale show?

3. Draw the scales, making each about 4 inches long.

4. Decide on a reasonable bar width. Then divide the horizontal scale and label each bar.

Now the vertical scale must be labeled and divided.

5. First, look at the information given. Find the greatest number in the "Calories Burned" column. The top of the vertical scale must be greater than this number. Why?

6. Erica decides to use 1000 as the top of the scale. Why would 1000 be easier to work with than a number such as 925?

The vertical scale must now be divided into equally spaced intervals. The scale covers 1000 units, from 0 to 1000. For Erica's graph, an interval of 100 is appropriate.

7. Divide and label the intervals of the vertical scale.

8. Why would an interval of 1 or 10 be inappropriate?

Finally, the bars must be drawn. Since the vertical scale has an interval of 100, the height of each bar must be estimated. Start with the bar for aerobic dance: 778 is about $\frac{3}{4}$ of the distance between 700 and 800.

9. On the vertical scale, find a point that is $\frac{3}{4}$ of the distance between 700 and 800. Draw a vertical bar for aerobic dance that goes as high as this point. Color the bar.

10. Playing basketball burns 564 calories per hour: 564 is slightly greater than $\frac{1}{2}$ the distance between 500 and 600. On the vertical scale, estimate where 564 lies. Now draw a vertical bar for basketball and color it.

11. Complete the graph.

▶ **Think and Discuss** (For more practice, see page 429.)

12. Describe the graph you constructed. Compare your graph to the list of numbers you started with. How are they different? How are they the same?

13. Refer to the introduction to this lesson. Erica organized her information by listing the activities in alphabetical order. How else might she have organized the information? Describe how the look of the graph would change if she reorganized the information.

14. How is the bar graph more helpful than the original table? Discuss.

▶ **Review** (Lessons 4.2, 4.5, 4.6)

Convert each measure.

15. 9.9 cm to m
16. 337 mL to L
17. 0.48 kg to g

18. 1650 m to km
19. 543 mm to cm
20. 75 g to kg

21. 43 L to mL
22. 9.7 km to m
23. 157 g to kg

24. 47.8 m to mm
25. 0.6 L to mL
26. 92 kg to g

9.5 Reading Line Graphs

Sallie is a meteorologist in Minneapolis. Part of her job involves graphing weather data. She often uses line graphs because they show trends, or changes over a period of time.

Examples

To read a line graph, first read the scale labels to see what is being studied. The line graph below shows the average monthly temperatures for Minneapolis, Minnesota.

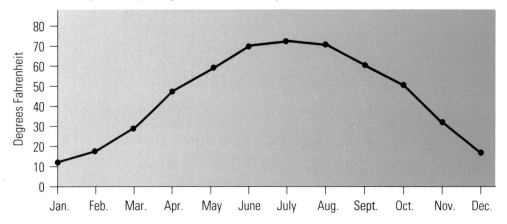

Average Monthly Temperatures in Minneapolis

Find the average temperature for May in Minneapolis.

Find May on the horizontal scale. Move straight up to the graph. Then trace a line straight across to the vertical scale.

The average temperature in May in Minneapolis is **about 59°**.

▶ Think and Discuss

1. Describe how to find the average temperature for February in Minneapolis.

2. What is the average temperature in Minneapolis in August?

3. Find the warmest month and the coldest month in Minneapolis. What is the difference in temperature?

4. How is the information on the horizontal scale organized? Would it be reasonable to alphabetize these labels? Why?

5. "That graph is wrong. I was in Minneapolis last March and it was 10° outside," one cynic said. Why is he mistaken?

Exercises (For more practice, see page 430.)

Use the graph to the right to answer the following questions.

6. Describe briefly what the graph shows.

7. Which month has the greatest number of rainy days?

8. According to this graph, when is the rainy season in San Francisco? When is the dry season?

9. About how many days a year does it rain in San Francisco?

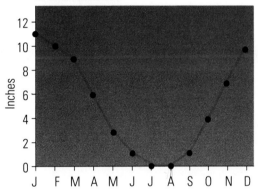

Average Number of Days with Rain in San Francisco

▶ **Applications**

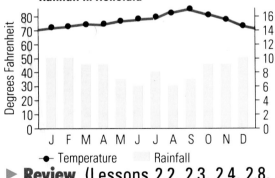

Average Monthly Temperature and Rainfall in Honolulu

-●- Temperature ▨ Rainfall

10. Use the graph to the left to describe the climate of Honolulu, Hawaii.

11. Compare the graphs for Honolulu and Minneapolis (opposite page). Describe how the two cities differ in temperature.

▶ **Review** (Lessons 2.2, 2.3, 2.4, 2.8, 2.10)

Multiply.

12.	327	13.	0.894	14.	88	15.	0.7	16.	606
	× 62		× 0.9		× 37		× 0.54		× 509

17.	9.07	18.	600	19.	6398	20.	0.004	21.	0.49
	× 7.6		× 80		× 5		× 0.05		× 3

9.6 Constructing Line Graphs

For a history project, you are studying immigration to the United States. You decide to present information for 1821–1980 in a line graph, to show the trend of immigration over time.

First you arrange your information in a table.

Immigration to the United States, 1821—1980

1821—1840	0.8 million
1841—1860	4.3 million
1861—1880	5.1 million
1881—1900	8.9 million
1901—1920	14.4 million
1921—1940	4.7 million
1941—1960	3.7 million
1961—1980	7.3 million

Line graphs and vertical bar graphs are constructed in similar ways. The previous lessons in this chapter can help you decide how to divide the scales for the immigration graph.

1. What information goes on the horizontal scale?

2. Describe how the scale should be divided. How many divisions are there? Explain how you decided.

Often in graphs, when the numbers are very large, the scales can be labeled with "abbreviated" numbers.

3. Look at the sample scale at the right. Write the numbers that the "abbreviated" numbers stand for.

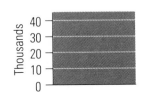

4. What should the greatest number on your vertical scale be? Write this number as an "abbreviated" number.

5. Describe how the vertical scale should be divided. What interval did you select? Why?

6. Draw, divide, and label the horizontal and vertical scales.

Constructing a line graph is similar to constructing a bar graph. For the period 1821–1840, for example, you start by finding the mark on the horizontal scale that is labeled 1821–1840. Lightly pencil in a line that rises vertically from this point.

Next find the point on the vertical scale that corresponds to 0.8 million. Lightly pencil in a line that extends horizontally from this point. Where the 2 penciled-in lines cross, place a point. This is the first point on your line graph.

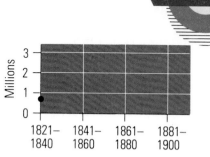

7. Continue to find the points for each 20-year period.

8. When you have found all the points, use a line to connect each point in order. Your line graph is now complete.

9. Describe in a few sentences the pattern of immigration to the United States since 1821.

Exercises (For more practice, see page 430.)

U.S. Life Expectancy
(average expected lifespan)

Year	
1920	54.1
1930	59.7
1940	62.9
1950	68.2
1960	69.7
1970	70.9
1980	73.7

10. Construct a line graph from the table at the right. Then describe the pattern in life expectancy since 1920.

▶ **Applications**

11. The two graphs below show the rise in a bank account balance over a 4-year period. Notice that the first graph gives an impression of a sharp increase, while the second graph gives an impression of a slow increase. Why? What is similar about the two graphs? What is different?

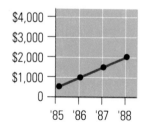
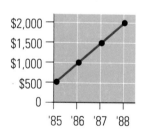

12. Draw a graph that gives an impression of a sharper increase in the bank account than the two graphs here do.

▶ **Review** (Lesson 1.7)

Round to the nearest hundred.

13. 9647   **14.** 11,557   **15.** 32,976   **16.** 48,339   **17.** 57,151

9.7 Constructing and Reading Pictographs

Neil works at the Computer Source, where a three-day sale has just ended. Neil must make a report on the sale to the store manager. He decided to construct a pictograph to show how many computers were sold.

Examples

To read a pictograph, determine what each symbol represents. An incomplete symbol represents a fractional part of the symbol's value.

Find the total number of computers sold at the sale.

Each computer symbol represents 10 computers sold. Count the number of whole computers pictured and multiply by 10.
$7 \times 10 = 70$

Count the number of half-computers pictured and multiply by 5. $2 \times 5 = 10$

Add. $70 + 10 = 80$

Approximately 80 computers were sold.

▶ Think and Discuss

1. How many computers does the symbol  represent?

2. How many model MT3 computers were sold?

3. How many Model INX/2S computers were sold?

4. If Neil made a graph showing total sales for the entire year, would he still have one picture equal 10 computers? Explain.

Exercises

(For more practice, see page 431.)

Refer to the introduction to this lesson. Use Neil's graph to answer the following questions.

5. Model SG2 costs $1500. What were the total sales in dollars for this model?

6. Model MT3 costs $2000. Model INX/2S costs $1000. What were the combined total sales in dollars for these models?

7. Using "$" to represent $5000, construct a pictograph to represent total sales in dollars for each computer model.

▶ Applications

Use the graph below to answer the following questions.

Rural and Urban United States Population, 1950 – 1980

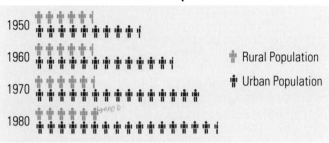

Each symbol represents 10 million people

8. What does the symbol 👤👤👤 represent? *urban population*

9. About how many people lived in cities in 1950? In 1980?

10. Approximately how much greater was the rural population in 1980 than in 1950?

11. Examine the growth of urban and rural populations over the period 1950–1980. Predict the 1990 populations. Explain your predictions.

12. Could this information be shown in a bar graph? Could it be shown in a line graph? Which method do you think would be most effective? Why?

▶ Review (Lessons 1.9, 1.10, 1.11)

Add or subtract.

13.	14.	15.	16.	17.
8.096	6	0.8	9007	27,375
+ 3.897	− 3.153	+ 12.4	− 3268	+ 46,786

9.8 Reading Circle Graphs

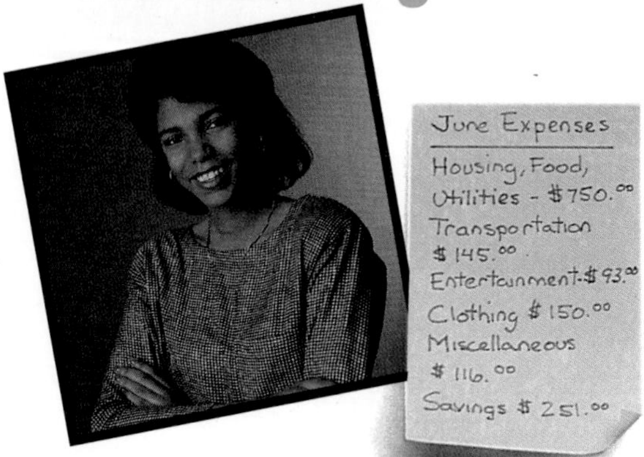
June Expenses
Housing, Food, Utilities - $750.00
Transportation $145.00
Entertainment-$93.00
Clothing $150.00
Miscellaneous $116.00
Savings $251.00

Carol sometimes wonders where all her money goes, so she decided to keep track of her expenditures. In June she wrote down her expenses in 6 categories. Then she made a circle graph to show the results.

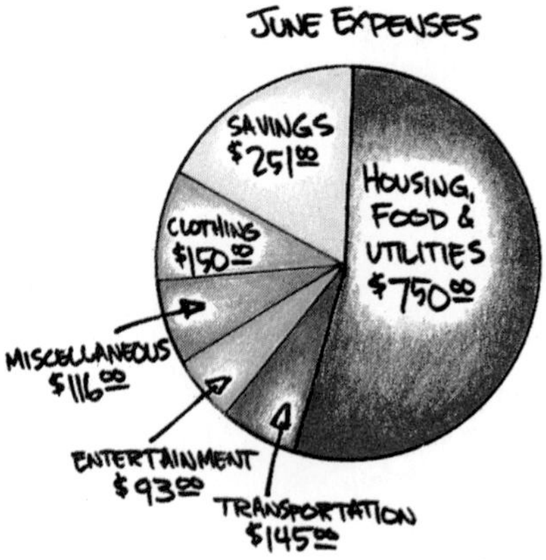
JUNE EXPENSES
SAVINGS $251.00
HOUSING, FOOD & UTILITIES $750.00
CLOTHING $150.00
MISCELLANEOUS $116.00
ENTERTAINMENT $93.00
TRANSPORTATION $145.00

1. In which category did Carol spend the most money? How did you find an answer?
2. In which category did Carol spend the least money?
3. List Carol's expense categories in order from most money spent to least money spent.
4. The sum of all the amounts in the circle graph equals Carol's total income for the month. What was her income?
5. About what fraction of Carol's income went into savings?
6. About what fraction of Carol's income went for transportation?
7. What might the Miscellaneous category include?

Exercises

(For more practice, see page 431.)

Thom decided to show in a graph how his time was divided up. Use his graph to answer the following questions.

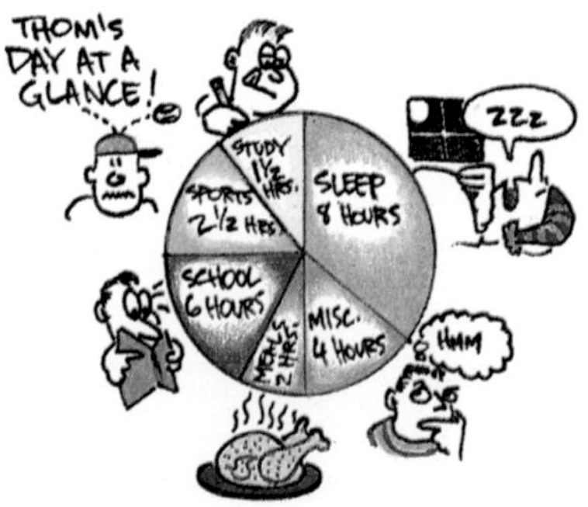
THOM'S DAY AT A GLANCE!
STUDY 1½ HRS.
SLEEP 8 HOURS
SPORTS 2½ HRS.
SCHOOL 6 HOURS
MEALS 2 HRS.
MISC. 4 HOURS
ZZZ
HMM

8. How many hours does Thom spend eating or studying each day?
9. Which activity is the second most time consuming?
10. What fractional part of his day does Thom spend sleeping?
11. What fractional part of his day does Thom spend in school?

12. How many hours does Thom spend playing sports?

13. What fractional part of Thom's school day is his study time equal to?

14. What might the Miscellaneous category include?

▶ Applications

The graph below shows the proposed U.S. budget for 1988.

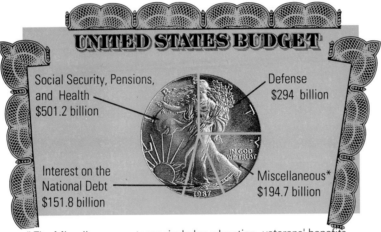

UNITED STATES BUDGET

Social Security, Pensions, and Health $501.2 billion

Defense $294 billion

Interest on the National Debt $151.8 billion

Miscellaneous* $194.7 billion

* The Miscellaneous category includes education, veterans' benefits, transportation, agriculture, natural resources, energy, science, etc.

15. What is the total U.S. budget in billions of dollars?

16. List the budget categories in order from least to greatest.

17. Estimate the fractional part of the total budget that goes to social security, pensions, and health.

18. Estimate the fractional part of the total budget that goes to interest on the national debt. How many billions of dollars is this?

19. Estimate the fractional part of the total budget that goes to defense. How many billions of dollars is this?

20. There are about 250 million U.S. citizens. Use a calculator to estimate how much is spent on defense per American.

▶ Review (Lessons 3.3, 3.5, 3.6, 3.9, 3.11, 3.12)

Divide.

21. $81\overline{)947}$ **22.** $3\overline{)5876}$ **23.** $500\overline{)6000}$ **24.** $1.2\overline{)10.8}$

25. $0.004\overline{)1.56}$ **26.** $90\overline{)4768}$ **27.** $0.006\overline{)16.98}$ **28.** $25\overline{)963}$

29. $42{,}287 \div 7$ **30.** $\$39.48 \div 2$ **31.** $1.176 \div 49$ **32.** $0.182 \div 2.8$

9.9 Choosing the Appropriate Graph

Shelby works at the Vistaville Teen Center raising money for the Center's programs. The Center has become so popular since it was founded in 1984 that Shelby thinks it's time to expand its programs. He is now preparing a presentation to convince the Center's board of directors of the need for more programs.

Shelby wants to use graphs in his presentation, but he's not sure just how. His goal is to show the rapid growth of the Center, especially in the past two years. He has experimented with one table, making four different graphs from it. The table and Shelby's graphs are shown below.

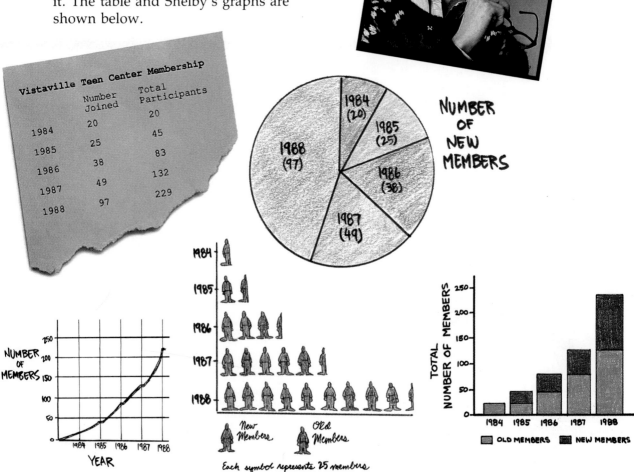

Vistaville Teen Center Membership		
	Number Joined	Total Participants
1984	20	20
1985	25	45
1986	38	83
1987	49	132
1988	97	229

NUMBER OF NEW MEMBERS

Each symbol represents 25 members

▶ Think and Discuss

1. In which graph is it easiest for you to see how fast total membership has grown? Why? Discuss.

2. In which graph is it easiest for you to see how fast new membership has grown? Why? Discuss.

3. In which graph is it most difficult for you to see how fast total membership has grown? Why? Discuss.

4. In which graph is it most difficult for you to see how fast new membership has grown? Why? Discuss.

5. In which graph is growth during 1988 made most clear? Why?

6. Reread the first part of this lesson. If you were Shelby, which graph would you use? How does it do the job better than the others? Discuss.

As treasurer of the Elevator Racing Society, you know that there is a crisis—the Society is running out of money! To convince the members that dues must be increased, you have made the table shown below. Now you want to make graphs from the table. Work with a classmate on the problems below.

7. Study the table and discuss ways to present your case to the Society. Draw the graph that will best convince the members.

8. Compare your graph with those drawn by others in your class. Discuss as a group the advantages and disadvantages of each.

Year 1987	Dues Income	Expenses	Year 1988	Dues Income	Expenses
Jan.–Mar.	$105	$ 97	Jan.–Mar.	$110	$132
Apr.–June	115	84	Apr.–June	100	147
July–Sept.	120	123	July–Sept.	125	156
Oct.–Dec.	100	104	Oct.–Dec.	105	148
1987 Total	**$440**	**$408**	**1988 Total**	**$440**	**$583**

▶ Review (Lesson 6.8)

9. Find the total cost for making 100 baseball pennants, based on the following table.

Materials and Costs for Making 2 Baseball Pennants

Materials	Unit Cost	Materials	Unit Cost
$\frac{1}{2}$ yard blue felt	$1.59/yd.	$1\frac{1}{4}$ yards yellow cord	$0.89/yd.
$\frac{1}{4}$ yard yellow satin	$8.99/yd.	$6\frac{1}{2}$ feet of $\frac{1}{4}$-inch doweling	$0.75/ft.

Chapter 9 Review

Use Graph A to answer Questions 1–4. (Lessons 9.2, 9.3, 9.4)

1. What country had the second highest number of winners?

2. What country had the highest number of female winners?

3. How many more American women than American men have won?

4. Construct a single bar graph based on this double graph.

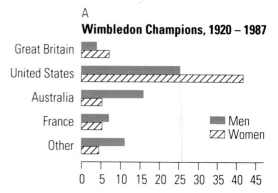

A

Wimbledon Champions, 1920 – 1987

Use Graph B to answer Questions 5–7. (Lessons 9.5, 9.6)

5. Estimate the number of students enrolled in grades Kindergarten–8 in 1930.

6. About what was the greatest number of students enrolled in grades 9–12 during the period shown on the graph? What year was that?

7. Construct a single line graph based on this double line graph.

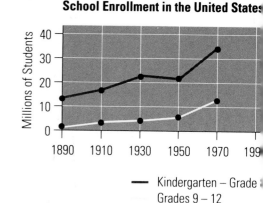

B

School Enrollment in the United States

— Kindergarten – Grade 8
Grades 9 – 12

Answer the following question. (Lesson 9.7)

8. If the symbol ▢ stands for 50 TV sets, what does the symbol ⌐ stand for?

Use Graph C to answer Questions 9–10. (Lesson 9.8)

9. Approximately what fractional part of the total budget goes to printed materials?

10. How much money is spent on salaries, building and maintenance costs, and miscellaneous expenses?

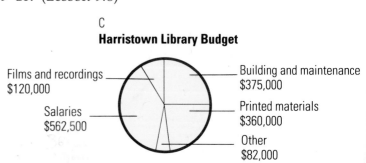

C

Harristown Library Budget

Films and recordings
$120,000

Salaries
$562,500

Building and maintenance
$375,000

Printed materials
$360,000

Other
$82,000

Chapter 9 Test

Use Graph A to answer Questions 1–3.

1. Which magazine has the most readers?

2. Which two magazines together have about the same number of readers as *Sports View*?

3. About how many readers do the sports, teen, and music magazines have all together?

A

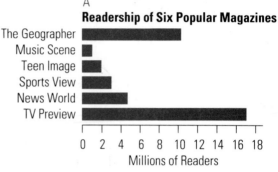

Readership of Six Popular Magazines

The Geographer · Music Scene · Teen Image · Sports View · News World · TV Preview

0 2 4 6 8 10 12 14 16 18
Millions of Readers

Use Graph B to answer Questions 4–7.

4. When did U.S. farms have the most beef cattle? Dairy cattle?

5. About how many more beef cattle were there in 1975 than in 1985?

6. Do you predict that in 1990 there will be fewer or more dairy cattle than in 1985? Why?

7. Construct a single line graph based on this graph.

B

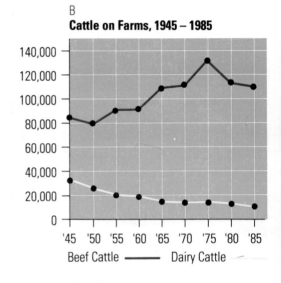

Cattle on Farms, 1945 – 1985

140,000
120,000
100,000
80,000
60,000
40,000
20,000
0

'45 '50 '55 '60 '65 '70 '75 '80 '85

Beef Cattle ——— Dairy Cattle

Use Graph C to answer Questions 8–10.

8. What does the symbol ☎ stand for?   ⚐ ?

9. How many telephones are there in Poland for every 100 people? Great Britain? Israel?

10. Compare the number of phones per 100 people in the United States and Mexico.

C

Telephones per 100 People

Mexico ☎ ☎

Poland ☎ ☎

Israel ☎ ☎ ☎ ☎ ☎

Great Britain ☎ ☎ ☎ ☎ ☎ ☎ ☎ ☎

United States ☎ ☎ ☎ ☎ ☎ ☎ ☎ ☎ ☎ ☎ ☎

☎ = 5 phones

"Now that our ball team has me to keep track of statistics like earned run averages, batting averages, and wins and losses, it's much easier to see where our strengths and weaknesses are. Who knows, maybe this year we'll win the pennant!"

Statistics

By using statistics, you can collect and analyze large amounts of numerical information. You use statistical information when you

- listen to a weather forecast.
- check the baseball standings.
- read about the popularity of a political candidate.
- buy the number-one-selling record album.

Can you think of other ways you use statistics?

Class Favorites

Each person should write his or her favorite color, sport, and musical instrument from the choices given below on separate pieces of paper. Working in one of the three categories (color, sport, or instrument), compile the information for that category. Then determine the class favorites and report your findings to the class.

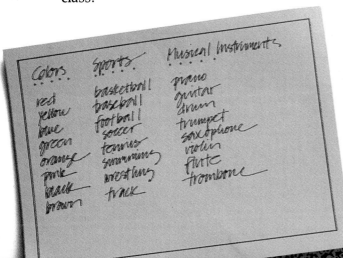

10.1 Introduction to Statistics

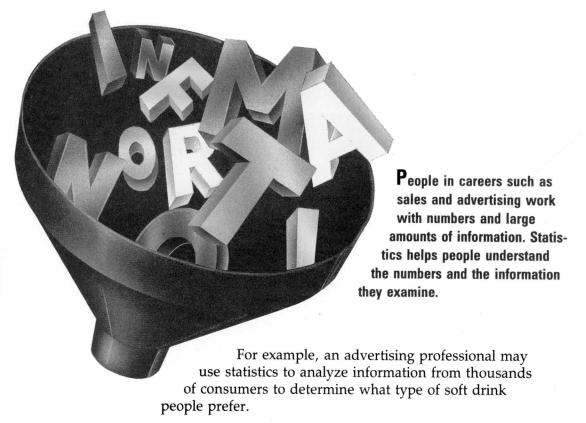

People in careers such as sales and advertising work with numbers and large amounts of information. Statistics helps people understand the numbers and the information they examine.

For example, an advertising professional may use statistics to analyze information from thousands of consumers to determine what type of soft drink people prefer.

In this chapter you will learn how to organize and display large amounts of information.

▶ Express Yourself

Many terms used in statistics are a part of everyday language, as you will see from the following exercises.

1. A teacher told her class that the mean score on last week's test was 48. "That's a pretty mean score," responded one student. What did the teacher mean by *mean*? Use a dictionary to help you identify the different definitions.

2. What is a highway *median*?

3. How is the word *average* used in everyday speech?

4. Look up each of the terms below in a dictionary. Write a definition for each word.
   average   **data**   **frequency**

THE RESULTS ARE IN!

Taste tests prove that Apple Sparkle is the best.

5. The phrases listed below are commonly used in advertising. For each phrase, write a short advertisement for a product you are familiar with.
"Four out of five"
"Laboratory tests prove"
"Number-one-selling"

6. Look in a newspaper for examples of news stories or reports that use statistics. The sports and business sections are good places to look. Bring in examples to show the class.

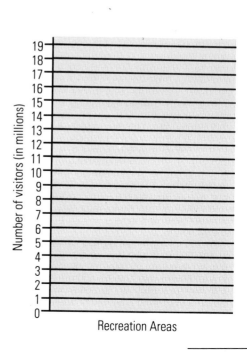

▶ **Practice What You Know**

Order each set of numbers from least to greatest.

7. 19   20   53   6   17   206   85   21   12

8. 3.5   18.0   35.0   28.6   1001.1   10.1

9. 30   7   1200   800   9   54   98   706

10. 112   88   16   91   57   47   31   306   80

11. $22,000 $18,500 $56,000 $2500 $10,600

12. $1465 $20,683 $108,675 $200,031 $18,563

Round each decimal to the nearest tenth.

13. 14.06   14. 106.531   15. 2.75   16. 47.31   17. 206.99

18. Complete a bar graph using the information shown here.

National Park Service Recreation Areas	Number of Visitors
Golden Gate National Recreation Area	18.4 million
Blue Ridge Parkway	16.7 million
Colonial National Historical Park	9.1 million
Hot Springs National Park	5.3 million
Rocky Mountain National Park	2.6 million

Number of visitors (in millions)

Recreation Areas

10.2 Making a Frequency Table

Rita is starting a business to care for the pets and plants of people who are on vacation. She is preparing a flyer to distribute around the neighborhood and she needs to figure out how much to charge for each visit.

Pet and Plant Care Fees
per visit
$ 7.50
$ 7.25
$ 5.00
$ 7.50
$ 9.00
$ 7.50
$ 8.00
$ 5.75
$ 7.25
$ 7.50
$ 6.00
$ 8.50
$ 7.00
$ 7.50
$ 9.00
$ 8.00

Rita decides to find out what other businesses charge for a similar service. She calls 16 services advertised in the yellow pages and makes a list of their fees.

Rita makes a plan for analyzing the fees. First she has to organize her information.

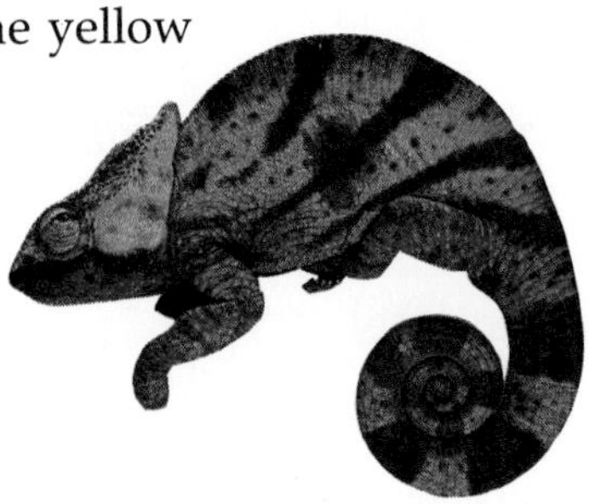

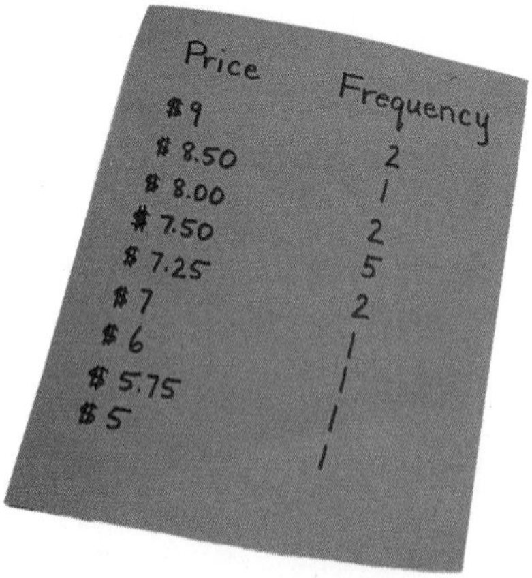

Price
Frequency
$9
$ 8.50
$ 8.00
$ 7.50
$ 7.25
$ 7
$ 6
$ 5.75
$ 5
2
1
2
5
2
1
1
1
1

Rita uses her data to make a **frequency table**. She writes the prices in order and counts how often each price appears. The number of times each price appears is called its **frequency**. Rita can see from her table that she should charge between $5 and $9 if she wants to be competitive. She decides to charge the price that appears most often in her list, $7.50.

▶ Think and Discuss

1. Check the frequencies in Rita's list. Does $7.50 seem like a reasonable price for Rita to charge? Discuss.
2. What price would you charge? Why? Discuss.
3. Write a paragraph explaining how organizing the data helped Rita determine her fee.

Exercises   (For more practice, see page 432.)

4. Make a frequency table for the following data on the size of men's shoes sold during one work day.

9   10   11   12   8   10   11   9   7   10   11

10   8   9   9   8   12   9   10   8   10   9

Use the frequency table you made for Question 4 to answer the following questions.

5. How many pairs of shoes were sold?

6. Which shoe size was sold most often?

7. How many pairs of shoes larger than size 10 were sold?

8. How could a shoe store use the kind of information shown here? Discuss.

9. Make a frequency table for the following scores on a 15-point quiz. Use your table to answer Questions 10, 11, and 12.

14   12   10   14   9   10   12   10   9   15   13   12

9   10   12   12   8   11   12   10   12   14   12   14

10. How many students took the quiz?

11. Which was the most common score?

12. How many scores were 12 or less?

▶ Applications

13. Conduct a survey of your class. Ask about such topics as favorite musical group or favorite brand of cereal. Construct a frequency table that shows the results of your survey.

14. Write a summary of the results of your survey. Questions 5–8 and 10–12 may help you analyze your table.

▶ Review (Lessons 1.10, 1.11)

Subtract.

15. $42 - 9.76$

16. $602 - 399$

17. $4.79 - 2.182$

18. $234 - 176$

19. $88.8 - 0.172$

20. $0.75 - 0.705$

21. $1472 - 595$

22. $2.93 - 1.09$

10.3 Finding the Range, Mode, and Median

Mr. Barnes has 21 students in his class. When he gives a test, he examines the class's performance by finding the range, mode, and median of the test scores.

Examples

To find the range, mode, and median of a set of numbers, first write the numbers in order.

range   the difference between the greatest and the least numbers

mode   the number or numbers that occur most often

median   the middle number in an ordered set of data

A Find the range of the test scores.

$$\begin{array}{r} 9\ 9 \\ -\ 7\ 4 \\ \hline 2\ 5 \end{array}$$ Subtract the least number from the greatest number in the set.

The range of the scores is **25**.

B Find the mode of the test scores.

Find the number or numbers that occur most often. 78 occurs 6 times.
The mode of the scores is **78**.

C Find the median of the test scores.

Find the middle number. There are 21 scores. 10 scores are above 83. 10 scores are below 83.
The median is **83**.

D Find the median.   76, 80, 81, 83, 87, 90

$83 + 81 = 164$ Add the two middle numbers.
$164 \div 2 = 82$ Divide by 2.
The median is **82**.

▶ Think and Discuss

1. Order the set from least to greatest.
   7 9 4 8 3 0 10 5 8 5

2. Find the mode: 1 2 3 1 2 1 2. Discuss.

3. Refer to the introduction to this lesson. If the score of 99 was erased from Mr. Barnes's list, would the range, mode, or median change? Explain.

4. Which of the following—range, mode, or median—would a frequency table be particularly useful for finding? Explain.

Exercises

Find the range. (See Example A.)

5. 3.2 2.7 4.1 6.5 3.8 4.9

6. 19 23 17 24 20 29 16 18

Find the mode. (See Example B.)

7. $1.25 75¢ $1.49 $2.35 $1.25
   $2.10 $1.10 80¢ $1.25 $1.89

8. 98 87 78 88 91 90 88 92
   91 88 78 91 87 86 88 82

Find the median. (See Example C.)

9. $12,500 $9600 $21,000 $5500
   $24,000 $14,700 $18,200

10. $24.20 $47.50 $31.74 $26.15
    $25.70 $35.90 $40.53

Find the median. (See Example D.)

11. 4 9 7 6 2 9 3 5

12. $4.50 $5.00 $6.60 $9.25

▶ Mixed Practice (For more practice, see page 432.)

Use this set of data for Questions 13–15.

76 73 80 82 77 77 69 82 75 79 70 71 80 77 76 77

13. Find the range.

14. Find the mode.

15. Find the median.

▶ Applications

16. Elroy's lunches for two weeks cost: $3.52, $2.06, $4.16, $3.52, $2.25, $3.05, $6.15, $2.06, $3.76, and $3.52. Find the range, the median, and the mode.

17. The prices and weights for running shoes are listed here. Find the median price. Find the median weight. Find the range of prices and of weights.

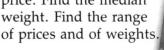

Price	Weight (oz.)
$53	24
$80	20
$59	22
$57	19

▶ Review (Lessons 3.2, 3.5)

Divide.

18. 36 ÷ 9

19. 150 ÷ 5

20. 372 ÷ 3

21. 1025 ÷ 5

22. 562 ÷ 18

10.4 Finding the Mean

CLASS ROSTER

SCOTT	75 in
DAVE	78
JUAN	79
PETE	78
SAMMY	63
PEDRO	79
JIM	62
MIKE	78
CARLOS	62
KEVIN	6
BOBBI	6
MARK	6
ANNE +	6

Southside High is building a new gym, and Coach Charles wants to determine the best height for the chin-up bar. To do this, he decides to find the mean, or average, height of his students and then add 11 inches.

Examples

To find the mean of a set of numbers, first find the sum of the numbers. Then divide the sum by the number of items.

Find the mean height in Coach Charles's survey.

Find the sum.

```
   7 5
   7 8
   7 9
   7 8
   6 3
   7 9
   6 2
   7 8
   6 2
   6 1
   6 3
   6 4
 + 6 1
   9 0 3
```

```
        6 9.4 6 1
   13)9 0 3.0 0 0
      7 8
      1 2 3
      1 1 7
            6 0
            5 2
              8 0
              7 8
                2 0
                1 3
```

Divide the sum by 13, the number of students.

You might want to use a calculator in this problem.

The mean is **about 69.46 inches.**

▶ Think and Discuss

1. Find the mean. 29 55 43 37 52.

2. Refer to the introduction to this lesson. If Coach Charles uses his original plan, how high will he position the chin-up bar?

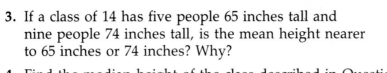

3. If a class of 14 has five people 65 inches tall and nine people 74 inches tall, is the mean height nearer to 65 inches or 74 inches? Why?

4. Find the median height of the class described in Question 3.

5. If a class of 12 has six people 65 inches tall and six people 74 inches tall, are the mean and the median equal? Explain.

6. Should you write a set of numbers in order before computing their mean? Explain.

Exercises

(For more practice, see page 433.)

Find the mean.

7. 12 16 4 25 10

8. 16 38 71 82 91 71

9. 375 902 434 182

10. 284 267 225 213

11. $15,000 $22,000 $18,500 $12,250 $24,400

12. 23 19 22 25 22 21 24 20 23 21

13. 78 67 93 82 88 56 84 86 90 79

14. $4.15 $3.75 $5.10 $4.50 $5.40 $3.80 $7.95 $6.25 $5.00

15. 176 181 167 202 195 186 158 198 211 170 169 205

▶ Applications

16. Ana's math grade is actually the mean of her test scores for the term. Ana's test scores were 89, 74, 67, 83, 73, 91, and 85. What will her final grade be?

17. Last week Morgan ran 6.5 miles, 7.25 miles, 8 miles, 4 miles, 12.5 miles, 14 miles, and 7.5 miles. What was his mean daily distance?

18. Greg wants a B in English class. He has three test scores of 86, 78, and 76. What score does he need on the final test to end up with a mean score of 80?

▶ Review (Lessons 6.7, 7.1, 7.3, 7.4)

Solve. Write the answers in lowest terms.

19. $1\frac{3}{4} \div 2$

20. $1\frac{1}{8} \div 3$

21. $2\frac{5}{8} \div 1\frac{3}{4}$

22. $2\frac{1}{2} \div 3\frac{3}{4}$

23. $3\frac{1}{5} \div 1\frac{1}{5}$

24. $\frac{4}{9} + \frac{7}{9}$

25. $\frac{5}{6} + \frac{4}{5}$

26. $\frac{9}{10} + \frac{7}{20}$

27. $3\frac{1}{8} + 6\frac{7}{8}$

28. $8\frac{3}{4} + 5\frac{11}{16}$

10.5 Investigating Averages

$13,600

$21,290

$14,090

Luisa just got a job. When she interviewed she was told that her division's average salary was a little over $21,290. Then a co-worker told her the average salary was $13,600. Luisa decided to do her own calculations. She found the average salary to be $14,090. Why do you think these figures are different?

Luisa's company consists of 3 managers and 12 clerks. Their salaries are shown below.

Salaries (in dollars)

50,000	50,000	50,000	15,325	15,060	15,000	14,125	14,090
14,000	13,750	13,600	13,600	13,600	13,600	13,600	

Find the mean salary.   Total payroll is $319,350.
Mean = $319,350 ÷ 15 = $21,290

Find the median salary.   There are 15 salaries: 7 are greater than $14,090, 7 are less than $14,090.
Median = $14,090

Find the mode of the salaries.   Five people earn $13,600.
Mode = $13,600

The mean, median, and mode can be different numbers. Yet each is sometimes called an average, just as in the above case.

1. Find the median and the mean salaries of the 12 clerks. Discuss why the mean of this group of salaries is different from the mean of the original group of salaries.

2. Suppose two of the managers in Luisa's department leave their jobs. What are the mean, median, and mode salaries now? Explain how changing one or two figures can affect the mean, median, and mode differently.

3. Now suppose one of the three managers earns $120,000 instead of $50,000. What are the mean, median, and mode salaries now? Discuss how each changes.

4. Which average—the mean, the median, or the mode—most accurately describes the typical salary in Luisa's division? Justify your answer.

5. Suppose you wanted to study average income in the United States as part of a history project. Would you use median or mean income in your study? Justify your choice.

Read the following example:

According to a recent study, half of Newark Community College's students are over 24 years old. The average age of the students is 29.

Based on the information in this lesson, use the word *mean* or *median* to complete each statement.

6. The ___ age of Newark Community College students is 29.

7. The ___ age of Newark Community College students is 24.

Sometimes finding the mean, median, or mode is not enough to help you understand a situation.

Planet	Approximate Length of a Day in Earth Hours
Mercury	1416
Venus	5832
Earth	24
Mars	24.6
Jupiter	9.9
Saturn	10.2
Uranus	22
Neptune	19
Pluto	144

8. What is the mean length of one day in our solar system?

9. What is the median length of one day in our solar system?

10. Is the use of mean or median helpful in understanding day length in our solar system? Discuss.

▶ **Review** (Lesson 9.4)

11. Use the following data to make a bar graph.

League Pennant Winners (through 1987)

Cubs—10 years   Cardinals—15 years
Tigers—9 years   Yankees—33 years

10.6 Constructing Histograms

How often do you go to the movies? A high school newspaper staff conducted a survey concerning freshman movie attendance. The results are shown here. The editors decided to present the results in a histogram to attract attention to the information.

Movies attended per month	0	1	2	3	4	5	6	7	8	9	10	11
Numbers of students	3	9	15	29	43	35	18	14	8	3	0	2

Histograms are bar graphs that show the frequency of events, prices, or scores. Histograms condense and organize large amounts of information.

Examples

To construct a histogram, follow these steps:

1. Draw and label the horizontal and vertical axes.
2. Label the numbers along the axes.
3. Draw the bars to their proper heights.
4. Write a title for the histogram.

Use the information above to construct a histogram.

Follow these steps:
1. Organize the movie survey results into 6 categories.
2. Find the frequency for each category.
3. Label the horizontal axis to show the 6 categories.
4. Label the vertical axis to show the frequencies.
5. Construct the histogram bars.

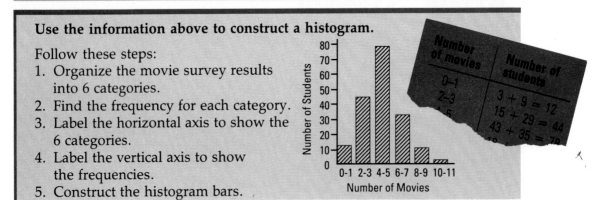

Number of movies	Number of students
0–1	3 + 9 = 12
2–3	15 + 29 = 44
4–5	43 + 35 = 78

▶ Think and Discuss

1. Use the movie survey results above to construct another histogram that shows 4 movie attendance categories.

2. Can you tell exact frequencies from a histogram? Explain.

3. Are all histograms bar graphs? Discuss.

Exercises

(For more practice, see page 433.)

Use the information in each table to construct a histogram.

4. Number of states visited by Miss Lal's math students (besides home state)

States visited	0	1	2	3	4	5	6	7	8	9	10	11+
Number of students	4	6	8	6	5	4	0	3	2	0	0	1

Divide the states visited into categories, 0–2, 3–5, and so on.

5. Torino Bicycle Shop sales (number of bicycles sold per month)

Month	Jan.	Feb.	Mar.	Apr.	May	June	July	Aug.	Sept.	Oct.	Nov.	Dec.
Number of bicycles	10	12	24	31	58	72	103	69	32	12	9	94

Divide the months into 6 categories.

6. Ages of Slocumville Symphony Orchestra members

Age	22	27	31	38	44	46	49	52	53	55	57	58	59	62	66	70
Number of members	1	1	2	1	3	4	8	10	14	9	18	15	20	14	7	3

Divide the ages into 20s, 30s, and so on.

▶ Applications

7. You work for a movie production studio. You want to design a graph showing the studio's income over the past year, using the data below.

   The studio president wants the information shown by quarters (3-month periods). Construct a histogram. Describe what the histogram tells about the studio's income.

Studio Income (in thousands of dollars)

Jan.	Feb.	Mar.	Apr.	May	June	July	Aug.	Sept.	Oct.	Nov.	Dec.
12	25	30	42	28	59	78	107	94	61	24	37

▶ Review (Lessons 7.1, 7.5, 7.7, 7.9)

Subtract. Write the answers in lowest terms.

8. $\frac{11}{20} - \frac{3}{20}$

9. $2\frac{2}{3} - 1\frac{1}{4}$

10. $3\frac{1}{2} - 1\frac{2}{3}$

11. $\frac{8}{9} - \frac{1}{3}$

12. $4\frac{5}{6} - 4\frac{1}{2}$

13. $\frac{7}{12} - \frac{1}{3}$

14. $9\frac{5}{9} - 7\frac{1}{6}$

15. $\frac{13}{16} - \frac{11}{16}$

16. $\frac{7}{8} - \frac{2}{5}$

17. $5 - 2\frac{3}{5}$

10.7 Statistics: A Career Application

Graphic artists illustrate statistics that apply to the nation, finances, and our lives. They organize data in ways that are interesting and informative.

The artist chooses a type of graph that is appropriate to the set of data. For example, the artist might choose a pictograph to illustrate the population of the five largest cities in the United States. To illustrate changes in the stock market, the artist might choose a line graph.

U.S. Motor Vehicle Fuel Consumption
(1983 estimate)

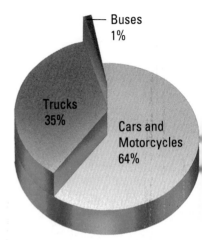

Buses
1%

Trucks
35%

Cars and
Motorcycles
64%

Source: Dept. of Transportation,
Federal Highway Administration

▶ Think and Discuss

1. What do you find interesting about the graphs on these pages? What things make a graph informative or interesting? Discuss.

2. What type of graph is used to report the fuel consumption of motor vehicles in the United States?

3. Why do you think a line graph is used for hurricanes instead of a circle graph?

4. How do you think the graphic artist selected the colors for the map of U.S. population density shown on the next page?

5. Why is a map a better choice than a graph to represent U.S. population density?

What type of graph would you choose if you were a media artist and you were reporting:

6. the amounts of corn and wheat grown in the United States?

7. the stray dog population in Tulsa over the last 10 years?

8. the number of men and women who worked outside the home in 1940, 1950, 1960, 1970, and 1980?

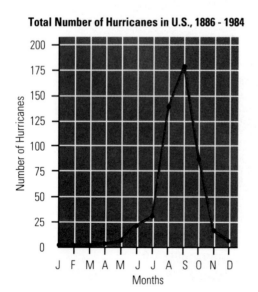

Total Number of Hurricanes in U.S., 1886 - 1984

Number of Hurricanes

Months

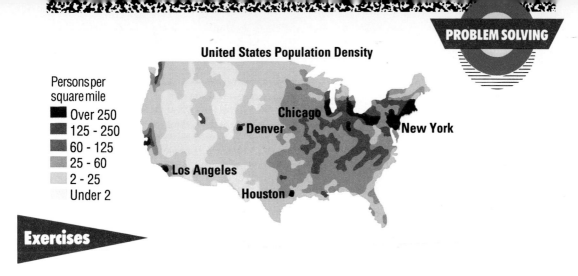

United States Population Density

Persons per square mile
■ Over 250
■ 125 - 250
■ 60 - 125
■ 25 - 60
■ 2 - 25
□ Under 2

Chicago
Denver
New York
Los Angeles
Houston

Exercises

A national teen magazine surveyed 600 employed students. The results are shown below:

Type of Job	Freshmen	Sophomores	Juniors	Seniors
Pizza delivery person	0	0	34	36
Waiter/waitress (fast food)	0	18	68	74
Sales clerk (department store)	0	5	26	38
Stock person	1	8	22	30
Babysitter	47	31	20	4
Lawn mower	42	27	12	6
Other	10	11	18	12

Now you are a graphic artist. Choose the type and style of graph that would best represent the information.

9. Draw a graph that shows the decline of babysitting employment throughout the four high school years.

10. Draw a graph that shows the types of jobs held by juniors.

11. Draw a graph that shows the findings of the entire survey.

▶ **Review** (Lesson 3.8)

Write which operation you would use to solve each problem. Then solve.

12. The Thompson family spends an average of $350 a month on food. Estimate the average cost of food for each week.

13. Jenny's car averages 24.7 miles per gallon. How much does Jenny spend to purchase 13.5 gallons of gas at $0.98 a gallon?

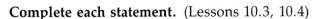

Chapter 10 Review

REVIEW

Complete each statement. (Lessons 10.3, 10.4)

1. Another word for mean is ▩ .

2. The ▩ is the middle number in a set of data when the numbers are listed in order.

3. The ▩ is the difference between the largest and the smallest numbers in a set of numbers.

Find the range, mean, median, and mode for each set of numbers. (Lessons 10.3, 10.4)

4. 84   68   81   76

5. 74   79   91   74   91

Use these salaries of Winsome baseball team members to answer the following questions. (Lesson 10.5)

6. What is the mean salary?

7. What is the mean salary of the first 4 players?

8. Which mean salary describes the salary situation better? Why?

Player	Salary
Spitball Harry (pitcher)	$230,000
Sturdy Al (shortstop)	$185,000
Traveling Troy (1st base)	$315,000
Lightning Sol (3rd base)	$195,000
Stop'em Joe (catcher)	$2,000,000

Make a frequency table for the set of data below. Then find the range, mean, median, and mode. (Lessons 10.2, 10.3, 10.4)

9. Cost of blank cassette tapes at local stores
   $2 $2 $2 $3 $4 $2 $2 $3 $4 $3 $4 $3 $2 $4 $3

Use the information below to construct a histogram. (Lesson 10.6)

10. Last year's sales of lawn furniture at Kelly's Patio Shoppe

Month	Jan.	Feb.	Mar.	Apr.	May	June	July	Aug.	Sept.	Oct.	Nov.	Dec.
Number of pieces	0	0	3	6	85	119	97	53	37	8	0	0

Divide the year into 3-month periods.

Answer the following questions. (Lesson 10.7)

11. Describe a situation when a graphic artist might choose a line graph to illustrate data.

12. What kind of graph would you choose to show a family's budget for one year?

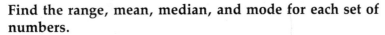

TEST

Find the range, mean, median, and mode for each set of numbers.

1. $25 $23 $25 $30 $32 $28   2. 4   9   3   8   5   8   6

Make a frequency table for the data below. Then find the range, mean, median, and mode.

3. Hourly charge of 20 child-care workers
   $1.50 $2 $1.75 $2 $1.50 $2.50 $1.25 $1.50 $1.75 $1.50
   $2.50 $1.75 $2 $1.50 $2.50 $1.50 $1.50 $2 $3.00 $1.25

Use the information in the table below to construct a histogram.

4. African-violet sales, Hooper Florists

Month	Jan.	Feb.	Mar.	Apr.	May	June	July	Aug.	Sept.	Oct.	Nov.	Dec.
Number of plants sold	17	25	15	12	8	10	6	4	9	18	28	33

Divide the year into 2-month periods.

Complete each statement.

5. The number that appears most often in a set of numbers is called the ____ .

6. A graph that shows the frequency of events, prices, or scores is called a(n) ____ .

7. The ____ , ____ , and ____ are all sometimes called the average.

Use the table below to answer the following questions.

8. What is the mean distance?

9. What is the median distance?

10. What is the range of distances?

Annual Driving Distances of Salespeople, Acme Distributors

Tom	21,700 mi.
Bette	33,900 mi.
Walt	95,500 mi.
Carmen	28,000 mi.
Bob	30,700 mi.

"I used ratios and proportions when my family decided to move. I made a scale drawing of my new bedroom so I could figure out how to arrange my furniture."

Ratio and Proportion

A ratio is a comparison of two numbers by division. You see ratios
  • in the supermarket when you buy 6 oranges for $1.00 (6/$1).
  • in a recipe that calls for 3 tablespoons oil to 2 tablespoons vinegar (3 to 2).
  • on a map with a scale where 1 inch is equivalent to 100 miles (1 inch : 100 miles).
  • when you paint and mix colors to find a certain shade.

Have you ever used a ratio in shopping, cooking, or traveling? Where else have you used or seen ratios?

Picturing Ratios

Materials: ruler, book or magazine

Work with 2 or 3 students. Choose 10 pictures from a book or magazine. Measure the height and width of each picture to the nearest centimeter. Record the measurements in a chart like the one below. In the last column, write the ratio of height to width as a fraction.

Height	Width	Height/Width

Are any of the fractions listed equal?

11.1 Writing Ratios and Rates

Bill stopped his friend Ravi while they were selling tee shirts on the Fourth of July. "You told me we'd sell just as many small as large shirts. I'm selling 3 large for every 2 small. What happened?"

Bill is using a **ratio**, which is a comparison of two numbers by division.

Examples

To write ratios, use the word *to*, a colon, or a fraction bar.
To find equivalent ratios, multiply or divide each term of the given ratio by the same number.

A Write the ratio of 100 miles to 4 gallons of gasoline in 3 different ways.

With the word *to*: 100 to 4

As a fraction: $\frac{100}{4}$

With a colon: 100 : 4

The ratio of 100 miles to 4 gallons of gasoline can be written as **100 to 4**, $\frac{100}{4}$, and **100 : 4**.

B Find 3 ratios equivalent to $\frac{6}{8}$.

$$\frac{6}{8} = \frac{6 \div 2}{8 \div 2} = \frac{3}{4}$$

$$\frac{6}{8} = \frac{6 \times 2}{8 \times 2} = \frac{12}{16}$$

$$\frac{6}{8} = \frac{6 \times 7}{8 \times 7} = \frac{42}{56}$$

Three ratios equivalent to $\frac{6}{8}$ are $\frac{3}{4}$, $\frac{12}{16}$, and $\frac{42}{56}$.

► Think and Discuss

1. Refer to the introduction to this lesson. Write the given ratio in three different ways.

2. Write three ratios equivalent to $\frac{7}{9}$.

3. Ten girls and twenty-four boys are enrolled in a class. Find the ratio of girls to boys and the ratio of boys to girls. Are these ratios equivalent? Explain.

4. Name two situations in which you could use ratios.

Exercises

Write each ratio in 3 different ways. (See Example A.)

5. 55 miles per hour

6. 12 laps in 15 minutes

7. 10-pound loss in 5 weeks

8. 8 quarters for 2 dollars

9. $1.56 for 4 pounds of bananas

10. $4.56 for 3 light bulbs

Find 3 ratios equivalent to each ratio. (See Example B)

11. 10 prizes for 5 games

12. 250 desks for 10 classrooms

13. 3 hours of homework in 4 classes

14. 500 miles traveled in 10 hours

▶ Mixed Practice (For more practice, see page 434.)

Write each ratio in 3 three different ways. Find 3 ratios equivalent to each ratio.

15. 7 records broken in 2 years

16. 52 passengers per bus

17. $10 for 8 gallons of gas

18. 5 hits in 8 times at bat

19. 7 games won out of 9 games played

20. $1.09 for 12 eggs

21. 200 seats for 10 rows

22. $5.99 for 2 tapes

▶ Applications

23. Marcy got three hits in five times at bat. Write her ratio of hits to times at bat and three ratios equivalent to it. If she continues to hit this way, how many hits will she have after thirty times at bat?

24. Ricardo took a 3-minute typing test. His results are shown here. Find the number of words he typed per minute.

▶ Review (Lessons 2.4, 2.7, 2.8, 2.10)

Multiply.

25. 5.7×3

26. $\begin{array}{r} 36 \\ \times\ 15 \end{array}$

27. 254×481

28. $\begin{array}{r} 0.83 \\ \times\ 1.6 \end{array}$

29. $\begin{array}{r} 420 \\ \times\ 0.02 \end{array}$

30. 9.11×0.32

31. $\begin{array}{r} 3.57 \\ \times\ 12.5 \end{array}$

32. 1.225×50

33. $\begin{array}{r} 683 \\ \times\ 475 \end{array}$

34. 5.39×4.2

35. 8.54×9.7

36. $\begin{array}{r} 294 \\ \times\ 0.78 \end{array}$

11.2 Finding Proportions

Marty usually charges $3 an hour to do odd jobs. After a bad snowstorm, one of the local merchants asked him to shovel the sidewalk around his store for $10. He finished the task in 3 hours. How did the money he earned on this job compare with his usual fee?

Marty could compare ratios by using a **proportion**, a mathematical statement that two ratios are equivalent.

Examples

To tell whether a statement is a proportion, find the cross products.

A Tell whether the statement is a proportion. Use = or ≠.

$\frac{2}{7}$ ※ $\frac{6}{21}$

$\frac{2}{7}$ ✕ $\frac{6}{21}$ Find the cross products.

$\begin{array}{l} 2 \times 21 = 42 \\ 7 \times 6 = 42 \end{array}$ Cross products

$42 = 42$, so $\frac{2}{7} = \frac{6}{21}$.

The statement is **a proportion.**

B Tell whether the statement is a proportion. Use = or ≠.

$\frac{12}{5}$ ※ $\frac{7}{3}$

$\frac{12}{5}$ ✕ $\frac{7}{3}$ Find the cross products.

$\begin{array}{l} 12 \times 3 = 36 \\ 5 \times 7 = 35 \end{array}$ Cross products

$36 \neq 35$, so $\frac{12}{5} \neq \frac{7}{3}$.

The statement is **not a proportion.**

▶ Think and Discuss

1. Tell whether the statement is a proportion. $\frac{6}{3}$ ※ $\frac{4}{2}$

2. Tell whether the statement is a proportion. $\frac{1}{3}$ ※ $\frac{6}{2}$
   Explain how you found your answer.

3. One student claimed that $\frac{2}{3} = \frac{6}{4}$ is a proportion. Do you agree? Why?

4. Refer to the introduction to this lesson. Write the two ratios involved. Do they form a proportion?

5. Write a proportion using the numbers 1, 18, 6, and 3. Compare your answers with other students' answers. Is there more than one correct answer?

6. Some ratios, such as 40 miles per hour, are called rates. What are some other rates you have seen?

Exercises

Tell whether each statement is a proportion. Use = or ≠.
(See Example A.)

7. $\frac{1}{2}$ ▦ $\frac{4}{8}$ 8. $\frac{2}{5}$ ▦ $\frac{4}{9}$ 9. $\frac{1}{11}$ ▦ $\frac{3}{32}$ 10. $\frac{4}{7}$ ▦ $\frac{3}{5}$

11. $\frac{7}{8}$ ▦ $\frac{21}{25}$ 12. $\frac{1}{24}$ ▦ $\frac{3}{72}$ 13. $\frac{3}{12}$ ▦ $\frac{12}{50}$ 14. $\frac{6}{9}$ ▦ $\frac{8}{12}$

15. $\frac{5}{7}$ ▦ $\frac{15}{21}$ 16. $\frac{4}{9}$ ▦ $\frac{8}{16}$ 17. $\frac{7}{12}$ ▦ $\frac{13}{24}$ 18. $\frac{9}{12}$ ▦ $\frac{15}{20}$

Tell whether each statement is a proportion. Use = or ≠.
(See Example B.)

19. $\frac{5}{2}$ ▦ $\frac{3}{1}$ 20. $\frac{7}{4}$ ▦ $\frac{21}{12}$ 21. $\frac{13}{2}$ ▦ $\frac{17}{3}$ 22. $\frac{14}{6}$ ▦ $\frac{7}{3}$

23. $\frac{8}{5}$ ▦ $\frac{12}{9}$ 24. $\frac{20}{3}$ ▦ $\frac{60}{9}$ 25. $\frac{15}{4}$ ▦ $\frac{50}{16}$ 26. $\frac{100}{7}$ ▦ $\frac{700}{56}$

27. $\frac{19}{2}$ ▦ $\frac{95}{10}$ 28. $\frac{37}{6}$ ▦ $\frac{187}{30}$ 29. $\frac{52}{8}$ ▦ $\frac{416}{64}$ 30. $\frac{125}{4}$ ▦ $\frac{350}{12}$

▶ Mixed Practice (For more practice, see page 434.)

Tell whether each statement is a proportion. Use = or ≠.

31. $\frac{6}{11}$ ▦ $\frac{18}{38}$ 32. $\frac{8}{2}$ ▦ $\frac{20}{5}$ 33. $\frac{4}{3}$ ▦ $\frac{60}{45}$ 34. $\frac{12}{15}$ ▦ $\frac{20}{25}$

35. $\frac{20}{9}$ ▦ $\frac{100}{45}$ 36. $\frac{10}{15}$ ▦ $\frac{12}{16}$ 37. $\frac{47}{50}$ ▦ $\frac{107}{120}$ 38. $\frac{12}{7}$ ▦ $\frac{84}{49}$

39. $\frac{89}{100}$ ▦ $\frac{44}{50}$ 40. $\frac{90}{5}$ ▦ $\frac{36}{2}$ 41. $\frac{168}{171}$ ▦ $\frac{56}{57}$ 42. $\frac{120}{7}$ ▦ $\frac{1200}{77}$

▶ Applications

43. Mati did 12 sit-ups in 15 seconds one day. Another day she did 36 sit-ups in 45 seconds. Are her rates the same?

44. Alice swam fifteen laps in ten minutes. Then she swam twenty laps in fifteen minutes. Are her rates the same?

▶ Review (Lesson 5.6)

Rewrite each fraction in lowest terms.

45. $\frac{18}{2}$ 46. $\frac{5}{7}$ 47. $\frac{24}{13}$ 48. $\frac{4}{12}$ 49. $\frac{42}{14}$

11.3 Solving Proportions

The art teacher wants to use Wesley's cartoon in the school paper. He asks Wesley to reduce the cartoon so that it is 6 inches tall.

How wide will the reduced cartoon be? You can use a proportion to find the width: $\frac{\text{▓}}{6} = \frac{10}{15}$.

Examples

To solve proportions, first find the cross products. Then set the cross products equal. Use division to solve.

A Solve the proportion. $\frac{\text{▓}}{6} = \frac{10}{15}$

$$\frac{\text{▓}}{6} \diagdown \frac{10}{15}$$ Cross products are __▓__ × 15 and 60.

$\underline{\text{▓}} \times 15 \qquad = 60$ Set the cross products equal.

$\underline{\text{▓}} \times 15 \div 15 = 60 \div 15$ Divide both sides by 15.

$\underline{\text{▓}} \qquad\qquad = 4$

B Solve the proportion. $\frac{3}{7} = \frac{\text{▓}}{10}$

$$\frac{3}{7} \diagdown \frac{\text{▓}}{10}$$ Cross products are 30 and 7 × __▓__.

$7 \times \underline{\text{▓}} \qquad = 30$ Set the cross products equal.

$7 \times \underline{\text{▓}} \div 7 = 30 \div 7$ Divide both sides by 7.

$\underline{\text{▓}} \qquad\qquad = \frac{30}{7}$, or $4\frac{2}{7}$

▶ Think and Discuss

1. Explain why $16 \times 3{,}147{,}878 \div 3{,}147{,}878 = 16$.

2. Solve the proportion $\frac{4}{9} = \frac{\text{▓}}{36}$.

3. Solve the proportion $\frac{1}{2} = \frac{\text{▓}}{6}$ mentally.

4. Write three different proportions involving $\frac{6}{7}$.

5. Can you solve the proportion $\frac{0}{3} = \frac{0}{\blacksquare}$? Explain.

6. When you have solved a proportion, how can you check to see that your answer is correct?

7. Refer to the introduction to this lesson. How wide will Wesley's cartoon be?

Exercises

Solve each proportion. (See Example A.)

8. $\frac{\blacksquare}{8} = \frac{9}{12}$ 9. $\frac{\blacksquare}{10} = \frac{4}{5}$ 10. $\frac{\blacksquare}{16} = \frac{15}{24}$ 11. $\frac{\blacksquare}{18} = \frac{20}{45}$ 12. $\frac{\blacksquare}{20} = \frac{21}{14}$

13. $\frac{18}{\blacksquare} = \frac{12}{9}$ 14. $\frac{24}{\blacksquare} = \frac{8}{3}$ 15. $\frac{78}{\blacksquare} = \frac{18}{12}$ 16. $\frac{40}{\blacksquare} = \frac{12}{21}$ 17. $\frac{\blacksquare}{12} = \frac{7}{5}$

Solve each proportion. (See Example B.)

18. $\frac{10}{15} = \frac{\blacksquare}{6}$ 19. $\frac{15}{3} = \frac{\blacksquare}{1}$ 20. $\frac{13}{39} = \frac{\blacksquare}{12}$ 21. $\frac{28}{4} = \frac{\blacksquare}{15}$

22. $\frac{9}{4} = \frac{36}{\blacksquare}$ 23. $\frac{15}{8} = \frac{60}{\blacksquare}$ 24. $\frac{15}{4} = \frac{75}{\blacksquare}$ 25. $\frac{4}{6} = \frac{57}{\blacksquare}$

▶ **Mixed Practice** (For more practice, see page 435.)

Solve each proportion.

26. $\frac{\blacksquare}{11} = \frac{12}{132}$ 27. $\frac{\blacksquare}{9} = \frac{12}{27}$ 28. $\frac{13}{10} = \frac{52}{\blacksquare}$ 29. $\frac{15}{9} = \frac{55}{\blacksquare}$

30. $\frac{18}{27} = \frac{\blacksquare}{9}$ 31. $\frac{14}{\blacksquare} = \frac{3}{20}$ 32. $\frac{12}{5} = \frac{30}{\blacksquare}$ 33. $\frac{\blacksquare}{21} = \frac{8}{12}$

34. $\frac{53}{100} = \frac{\blacksquare}{600}$ 35. $\frac{\blacksquare}{12} = \frac{5}{20}$ 36. $\frac{41}{5} = \frac{369}{\blacksquare}$ 37. $\frac{7}{16} = \frac{\blacksquare}{192}$

▶ **Applications**

38. Jerome drove an average of fifty-two miles per hour on a trip. About how many miles did he go in three hours of nonstop driving?

39. Ruth missed 12 problems on a 75-question test. If each problem was worth the same amount, how many points did she earn out of 100 points?

▶ **Review** (Lessons 1.7, 1.8)

Round to the nearest tenth.

40. 4.871   41. 0.135   42. 2.092   43. 19.9842   44. 25.64

Round to the nearest hundred.

45. 1879   46. 58   47. 44,326   48. 27,647   49. 39,961

11.4 Unit Pricing

Tamara wanted to try a new type of shampoo. She selected the brand shown here. Which bottle is the better value?

$1.98
8 oz.

$2.75
12 oz.

Unit pricing is a way to compare values. The **unit price** tells you how much a product costs for a given unit. The unit could be an ounce, a pound, a gallon, a quart, or some other unit.

Examples

To find the **unit price of an item**, first identify the appropriate unit. Then set up a proportion and solve.

$$\frac{\text{Price paid}}{\text{Quantity bought in units}} = \frac{\text{Unit price}}{\text{Unit}}$$

A Find the unit price of the smaller bottle of shampoo.

$\frac{\$1.98}{8 \text{ oz.}} = \frac{\text{\textcolor{gray}{▨}}}{1 \text{ oz.}}$ Let the unit be an ounce. Set up the proportion.

$8 \times \text{▨} = 1.98 \times 1$

$8 \times \text{▨} \div 8 = 1.98 \div 8$

$\text{▨} = \$0.2475$, or $0.25 rounded to the nearest cent.

The unit price is **$0.25 per ounce**.

B Determine the better buy: $9 for 6 tapes or $8 for 5 tapes.

The appropriate unit is 1 tape.

$\frac{\text{▨}}{1 \text{ tape}} = \frac{\$9}{6 \text{ tapes}}$

$\text{▨} = \frac{9}{6}$

$\text{▨} = \$1.50$ per tape

$\frac{\text{▨}}{1 \text{ tape}} = \frac{\$8}{5 \text{ tapes}}$

$\text{▨} = \frac{8}{5}$

$\text{▨} = \$1.60$ per tape

Since its unit price is less, **$9 for 6 tapes is the better buy.**

▶ Think and Discuss

1. Solve the proportion. $\frac{3}{4} = \frac{\text{▨}}{1}$

2. Find the unit price if 10 tapes cost $19.99. Solve mentally.

3. Refer to the introduction to this lesson. Which size bottle is the better buy?

4. Is buying a larger quantity at a lower unit price always the better buy? Explain your answer.

5. When could you use unit pricing? Discuss.

Exercises

Find the unit price. (See Example A.)

6. $1 for 5 light bulbs

7. $3 for 30 trash bags

8. $2 for a pack of 24 pens

9. $1 for 20 pencils

10. $3 for 12 cans of soda

11. $1.50 for 200 sheets of paper

Determine the better buy. (See Example B.)

12. $9.54 for 6 tapes or $8.25 for 5 tapes

13. $7.12 for 8 batteries or $5.70 for 6 batteries

14. $2.20 for 12 prints or $6 for 36 prints

▶ **Mixed Practice** (For more practice, see page 435.)

Find the better buy.

15. $4 for 6 pairs of socks or $3.50 for 5 pairs of socks

16. $0.89 for a quart of milk or $3.16 for a gallon of milk

17. $1.08 for a dozen eggs or $1.71 for 18 eggs

18. $1.26 for an 18-ounce bottle of detergent or $2.08 for a 32-ounce bottle of detergent

▶ **Applications**

19. Which pack of film shown is the better buy?

20. Mr. Blaine buys a woodworking magazine every month for $2.25. A one-year subscription to the magazine costs $19.20. How much would he save each month by subscribing?

▶ **Review** (Lessons 2.3, 2.7)

Multiply mentally.

21. 60×700

22. 8.4×1000

23. 90×9000

24. 500×500

25. 0.053×100

26. 10×0.967

27. 100×4.486

28. 6000×400

11.5 Using Scale Drawings

A *fashion designer uses scale drawings like the one shown to illustrate dress designs. The designer then works from scale drawings to make the designs into actual-size clothing. Proportions are used when working with scale drawings.*

Examples

To find lengths from scale drawings, set up and solve proportions.

A Find the actual length of side *b*.

scale: $\frac{1}{2}$ in. = 4 ft.

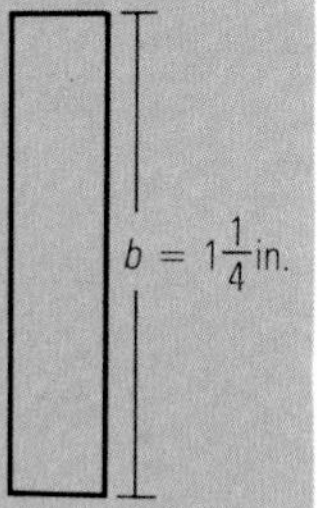

$$\frac{\frac{1}{2}\text{ in.}}{4\text{ ft.}} = \frac{1\frac{1}{4}\text{ in.}}{\square}$$

$$\frac{1}{2} \times \square = 4 \times 1\frac{1}{4}$$

$$\frac{1}{2} \times \square \div \frac{1}{2} = 5 \div \frac{1}{2}$$

$$\square = 10$$

The actual length of side *b* is **10 feet.**

B Find the scale length for side *f*.

scale: $\frac{1}{4}$ in. = 8 ft.

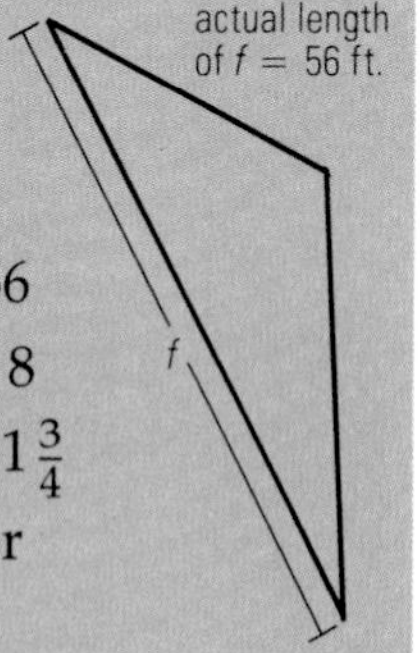

$$\frac{\frac{1}{4}\text{ in.}}{8\text{ ft.}} = \frac{\square}{56\text{ ft.}}$$

$$8 \times \square = \frac{1}{4} \times 56$$

$$8 \times \square \div 8 = 14 \div 8$$

$$\square = \frac{14}{8} \text{ or } 1\frac{3}{4}$$

The scale length for side *f* is **$1\frac{3}{4}$ inches.**

▶ Think and Discuss

1. Solve. $\frac{5}{8} = \frac{10}{\square}$

2. Solve. $\frac{1}{2} \div 6 = \frac{\square}{3}$

3. Find the actual length of a side if its scale length is 2 inches.
   scale: $\frac{1}{8}$ in. = 3 ft.

4. Find the scale length of a side if the actual length is 10 yards. scale: $\frac{1}{2}$ in. = 4 yd.

5. Name some other professions that might use scale drawings. How are scale drawings helpful?

Exercises

Find the actual length for each scale length given.
(See Example A.)

6. 3 in.   scale: 1 in. = 25 mi.

7. $1\frac{1}{4}$ in.   scale: $\frac{1}{4}$ in. = 12 ft.

8. $5\frac{1}{2}$ in.   scale: $\frac{1}{2}$ in. = 5 yd.

9. $\frac{3}{4}$ in.   scale: $\frac{1}{2}$ in. = 2 ft.

Find the scale length for each actual length given.
(See Example B.)

10. 45 ft.   scale: 1 in. = 12 ft.

11. 120 mi.   scale: 1 in. = 24 mi.

12. 75 yd.   scale: $\frac{1}{2}$ in. = 15 yd.

13. $6\frac{1}{2}$ ft.   scale: $\frac{1}{2}$ in. = 1 ft.

▶ **Mixed Practice** (For more practice, see page 436.)

Find the unknown length.

14. scale length = $3\frac{1}{2}$ in.
    scale: $\frac{1}{4}$ in. = 9 ft.

15. actual length = 90 ft.
    scale: 1 in. = 12 ft.

16. actual length = 144 mi.
    scale: $\frac{3}{4}$ in. = 12 mi.

17. scale length = 9 in.
    scale: $\frac{1}{2}$ in. = 5 ft.

18. scale length = $4\frac{1}{4}$ in.
    scale: $\frac{1}{2}$ in. = 5 mi.

19. actual length = 100 yd.
    scale: $\frac{3}{4}$ in. = 15 yd.

▶ **Applications**

20. The actual dimensions of a stage are twelve feet by twenty feet. Using a scale of three inches equals two feet, what will be the dimensions of a scale drawing of the stage?

21. Melanie starts a trip with a full tank of gas. Her car gets 375 miles to a tank of gas. On the map, her trip covers $3\frac{3}{4}$ inches. Does she have enough gas?

▶ **Review** (Lesson 3.9)

Simplify.

22. $105 \div 5 \times 4$

23. $(8 + 4) \times 8 + 11$

24. $50 - 7 \times 4 + 16$

11.6 Planning a House

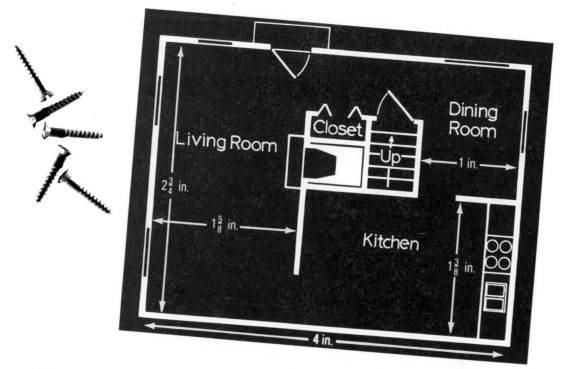

The Brandjords used the drawing shown here to get an idea of the amount of space they might have when their new house is built. According to the architect, 1 inch represents 10 feet.

If the depth of a closet measures $\frac{3}{8}$ inch on the diagram, how deep will the actual closet be?

To find the actual depth of the closet, set up a proportion.

$$\frac{\text{Drawing in inches} \rightarrow}{\text{Actual size in feet} \rightarrow} \quad \frac{1}{10} = \frac{\frac{3}{8}}{\text{\tiny\boxplus}} \quad \frac{\leftarrow \text{Closet (drawing)}}{\leftarrow \text{Closet (actual size)}}$$

$$1 \times \text{\tiny\boxplus} = 10 \times \tfrac{3}{8} \qquad \text{Find cross products.}$$

$$1 \times \text{\tiny\boxplus} \div 1 = \tfrac{30}{8} \div 1$$

$$\text{\tiny\boxplus} = \tfrac{30}{8} = \tfrac{15}{4} = 3\tfrac{3}{4} \textbf{ feet, or 45 inches}$$

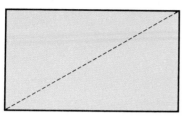

PROBLEM SOLVING

Refer to the introduction to this lesson.

1. What is the actual width (the shorter dimension) of the dining room in feet?

2. What is the actual width of the kitchen in feet?

3. What is the actual length of the living room in feet?

4. If the living room, from one corner diagonally to another, measures about $3\frac{1}{4}$ inches on the drawing, what proportion would you use to find the actual measurement in feet?

5. Find the actual measure of the diagonal of the living room.

6. If the diagonal of the kitchen measures $2\frac{5}{8}$ inches on the drawing, what is the actual measurement in feet?

7. Will a 91-inch couch fit in the living room?

8. A kitchen wall that is $1\frac{1}{4}$ inches long on the drawing is to be covered with cabinets 18 inches wide. How many cabinets will fit across?

The Brandjords would like to use the following furniture in their living room: an 88-inch sofa that is 36 inches deep, two chairs each 32 inches wide and 34 inches deep, a table that is 42 inches wide and $11\frac{1}{4}$ inches deep, a recliner that is $41\frac{1}{2}$ inches wide and 39 inches deep, a cabinet that is 25 inches wide and 13 inches deep, and a television set that is 43 inches wide and 22 inches deep.

9. Make a scale drawing of each piece of furniture and the living room. Show a possible arrangement of the furniture.

▶ **Review** (Lesson 8.9)

Copy and complete the table below.

	Starting Time	Finishing Time	Elapsed Time
10.	6:15 p.m.	12:01 a.m.	▨
11.	▨	9:25 a.m.	15 hours
12.	11:47 a.m.	▨	6 hours, 35 minutes

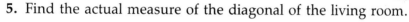

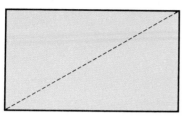

11.7 Using Similar Figures

Marius wants to enlarge a picture to fit the frame shown here. He brought the picture and the negative to a photo processor and told the clerk the dimensions of the frame.

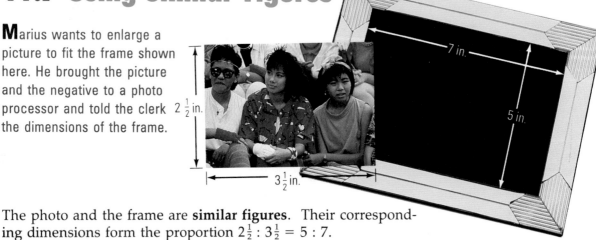

$2\frac{1}{2}$ in.

$3\frac{1}{2}$ in.

7 in.

5 in.

The photo and the frame are **similar figures**. Their corresponding dimensions form the proportion $2\frac{1}{2} : 3\frac{1}{2} = 5 : 7$.

Examples

To find **unknown lengths for similar figures**, set up and solve a proportion.

Find the length of the unknown side for these similar triangles.

$$\frac{3 \text{ in.}}{4 \text{ in.}} = \frac{7 \text{ in.}}{※}$$
$$3 \times ※ = 4 \times 7$$
$$3 \times ※ \div 3 = 28 \div 3$$
$$※ = \frac{28}{3}, \text{ or } 9\frac{1}{3}$$

3 in.

5 in.

4 in.

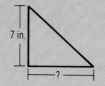

7 in.

?

The length of the unknown side is $9\frac{1}{3}$ **inches**.

▶ Think and Discuss

1. Refer to the introduction to this lesson. Could the photo be enlarged to fit a 9 × 12 frame? Explain.

2. The pentagons shown to the right are similar. Find the length of the unknown side.

3. One rectangle measures 12 inches by 15 inches. Is it similar to a 4-inch by 5-inch rectangle?

4. Are all squares similar? Explain your answer.

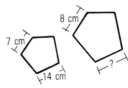

8 cm

7 cm

14 cm

?

Exercises (For more practice, see page 436.)

Each pair of polygons is similar. Find the length of the unknown sides.

5.
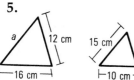

6.
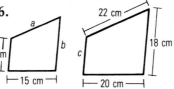

7.
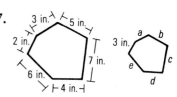

8. Rectangles
   $w = 2$ in.; $l = 7$ in.
   $w = 5$ in.; $l =$ ___

9. Rectangles
   $w = 7$ cm; $l =$ ___
   $w = 25.2$ cm; $l = 32.4$ cm

10. Rectangles
   $w = 5\frac{1}{4}$ in.; $l = 6\frac{1}{2}$ in.
   $w =$ ___ in.; $l = 13$

11.
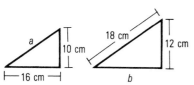

12.
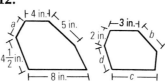

13.
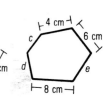

14. Rectangles
   $w = 5$ in.; $l = 6$ in.
   $w =$ ___ ; $l = 18$ in.

15. Rectangles
   $w =$ ___ ; $l = 4$ cm
   $w = 19.5$ cm; $l = 3$ cm

16. Rectangles
   $w = 20$ in.; $l =$ ___
   $w = 12\frac{1}{2}$ in.; $l = 10$ in.

▶ **Applications**

17. Kevin wanted to enlarge a six-inch by four-inch picture to fit a fourteen-inch by ten-inch frame. Is this possible? Explain.

18. Mandisa made a kite like the one shown. Her next kite had dimensions $1\frac{1}{2}$ times as large. What were the second kite's dimensions?

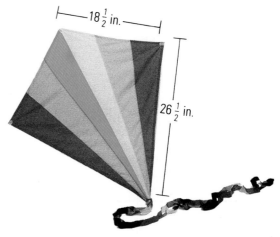

▶ **Review** (Lesson 10.4)

Find the mean of each set of numbers.

19. 32   25   10   41   **20.** 121   115   136   128   **21.** 92   88   79   95   86

11.8 Using Proportions to Solve Problems

As part of a geography project, you are to paint a scale map of the United States on one of the school walls. You will need mathematics to figure out how far apart cities should be placed.

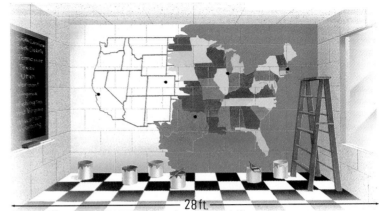

— 28 ft. —

Start with the cities shown in the table and use rounded distances as shown. The wall is 28 feet wide, so make the distance between Boston and San Francisco 25 feet. The real distance between these two cities is 3100 miles.

	Boston	Chicago	Dallas	Denver	San Francisco
Boston	——	1000	1750	2000	3100
Chicago		——	900	1000	2200
Dallas			——	800	1750
Denver				——	1250

How many feet should the distance between Boston and Chicago be on your map? To make the map look right, you have to consider the distances you used for Boston and San Francisco. Since these cities are 3100 miles apart, and you chose to draw them 25 feet apart, the scale of $\frac{25 \text{ feet}}{3100 \text{ miles}}$ must be used for the rest of the map.

To find the distance from Boston to Chicago, use a proportion.

$$\frac{\text{Distance on map}}{\text{Actual distance}} = \frac{\text{Boston to San Francisco on map}}{\text{Actual Boston to San Francisco distance}}$$

$$\frac{\text{Distance on map}}{\text{Actual distance}} = \frac{25}{3100}$$

$$\frac{\text{Distance on map}}{1000} = \frac{25}{3100} \quad \text{Find Boston to Chicago on the table above.}$$

$$3100 \times \text{Distance on map} = 1000 \times 25$$
$$\text{Distance on map} = 25{,}000 \div 3100$$
$$\text{Distance on map} = 8\tfrac{2}{31} \text{ feet} \approx 8 \text{ feet}$$

Draw Boston and Chicago about 8 feet apart on your map.

Exercises

1. What is the distance from Chicago to Denver?

2. How many feet apart will Chicago and Denver be on the map?

3. What is the distance from Denver to Boston?

4. How many feet apart will Denver and Boston be on the map?

5. Set up a ratio comparing map measurements to real distances for Chicago/Dallas and Boston/Dallas. Solve a proportion for each pair of cities using $\frac{25}{3100}$.

6. Use proportions to find the four map measurements that you have not yet found: Chicago/San Francisco, Denver/Dallas, Dallas/San Francisco, and Denver/San Francisco.

Write a proportion for each problem. Then solve.

7. There were 400 people at last week's football game. They ate 250 frankfurters. This week 650 people are expected. How many frankfurters should be prepared?

8. A car uses 5 gallons of gas to travel 115 miles. How far can it travel on 12 gallons of gas?

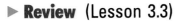

9. A plane travels 900 miles in $1\frac{1}{2}$ hours. How far will it travel in 5 hours if it continues at the same rate?

▶ Applications

For the following exercises, state whether a proportion can be used to solve the problem. If it can, use it to solve the problem. If a proportion cannot be used, explain why not.

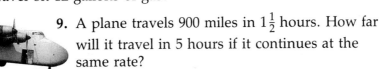

10. If a child is 3 feet tall at age 2, how tall will she be at age 10?

11. If a car travels 160 miles in 3 hours on the open road, about how far will it travel in 7 hours on the open road?

12. If a car travels 1 mile in 30 minutes during rush hour, how long will it take to go 5 miles?

▶ Review (Lesson 3.3)

Divide.

13. 95 ÷ 100   14. 600 ÷ 20   15. 49,000 ÷ 70   16. 8.431 ÷ 10

17. 64,000 ÷ 800   18. 4500 ÷ 50   19. 3.13 ÷ 100   20. 2067 ÷ 1000

Chapter 11 Review

REVIEW

Write each ratio in 3 different ways. (Lesson 11.1)

1. 60 pins in 7 pincushions

2. 10 shoes for 5 feet

3. 63 bills for 18 wallets

4. 45 students to 5 chaperones

Find 3 ratios equivalent to each ratio. (Lesson 11.1)

5. $1.25 for 3 stamps

6. 200 miles in 4 hours

7. $25 for 2 shirts

Tell whether each statement is a proportion. Use = or ≠.
(Lesson 11.2)

8. $\frac{4}{99}$ ___ $\frac{8}{198}$

9. $\frac{30}{50}$ ___ $\frac{5}{3}$

10. $\frac{19}{20}$ ___ $\frac{39}{40}$

11. $\frac{5}{9}$ ___ $\frac{6}{9}$

Solve each proportion. (Lesson 11.3)

12. $\frac{▧}{5} = \frac{9}{15}$

13. $\frac{8}{12} = \frac{▧}{3}$

14. $\frac{13}{▧} = \frac{52}{84}$

15. $\frac{600}{50} = \frac{144}{▧}$

Find the unit price. (Lesson 11.4)

16. $28.74 for 3 pairs of earrings

17. $19.95 for 5 tapes

Determine the better buy. (Lesson 11.4)

18. $1.09 for 3 pounds of apples or $1.55 for 5 pounds of apples

19. $34.50 for 6 ounces of cologne or $9.95 for 1.5 ounces of cologne

Write a proportion for each problem. Then solve.
(Lessons 11.5, 11.6, 11.7, 11.8)

20. A roll of film contains 36 exposures. How many rolls of film must you buy to have 324 exposures?

21. On a map, 1 inch represents 64 miles. Find the actual distance that is represented by 4.5 inches on the map.

22. The distance between two cities in Europe is 840 kilometers. On a map, the distance is 6 centimeters. How many kilometers does each centimeter represent on the map?

23. Mr. Burbank is making a sketch of his garden on paper. On his scale, one inch equals $1\frac{1}{2}$ feet. The sketch is 9 inches by 7 inches. What are the actual dimensions of his garden?

Chapter 11 Test

Write a proportion for each problem. Then solve.

1. There are 993 students at Highland High School. Two out of every three students are on a sports team. How many students are on sports teams?

2. Fourteen girls made the softball team. One out of every six girls made the team. How many girls tried out?

Find the unit price.

3. $29.95 for 2 sweaters

4. $19.88 for 8 pairs of socks

Write each ratio in 3 different ways.

5. 9 miles in 4 hours

6. 1 egg to 70 calories

7. 9 pies for 72 pieces

Tell whether each statement is a proportion. Use = or ≠.

8. $\frac{6}{4}$ ▧ $\frac{12}{10}$

9. $\frac{1}{7}$ ▧ $\frac{7}{56}$

10. $\frac{96}{16}$ ▧ $\frac{60}{10}$

11. $\frac{5}{65}$ ▧ $\frac{1}{13}$

Find 3 ratios equivalent to each ratio.

12. 5 boxes for 50 cans

13. 70 miles in 9 days

14. $15 for 4 books

Determine the better buy.

15. $1.54 for 9 bars of soap or $0.90 for 5 bars of soap

16. $34.45 for 5 pairs of gloves or $41.40 for 6 pairs of gloves

Solve.

17. The drawing of a room is 3 inches by $2\frac{1}{2}$ inches. One inch represents 12 feet. What are the actual dimensions of the room?

18. Sharon shot twelve rolls of film and took 288 photos. How many photos were there from each roll of film?

19. Jenny drew a plan for a bookcase. The plan uses $1\frac{1}{2}$ inches to represent 9 inches. The bookcase is 6 inches by $2\frac{1}{2}$ inches on the plan. What are the actual measurements of the bookcase?

20. Mrs. Smith's garden measures 45 feet by 39 feet. What is the size of her garden on paper if she uses a scale where $\frac{1}{2}$ inch represents 3 feet?

"I had been saving for a stereo for nearly 6 months. I didn't think I would ever have enough money. But then the one I wanted went on sale for 15% off. It was great. One quick calculation and I saw that I had enough money."

Percents

"With 30% of the vote counted, we predict Smith will recieve 62% of the vote."

Percents are used every day. Percents are involved when
- you figure how much to leave for a tip.
- a salesperson computes a sale price.
- a news broadcaster predicts the results of an election before the polls close.
- you compare banks to decide which one will give you the greatest amount of interest on your savings account.

Can you think of other situations where percents are used?

Picturing Percents

Materials: grid paper, scissors

Work with 2 or 3 classmates. Cut the grid paper into 6 large squares. Each square should be 10 units on each side. For each square, shade the number of small squares that is represented by the decimal, ratio, or fraction given below.

$\frac{17}{100}$	43 out of 100	0.29
$\frac{1}{2}$	0.8	$\frac{1}{4}$

12.1 Introduction to Percents

The sale sign shown and others like it are a common sight in most shopping areas. The signs are meaningful only if you understand the meaning of percent and how to use the % symbol.

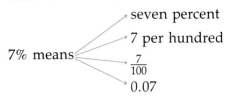

▶ Express Yourself

Here are some terms that will help you in this chapter.

%   symbol for percent.

7% means
- seven percent
- 7 per hundred
- $\frac{7}{100}$
- 0.07

of   in percent statements, another word for "times"

For example, 8% of 200 means 8% × 200.

is   in mathematics, another word for "equals"

You already know a lot about percents just by being a consumer.

1. A sports store is having a 30%-off sale. The price of a tennis racket is usually $60. If you buy the tennis racket on sale, will you pay more or less than $60? Can you find how much less? Discuss.

2. Look at the sign in the store window above. What percent off are the tennis shoes? How much would you have to pay for the shoes? How did you arrive at your answer? Explain.

3. Give an example of a percent that you have come across outside of school.

4. What is the meaning of the word *of* in the following sentence? **Ten percent of 60 is 6**.

5. What is the meaning of the word *is* in the sentence above?

6. You work with percents every time you pay sales tax. If the sales tax is 6% on every dollar, how much tax would you pay on a $5 purchase? Explain how you would compute the answer.

Percents can be illustrated in numerous ways. Here are some ways to represent 25%.

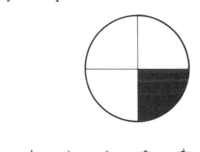

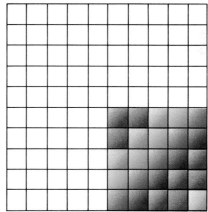

▶ Practice What You Know

As you work with percents in this chapter, you will be multiplying and dividing by 10 and by 100. You already know that to multiply by 10, you move the decimal point 1 place to the right. To multiply by 100, you move the decimal point 2 places to the right.

$$6.2 \times 10 = 62 \qquad\qquad 6.2 \times 100 = 620$$

Remember that to divide by 10, you move the decimal point 1 place to the left. To divide by 100, you move the decimal point 2 places to the left.

$$6.2 \div 10 = 0.62 \qquad\qquad 6.2 \div 100 = 0.062$$

Multiply or divide mentally.

7. 0.37×10
8. $13 \div 10$
9. $47.2 \div 100$
10. 4.5×100

11. $8.7 \div 10$
12. 0.276×100
13. $343 \div 100$
14. 56.3×10

12.2 Decimals and Percents

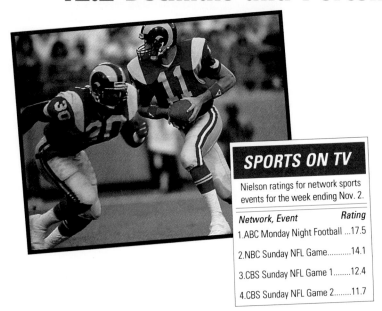

SPORTS ON TV

Nielson ratings for network sports events for the week ending Nov. 2.

Network, Event	Rating
1.ABC Monday Night Football	17.5
2.NBC Sunday NFL Game	14.1
3.CBS Sunday NFL Game 1	12.4
4.CBS Sunday NFL Game 2	11.7

Nielsen reports tell us about television viewing. A Nielsen report is shown here. Each rating point is equivalent to 1% of the television sets in the United States. You can express the rating of each program as a percent or as a decimal.

Examples

To write a decimal as a percent, multiply the decimal by 100 and write a percent sign after it.

To write a percent as a decimal, divide the percent by 100 and omit the percent sign.

A Write 0.375 as a percent.

$0.375 = 0.37.5$ Move the decimal point two places to the right.
$\quad\ = \mathbf{37.5\%}$ Write a % sign.

B Write 4% as a decimal.

$4\% = .04.$ Move the decimal point two places to the left.
$\quad\ = \mathbf{0.04}$ Add zeros as placeholders and omit the percent sign.

▶ Think and Discuss

1. Write 13% as a decimal. How many places and in what direction did you move the decimal point?

2. Refer to the introduction to this lesson. Write the rating of the fourth-rated program as a percent and as a decimal.

3. Write 67.3 as a percent.

4. Refer to the introduction to this lesson. What percent of the television sets were not tuned to Monday Night Football? Explain how you found your answer.

5. What does 0% mean? How do you change it to a decimal?

6. Jason bought a used car for $1250 and had to pay 7% sales tax. He decided to figure the tax on his calculator, but it doesn't have a percent key. How would he write 7% as a decimal so that he could calculate the answer?

Exercises

Write each decimal as a percent. (See Example A.)

7. 0.65 **8.** 0.05 **9.** 0.5 **10.** 0.73 **11.** 1.00

12. 0.09 **13.** 0.215 **14.** 0.9 **15.** 0.075 **16.** 0.11

Write each percent as a decimal. (See Example B.)

17. 8% **18.** 32% **19.** 11.4% **20.** 15% **21.** 13.5%

22. 25% **23.** 44.5% **24.** 81% **25.** 66.5% **26.** 100%

Mixed Practice (For more practice, see page 437.)

Write each percent as a decimal and each decimal as a percent.

27. 0.2 **28.** 0.16 **29.** 75% **30.** 1.25% **31.** 16.5%

32. 0.08 **33.** 0.135 **34.** 9.6% **35.** 0.505 **36.** 3.6%

37. 0.01 **38.** 99% **39.** 78% **40.** 0.12 **41.** 0.021

Applications

42. Outfielder Ted Williams, the "Splendid Splinter," hit 0.406 in 1941. Write his batting average as a percent.

43. A bank pays the interest advertised below. Write the interest rate as a decimal.

7.8% ON LONG TERM DEPOSITS

Review (Lesson 5.7)

Convert each fraction to a decimal.

44. $\frac{4}{5}$ **45.** $\frac{1}{2}$ **46.** $\frac{2}{3}$ **47.** $\frac{15}{20}$ **48.** $\frac{3}{4}$ **49.** $\frac{7}{8}$

Convert each decimal to a fraction in lowest terms.

50. 0.4 **51.** 0.33 **52.** 0.95 **53.** 0.056 **54.** 0.04 **55.** 0.125

12.3 Writing Fractions as Percents

Lou

Freethrows	
Attempted	5
Made	3

George and Lou play on a basketball team. After the first game, Lou said, "Let's see who has the better free-throw percent."

George

Freethrows	
Attempted	3
Made	2

To find who had the better free-throw percent, you can write each fraction as a percent and then compare.

Examples

To write a fraction as a percent, first convert the fraction to a decimal. Then write the decimal as a percent.

A Write $\frac{3}{5}$ as a percent.

$$\frac{3}{5} = 0.6$$
$$= 60\%$$

$$\begin{array}{r} 0.6 \\ 5{\overline{\smash{\big)}\,3.0}} \\ \underline{-3\ 0} \\ 0 \end{array}$$

B Write $\frac{2}{3}$ as a percent.

$$\frac{2}{3} = 0.66\tfrac{2}{3}$$
$$= 66\tfrac{2}{3}\%$$

$$\begin{array}{r} 0.6\ 6\ \tfrac{2}{3} \\ 3{\overline{\smash{\big)}\,2.0\ 0}} \\ \underline{-1\ 8} \\ 2\ 0 \\ \underline{-1\ 8} \\ 2 \end{array}$$

▶ Think and Discuss

1. Which part of the fraction is the divisor when you convert a fraction to a decimal?

2. Write $\frac{3}{4}$ as a percent.

3. Refer to the introduction to this lesson. Which player had the higher percent of successful free throws?

4. Can you determine if $\frac{5}{6}$ is larger than $\frac{2}{3}$ without converting to percents?

Exercises

Write as a percent. (See Example A.)

5. $\frac{1}{5}$ 6. $\frac{1}{2}$ 7. $\frac{3}{4}$ 8. $\frac{3}{10}$ 9. $\frac{17}{25}$ 10. $\frac{23}{50}$

11. $\frac{3}{100}$ 12. $\frac{9}{36}$ 13. $\frac{4}{20}$ 14. $\frac{12}{12}$ 15. $\frac{3}{20}$ 16. $\frac{9}{15}$

Write as a percent. (See Example B.)

17. $\frac{1}{3}$ 18. $\frac{1}{6}$ 19. $\frac{5}{6}$ 20. $\frac{17}{40}$ 21. $\frac{1}{8}$ 22. $\frac{3}{16}$

23. $\frac{11}{16}$ 24. $\frac{9}{16}$ 25. $\frac{7}{21}$ 26. $\frac{8}{12}$ 27. $\frac{1}{15}$ 28. $\frac{7}{56}$

▶ Mixed Practice (For more practice, see page 437.)

Write as a percent.

29. $\frac{3}{8}$ 30. $\frac{7}{8}$ 31. $\frac{2}{5}$ 32. $\frac{7}{10}$ 33. $\frac{1}{16}$ 34. $\frac{15}{75}$

35. $\frac{9}{12}$ 36. $\frac{19}{20}$ 37. $\frac{11}{66}$ 38. $\frac{37}{50}$ 39. $\frac{15}{16}$ 40. $\frac{4}{15}$

▶ Applications

41. Seven-tenths of the earth is covered by water. Write this as a percent.

42. Two out of every nine people in the world are Chinese. Write this as a percent.

43. Jeremy attended seven of his school's eleven home games. What percent of the home games did Jeremy attend?

44. Refer to the introduction to this lesson. Together Lou and George attempted 8 free throws in the first game. What was their combined free-throw percent?

45. A 1-hour aerobic exercise class is divided into the categories shown here. Write each as a percent.

Warm-up exercises	15 minutes
Abdominal exercises	10 minutes
Leg exercises	9 minutes
Hip exercises	6 minutes
Arm exercises	5 minutes
Aerobic exercises	12 minutes
Cool-down exercises	3 minutes

▶ Review (Lesson 5.6)

Rewrite each fraction in lowest terms.

46. $\frac{8}{100}$ 47. $\frac{12}{200}$ 48. $\frac{355}{1000}$ 49. $\frac{57}{300}$ 50. $\frac{12.5}{100}$ 51. $\frac{31}{200}$ 52. $\frac{82}{300}$

12.4 Writing Percents as Fractions

Torrez wins election with 52.5% of vote!

In St. Louis, 12% of the families subscribe to cable television.

62% of all downtown commuters *use* **PUB-TRANS**

90% of all dentists polled recommended Happy Smile toothpaste to help prevent cavities.

SALE 33⅓% OFF!

Percents can be found in newspapers, television advertisements, and magazines. Often it is easier to understand the situation if you write the percent as a fraction. Remember, percent means *per hundred*.

Examples

To write a percent as a fraction, first write the percent as the numerator without the percent sign. Use 100 as the denominator. Then write the fraction in lowest terms.

A Write 30% as a fraction in lowest terms.

$$30\% = \frac{30}{100} = \frac{3}{10}$$

B Write $12\frac{1}{2}\%$ as a fraction in lowest terms.

$$12\frac{1}{2}\% = \frac{12\frac{1}{2}}{100} = \frac{\frac{25}{2}}{\frac{100}{1}} = \frac{25}{2} \times \frac{1}{100} = \frac{25}{200} = \frac{1}{8}$$

First rewrite as fractions. Then multiply by the reciprocal of $\frac{100}{1}$.

$$12\frac{1}{2}\% = \frac{1}{8}$$

C Write 20.5% as a fraction in lowest terms.

$$20.5\% = \frac{20.5}{100} = \frac{20.5 \times 10}{100 \times 10} = \frac{205}{1000} = \frac{41}{200}$$

Multiply the numerator and denominator by 10 to move the decimal point.

$$20.5\% = \frac{41}{200}$$

▶ Think and Discuss

1. Write $\frac{6}{1000}$, $\frac{20}{100}$, and $\frac{50}{100}$ in lowest terms.

2. Write 58% as a fraction in lowest terms.

3. How could 40.5% be written as a percent in another way?

4. Refer to the introduction to this lesson. If 50 different dentists are polled tomorrow, how many do you think will recommend Happy Smile toothpaste? Discuss.

5. Why would a store choose "$33\frac{1}{3}$% off" for a sale?

Exercises

Write as a fraction in lowest terms. (See Example A.)

6. 26%   7. 75%   8. 48%   9. 99%   10. 5%   11. 4%

12. 40%   13. 24%   14. 57%   15. 73%   16. 85%   17. 10%

18. 12%   19. 55%   20. 2%   21. 64%   22. 3%   23. 25%

Write as a fraction in lowest terms. (See Example B.)

24. $7\frac{1}{2}$%   25. $37\frac{1}{2}$%   26. $6\frac{1}{4}$%   27. $41\frac{2}{3}$%   28. $3\frac{1}{2}$%   29. $50\frac{1}{3}$%

Write as a fraction in lowest terms. (See Example C.)

30. 1.25%   31. 19.5%   32. 22.75%   33. 12.5%   34. 6.25%   35. 9.75%

▶ Mixed Practice (For more practice, see page 438.)

Write as a fraction in lowest terms.

36. 25%   37. $33\frac{1}{3}$%   38. 22.5%   39. 44%   40. 9.5%   41. 57.5%

42. 1%   43. 14%   44. 95%   45. $16\frac{2}{3}$%   46. 30.5%   47. 87.5%

48. 66%   49. 54%   50. 5.75%   51. $6\frac{1}{2}$%   52. 11%   53. 62%

▶ Applications

54. A shirt is 60% cotton and 40% polyester. Write the percents as fractions.

55. If the sales tax is five percent, what fraction of an item's price must you add to the price?

▶ Review (Lesson 11.3)

Solve each proportion.

56. $\frac{6}{9} = \frac{12}{\text{▓}}$

57. $\frac{2}{1} = \frac{1}{\text{▓}}$

58. $\frac{8}{3} = \frac{\text{▓}}{24}$

12.5 Using Percents and Circle Graphs

You probably have heard someone say, "Back in the good old days I used to walk to school 3 miles through rain, sleet, and snow . . ."

You can use circle graphs like the ones shown here to compare how high school students traveled to school 30 years ago with how they get to school today. The circle in a circle graph represents the whole group, or 100%. Each slice of the circle represents a portion of the whole.

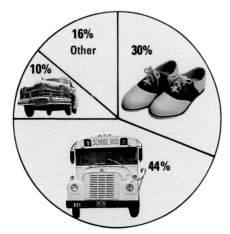

Methods of transportation to high schools in 1950's

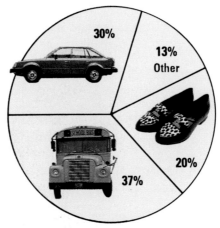

Methods of transportation to high schools in 1980's

1. What percent of students walked to school 30 years ago? What percent walk now?

2. Which mode of transportation did students use most 30 years ago? Which do they use most now?

3. What percent of students drive now? What fraction does that represent? How much of the circle is this slice?

4. Suppose there were no numbers on the circle graphs. Could you still get any information from them? Explain.

5. What is the sum of the percents in each circle graph?

6. Discuss why the "Other" category is important, even though you may not know exactly what it includes. What are some modes of transportation that might be included in the category "Other"?

A survey was given to people in three age groups: those under 21, those 21 to 50, and those over 50. Each group was asked to choose its favorite type of car. The results are shown below:

Vehicle	Under 21	21–50	Over 50
Sports car	200	12	10
Van	125	12	8
Economy	100	60	20
Full size	50	30	40
Other	25	6	2
Total number surveyed	**500**	**120**	**80**

7. What fraction of persons under 21 prefers sports cars? How would you write this as a percent?

8. What fraction of persons 21 to 50 prefers vans? How would you write this as a percent?

9. Convert the figures in the table above for each group to fractions and then to percents.

10. Use your answers from Question 9 to select the graph that represents each group. Copy the graphs and label each one. Then label each slice with the type of car it represents and its percent.

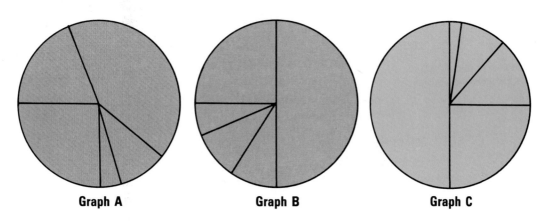

| Graph A | Graph B | Graph C |

▶ **Review** (Lesson 1.12)

Estimate.

11.   335
   + 467

12.   1119
   + 5880

13.   3852
   + 4439

14.   523
   + 1991

12.6 Percents Less Than One and Greater Than One Hundred

At the right are the prizes and chances of winning in a $1 instant state lottery game. The chances of winning $1000 are 1 in 3,480. Do you realize that this is less than a 1% chance of winning?

LOTTERY

Prize	* Chance
$1,000	1 in 7,692
$500	1 in 4,170
$100	1 in 912
$50	1 in 496

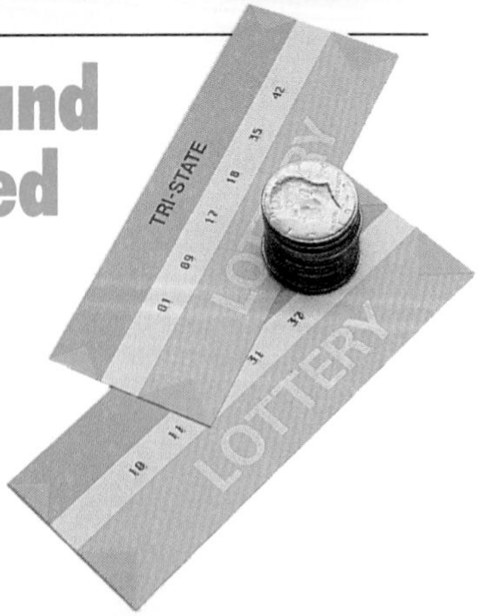

Examples

To work with percents less than one or greater than one hundred, follow the rules you have already learned about percents in this chapter.

A Write 9.8 as a percent.

$9.8 = \mathbf{980\%}$ Move the decimal point 2 places to the right.

B Write $\frac{1}{400}$ as a percent.

$\frac{1}{400} = 0.0025$ Change the fraction to a decimal.

$= \mathbf{0.25\%}$ Move the decimal point 2 places to the right.

C Write $\frac{1}{2}\%$ as a fraction and as a decimal.

$$\frac{1}{2}\% = \frac{\frac{1}{2}}{\frac{100}{1}} = \frac{1}{2} \times \frac{1}{100} = \frac{1}{200} = 0.005$$

Multiply by the reciprocal. Fraction Decimal

$\frac{1}{2}\% = \mathbf{\frac{1}{200}} = \mathbf{0.005}$

▶ Think and Discuss

1. Divide $\frac{3}{4}$ by 100.
2. Write 0.008 as a percent.
3. Convert the following percents to decimals. 100% 125%
4. Convert the following to percents. $\frac{1}{10}$ $\frac{1}{1000}$ 1 $1\frac{1}{4}$

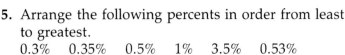

SKILLS

5. Arrange the following percents in order from least to greatest.
   0.3%   0.35%   0.5%   1%   3.5%   0.53%

6. A sign reads "SALE $\frac{1}{2}$% off!" Is the sign accurate? Explain.

Exercises

Write each decimal as a percent. (See Example A.)

7. 1.7   8. 3.6   9. 4   10. 5.75   11. 0.7

12. 0.009   13. 0.001   14. 0.0052   15. 0.0075   16. 0.0019

Write each fraction as a percent. (See Example B.)

17. $\frac{1}{200}$ 18. $\frac{11}{10}$ 19. $\frac{7}{1000}$ 20. $\frac{29}{20}$ 21. $\frac{17}{4}$

22. $\frac{5}{2000}$ 23. $\frac{4000}{8}$ 24. $\frac{8}{8000}$ 25. $\frac{900}{4}$ 26. $\frac{12}{2400}$

Write each percent as a fraction and as a decimal. (See Example C.)

27. $\frac{1}{4}$%   28. 175%   29. $\frac{5}{8}$%   30. $\frac{9}{10}$%   31. 324%

32. 156%   33. $220\frac{1}{2}$%   34. 433%   35. $\frac{5}{16}$%   36. $\frac{3}{8}$%

▶ **Mixed Practice** (For more practice, see page 438.)

Write each fraction or decimal as a percent.

37. $\frac{3}{400}$ 38. $1\frac{1}{5}$ 39. 4.25   40. $\frac{13}{6}$ 41. $\frac{1}{250}$

42. 5.3   43. $\frac{3}{1000}$ 44. 9   45. $3\frac{7}{8}$ 46. $\frac{4}{3}$

47. 7.08   48. $\frac{5}{8}$ 49. $6\frac{1}{4}$ 50. 0.085   51. $7\frac{5}{12}$

52. $\frac{11}{12}$ 53. 0.0003   54. $3\frac{8}{9}$ 55. 10   56. $11\frac{19}{20}$

▶ **Applications**

57. Refer to the introduction to this lesson. Express your chances of winning $50 as a percent. Round to the nearest hundredth of a percent.

58. The price of mailing a letter rose 1150% from 1920 to 1988. Express the price rise as a fraction and a decimal.

▶ **Review** (Lessons 2.8, 6.1, 6.2)

Multiply.

59. 0.7×8 60. $\frac{3}{4} \times 16$ 61. 5.83×17 62. $\frac{19}{20} \times 55$

12.7 Finding the Percent of a Number

Wilma works as a volunteer for the election campaign of a local politician. Her job is to see that all citizens are registered to vote. So far, 80% of the 16,000 citizens of voting age are registered. How many people are registered?

Examples

To find the percent of a number, first convert the percent to a decimal or fraction. Then multiply.

A Find 80% of 16,000.

80% = 0.80
80% of 16,000 = 0.80 × 16,000
80% of 16,000 = 12,800
80% of 16,000 is **12,800**.

B Find $33\frac{1}{3}\%$ of $75.

$33\frac{1}{3}\% = \frac{1}{3}$

$33\frac{1}{3}\%$ of $75 = \frac{1}{3} \times 75$

$33\frac{1}{3}\%$ of $75 = \frac{75}{3} = 25$

$33\frac{1}{3}\%$ of $75 is **$25**.

▶ Think and Discuss

1. What do *of* and *is* mean in "30% of 200 is 60"?

2. Find 20% of 40.

3. Find 1% of 800. Can you do it mentally?

4. Refer to the introduction to this lesson. How many of the 16,000 citizens are registered?

5. Suppose $12\frac{1}{2}\%$ of a class of 120 students went to a soccer game. How many students did not attend?

6. When might you need to find the percent of a number?

7. Write $33\frac{1}{3}\%$ as a fraction. How would you enter $33\frac{1}{3}\%$ on your calculator? Discuss.

Exercises

Find each number. (See Example A.)

8. 12% of $75   **9.** 32% of $110   **10.** 9% of $80   **11.** 125% of 16

12. 50% of 120   **13.** 5% of 15.6   **14.** 0.5% of 126   **15.** 275% of 56

Find each number. (See Example B.)

16. $12\frac{1}{2}$% of 24   **17.** $66\frac{2}{3}$% of 36.6   **18.** $87\frac{1}{2}$% of 64   **19.** $66\frac{2}{3}$% of 90

20. $33\frac{1}{3}$% of 900   **21.** $37\frac{1}{2}$% of 40   **22.** $33\frac{1}{3}$% of 96   **23.** $62\frac{1}{2}$% of 700

▶ Mixed Practice (For more practice, see page 439.)

Find each number.

24. 40% of 250   **25.** 35.5% of 100   **26.** 50% of 51   **27.** $33\frac{1}{3}$% of 54

28. 75% of 40   **29.** 35% of 60   **30.** 0.4% of 85   **31.** 10% of $70

32. $37\frac{1}{2}$% of 8.8   **33.** 15% of $25   **34.** $16\frac{2}{3}$% of 72   **35.** 0.1% of 210

36. 175% of 60   **37.** 95% of 1122   **38.** 19.5% of 46   **39.** 2.5% of 17

▶ Applications

40. During the Great Depression, the unemployment rate went as high as 25%. The work pool was 52 million people. Find 25% of 52 million.

41. The sales tax in New York City is $8\frac{1}{4}$%. How much sales tax must New Yorkers pay when purchasing the automobile below?

42. Dinah ordered a hamburger for $3.75 and iced tea for $0.75. The tax was 6%. If she gave the waitress 20% of the bill (before taxes) as a tip, how much did she spend altogether?

$15,000

▶ Review (Lesson 1.8)

Round to the nearest tenth.

43. 7.19   **44.** 18.64   **45.** 99.92   **46.** 3.0518   **47.** 2.1745   **48.** 6.268

Round to the nearest hundredth.

49. 0.873   **50.** 4.285   **51.** 1.964   **52.** 2.996   **53.** 8.3435   **54.** 9.0572

12.8 Finding What Percent One Number Is of Another

RECEIPT

Time *7:00pm*

Received from *Henry E. Smith*

for Cab Fare *16⁰⁰ + 2⁰⁰ tip*

from *202 Lake* to *86 W. 57ᵀᴴ*

Driver *Daniel O'Brien*

Cab No. *75067* Lease No. *SA2098*

"Ride like a gem"

Sheri's boss gave her cab fare so that she could make a quick delivery to a customer's store. "The fare is $16, and give the driver a $2 tip," the boss said. Sheri knew that her parents left a 15% tip in restaurants and she wondered what percent she tipped the driver.

Sheri must answer the question "What percent of $16 is $2?"

To find what percent one number is of another, use division to get a decimal. Then convert the decimal to a percent.

A 12 is what percent of 60?

$\frac{12}{60}$ Write the number that represents the whole as the denominator.
Write the number that represents part of the whole as the numerator.

$\frac{12}{60} = 0.20$ Convert the fraction to a decimal.

$0.20 = 20\%$ Write the decimal as a percent.

12 is **20%** of 60.

B What percent of $16 is $2?

$\frac{2}{16} = 0.125$ Convert the fraction to a decimal.

$0.125 = 12.5\%$ Write the decimal as a percent.

$2 is **12.5%** of $16.

$$\begin{array}{r} 0.125 \\ 16\overline{)2.000} \\ \underline{1\,6} \\ 40 \\ \underline{32} \\ 80 \\ \underline{80} \end{array}$$

▶ Think and Discuss

1. Find 20% of 60. Explain how this is a check for Example A.

2. Twelve is what percent of 120?

3. What is 100% of 7?

4. Refer to the introduction to this lesson. What percent of the fare was Sheri's tip? Is it more or less than 15%?

5. Name 2 situations in which you might want to find what percent one number is of another.

Exercises

Find each percent. (See Example A.)

6. 14 is what percent of 25?

7. 33 is what percent of 44?

8. 12 is what percent of 75?

9. 31 is what percent of 62?

10. 1 is what percent of 50?

11. 129 is what percent of 300?

Find each percent. (See Example B.)

12. What percent of 200 is 65?

13. What percent of 400 is 170?

14. What percent of 15 is 18?

15. What percent of 82 is 10.25?

16. What percent of 84 is 7?

17. What percent of 150 is 2?

▶ Mixed Practice (For more practice, see page 439.)

Find each percent.

18. 7 is what percent of 70?

19. 25 is what percent of 75?

20. 3 is what percent of 4?

21. 28 is what percent of 80?

22. What percent of 25 is 5?

23. What percent of 240 is 6?

24. 7 is what percent of 35?

25. 56 is what percent of 128?

26. What percent of 3 is 7?

27. 62 is what percent of 62?

▶ Applications

28. Jeanne wanted a one-hundred-fifty-dollar coat and put down a thirty-dollar deposit in order to place the coat on layaway. What percent did she put down?

29. The Recommended Daily Allowance of calcium for young people is 1200 milligrams. What percent of the RDA is 1800 milligrams?

▶ Review (Lessons 2.3, 2.7, 3.3)

Multiply or divide.

30. 6.5×1000

31. $98.7 \div 1000$

32. 400×500

33. $900 \div 30$

34. $0.34 \div 100$

35. 60×9000

36. $0.017 \div 100$

37. 0.671×100

12.9 Finding a Number When a Percent of It Is Known

Fifty freshmen were asked about their favorite books. If 8% of the freshmen chose George Orwell's *1984*, how many students chose this book? You can write this as

<u> 8% </u> of <u> 50 </u> is <u> ? </u> .

What percent of the 50 students chose *1984* if 4 students chose this book? You can write this as

<u> ?% </u> of <u> 50 </u> is <u> 4 </u> .

These types of questions were solved in Lessons 12.7 and 12.8.

How would you solve the following problem: how many students were asked about their favorite books, if 4 students are 8% of the total?

Examples

To find a number when a percent of it is known, first convert the percent to a decimal. Then divide.

A 8% of what number is 4?

This sentence translates into
8% of ? is 4, which becomes
$8\% \times ? = 4.$
$0.08 \times ? = 4$ Write the percent as a
$\frac{0.08}{0.08} \times ? = \frac{4}{0.08}$ decimal. Divide.
$? = 50$

8% of **50** is 4.

B 15 is 6% of what number?

This sentence translates into
15 = 6% of ?, which becomes
$15 = 6\% \times ?.$
$15 = 0.06 \times ?$ Write the percent as
$\frac{15}{0.06} = \frac{0.06}{0.06} \times ?$ a decimal. Divide.
$250 = ?$

15 is 6% of **250**.

▶ Think and Discuss

1. What is 10% of 170?

2. 17 is 10% of what number?

3. Refer to the introduction to this lesson. Describe the pattern for the three types of percent problems. What are some ways you can remember how to solve each type of percent problem?

4. Look at Example B. Convert 6% to a fraction instead of a decimal. Then solve the problem. Is your answer the same?

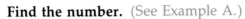

Exercises

Find the number. (See Example A.)

5. 40% of what number is 22?

6. 95% of what number is 38?

7. $33\frac{1}{3}\%$ of what number is 15?

8. 0.8% of what number is 2?

9. 5% of what number is 25.75?

10. $166\frac{2}{3}\%$ of what number is 65?

Find the number. (See Example B.)

11. 16 is 4% of what number?

12. 66 is 75% of what number?

13. 36 is 50% of what number?

14. 0.8 is 0.4% of what number?

15. 125 is 12.5% of what number?

16. 9 is $1\frac{1}{3}\%$ of what number?

▶ Mixed Practice (For more practice, see page 440.)

17. 30% of what number is 81?

18. 13 is 2% of what number?

19. 17 is $33\frac{1}{3}\%$ of what number?

20. 25% of what number is 30?

21. 200% of what number is 50?

22. 40 is 20% of what number?

23. 2% of what number is 40?

24. 40 is 1% of what number?

▶ Applications

25. A box of cereal states that a 1-ounce serving contains 0.36 milligram of iron. This is 2% of the Minimum Daily Requirement for teenagers. What is the Minimum Daily Requirement?

26. Workers at a telethon for a local charity raised the amount shown. One hundred twenty-five percent of the goal was raised. What was the goal?

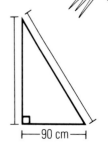

▶ Review (Lesson 11.7)

Find the length of the missing sides in these similar drawings.

27.

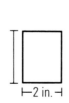

5 in.

4 in.

2 in.

28.

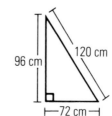

96 cm

120 cm

72 cm

90 cm

12.10 Finding the Percent of Change

Anderson's Department Store is offering $15 shirts for $12. What percent off is that?

SALE

Originally $15.00
Now only $12.00

Examples

To find the percent of change, first subtract to find the amount of change. Then use division to find the percent of change.

A Find the percent of increase from 24 to 30.

$30 - 24 = 6$ Subtract to find the amount of increase.

$\frac{6}{24} = 0.25$ Divide the amount of increase by the original amount.

$0.25 = 25\%$ Write the decimal as a percent.

The percent of increase is **25%**.

B Find the percent of decrease from $15 to $12.

$\$15 - \$12 = \$3$

$\frac{3}{15} = 0.20$

$0.20 = 20\%$

The percent of decrease is **20%**.

▶ Think and Discuss

1. An item's price rose from $40 to $50. To find the percent of change, would you divide 10 by 40 or by 50?

2. A price fell from $50 to $40. To find the percent of change, would you divide 10 by 40 or by 50?

3. Refer to the introduction to this lesson. What was the percent off?

4. The percent of change from 10 to 8 is 20%. What is the percent of change from 8 to 10?

5. Explain why the two percents in Question 4 are not the same.

6. What is the percent of change from 40 to 60? Solve mentally.

Exercises

Find the percent of increase. (See Example A.)

7. $4 to $5 **8.** $3 to $6 **9.** $2 to $3 **10.** $4.50 to $8.50

11. 4000 to 10,000 **12.** 1000 to 1003 **13.** 30 to 90 **14.** $32 to $38

Find the percent of decrease. (See Example B.)

15. 5 to 3 **16.** $25 to $7.50 **17.** $235 to $205 **18.** $45 to $36

19. 12 to 11 **20.** $600 to $450 **21.** 38 to 19 **22.** 21 to 14

▶ **Mixed Practice** (For more practice, see page 440.)

Find the percent of change.

23. 50 to 57 **24.** 48 to 36 **25.** 60 to 36 **26.** $10.00 to $9.30

27. 1000 to 1001 **28.** 575 to 690 **29.** 18 to 15 **30.** $125 to $100

31. $3.90 to $2.73 **32.** 216 to 189 **33.** $25 to $25.50 **34.** $50 to $200

▶ **Applications**

35. After taking a walk, Andrea's pulse rate rose from 70 beats per minute to 77 beats per minute. What percent of increase is this?

36. A house was originally listed as shown here. After six months, the seller dropped the price to $175,000. By what percent did the seller reduce the price?

37. Over a 30-year period the number of people living in New York City dropped from 7.9 million to 7.1 million. What is the percent of change?

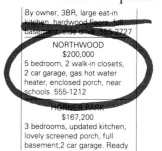

By owner, 3BR, large eat-in kitchen, hardwood floors, full basement, side drive. 555-2727

NORTHWOOD
$200,000
5 bedroom, 2 walk-in closets, 2 car garage, gas hot water heater, enclosed porch, near schools. 555-1212

HORNER PARK
$167,200
3 bedrooms, updated kitchen, lovely screened porch, full basement, 2 car garage. Ready

▶ **Review** (Lesson 1.12)

Estimate.

38. $119.33 − $79.80 **39.** $66.78 + $39.15 **40.** $243.88 − $96.25

41. $512.51 + $385.95 **42.** $489.50 + $103.10 **43.** $931.98 − $629.57

12.11 Applying Percents

Charlotte has a job at Grand Canyon National Park. She hopes to find time next summer for some hiking and camping. She has been looking at catalogs for clothes and equipment. Her selections are shown below.

Sun Visor
Reg. $6
10% off

Item No.: A2847

Hiking Boots
Reg. $80
20% off

Item No.: B1134

Thermos Bottle
Reg. $19.90
10% off

Item No.: C3098

Binoculars
Reg. $36
15% off

Item No.: D1952

Flashlight
Reg. $5.55
20% off

Item No.: E4204

Knee-Length Socks
Reg. $5.00
5% off

Item No.: F8866

The sale price of Item No. A2847 can be found in two steps:

Step 1. Compute the discount.
10% of $6 = 0.10 × 6 = 0.60 = $0.60

Step 2. Subtract the discount price from the regular price.
$6.00 − $0.60 = $5.40
The sale price of the sun visor is $5.40.

Another way to solve this problem is as follows:
Compute the sale price as a percent of the regular price (100% − 10% = 90%).
90% of $6 = 0.90 × 6 = 5.40 = $5.40

▶ Think and Discuss

1. What percent of the regular price is the discount for hiking boots? What percent do you actually pay?

2. What is the sale price of the hiking boots? Use both methods shown to solve.

3. Which of the two methods would you use to estimate? Why would the method you chose help you check your answer?

4. How much sales tax would be charged on the sale price of the hiking boots if the tax rate were 6%?

5. Look at the shipping and handling chart below. What would be the final total if the subtotal (the total of all items before tax and shipping and handling are added) were $25?

Exercises

Copy this order blank and complete it. Order one of each item. You will need the information on shipping and handling that appears below the order blank.

Catalog #	Qty.	Description	Reg. Price	Sale Price Each	Total
A2847	1	Sun visor	6.	7.	8.
B1134	1	Hiking boots	9.	10.	11.
C3098	1	Thermos bottle	12.	13.	14.
D1952	1	Binoculars	15.	16.	17.
E4204	1	Flashlight	18.	19.	20.
F8866	1	Socks	21.	22.	23.

24. Subtotal _____

25. Sales Tax: add 6% _____

26. Shipping and Handling _____

27. Total _____

Shipping and Handling

Subtotal

Up to $15	$2.75	$75.01 – $95	$10.00
$15.01 – $35	$4.50	$95.01 – $120	$20.00
$35.01 – $55	$6.25	Over $120	$20.00
$55.01 – $75	$8.00		

▶ Review (Lesson 11.1)

Write 3 ratios equivalent to each ratio.

28. 25 students to 1 teacher

29. 150 freshmen to 140 sophomores

30. 75 campers to 10 counselors

31. 52 cards to 4 players _____

Chapter 12 Review

Complete each statement. (Lesson 12.1)

1. Percent means per _____.

2. Six percent _____ 50 means 6% × 50.

Write each decimal as a percent and each percent as a decimal. (Lesson 12.2)

3. 0.175   4. $33\frac{1}{3}\%$ 5. 0.4   6. 7%   7. 0.875

Write each fraction as a percent. (Lesson 12.3)

8. $\frac{3}{10}$ 9. $\frac{5}{8}$ 10. $\frac{1}{4}$ 11. $\frac{1}{20}$ 12. $\frac{2}{3}$

Write each percent as a fraction in lowest terms. (Lesson 12.4)

13. 48%   14. 50%   15. 37.5%   16. 2%   17. $41\frac{2}{3}\%$

Write each fraction or decimal as a percent. (Lesson 12.6)

18. $\frac{1}{400}$ 19. 5   20. 8.2   21. $\frac{7}{1000}$ 22. $\frac{4}{3}$

Write each percent as a fraction and as a decimal. (Lesson 12.6)

23. $\frac{3}{4}\%$ 24. $\frac{1}{10}\%$ 25. $\frac{7}{8}\%$ 26. $\frac{7}{4}\%$ 27. $\frac{3}{5}\%$

Find each number or percent. (Lessons 12.7, 12.8, 12.9)

28. 65% of 480   29. 2% of $56   30. 580% of 25   31. $\frac{1}{2}\%$ of 44

32. 77.5 is what percent of 775?   33. 15 is what percent of 300?

34. 594 is what percent of 600?   35. 224 is what percent of 28?

36. 40% of what number is 24?   37. 3% of what number is 27?

38. 225% of what number is 72?   39. 50.5% of what number is 6.06?

Find the percent of change. (Lesson 12.11)

40. 75 to 100   41. 30 to 24

Find the sale price of each item.

42. Mittens, regularly $25, now 10% off   43. Camera, was $278, marked down 35%

Chapter 12 Test

Find each number or percent.

1. 2 is what percent of 8?

2. 2% of 50

3. $33\frac{1}{3}\%$ of what number is 284?

4. 98% of what number is 343?

5. 78% of 650

6. 195 is what percent of 39?

7. 145 is 50% of what number?

8. $12\frac{1}{2}\%$ of 86

Find the percent of change.

9. 8 to 24

10. 48 to 60

11. 300 to 270

12. What is the total of all the percents in a circle graph?

Write each decimal as a fraction and as a percent.

13. 0.06

14. 0.625

15. 0.001

Write each percent as a fraction and as a decimal.

16. 80%

17. 450%

18. 4%

Write each fraction as a decimal and as a percent.

19. $\frac{2}{5}$

20. $\frac{3}{20}$

21. $\frac{81}{9}$

Solve.

22. A watch is marked down 25% from $19.96. How much is the sale price?

23. Jose left a 15% tip. The bill came to $29. How much was his tip?

"You wouldn't believe the number of people who ask me what their chances are of winning first prize in the raffle. I must have explained probability to a dozen different people."

Probability

Many events in everyday life involve chance, or probability.
Probability is involved when
- you play a board game.
- a football referee tosses a coin.
- a contestant on a television show spins a wheel to determine the prize won.
- a weatherman predicts the chance of rain.

What other situations do you know of that involve probability?

Rolling Cubes

Materials: a number cube for each group

Work with 1 or 2 classmates. Write the words *odd* and *even* on a sheet of paper.

1. How many faces of the number cube are odd?

2. How many faces of the number cube are even?

3. Predict the number of even and odd rolls, if the number cube is rolled 10 times.

Now roll the number cube 10 times. After each roll, write a tally mark under *odd* or *even*, as appropriate. Total the marks.

4. Out of the first 10 tries, how many were odd? Even?

5. Predict the number of even and odd rolls, if the number cube is rolled 50 times.

Repeat the procedure until you have a total of 50 marks.

6. Out of the 50 tries, how many were odd? Even?

7. How do the results compare with your prediction?

13.1 Introduction to Probability

If you flip a coin into the air, the chance of it landing heads is 50%. The chance of it landing tails is also 50%. In mathematics, chance is called probability. You could say, then, that the probability of a tossed coin landing heads is 50%. Fifty percent is the same as 0.5, $\frac{1}{2}$, and 1 out of 2.

▶ **Express Yourself**

Here are some terms you will become familiar with:

chance or **probability**   the likelihood of a particular result

outcome   the result of one experiment

event   an outcome or a group of outcomes

favorable outcome   the desired outcome

Complete each statement.

1. The desired outcome is a(n) _____ .

2. Another word for chance is _____ .

3. A group of outcomes is a(n) _____ .

Today's Weather Outlook

Partly cloudy
30% chance of rain
High **77°** F /Low **55°** F

▶ Practice What You Know

Probability involves using fractions or ratios. You are familiar with fractions, so you already know something about probability.

Look at the concert tickets below and find the following.

4. the ratio of the number of Row A tickets to the total number of tickets

5. the ratio of the number of Row A tickets to the number of Row B tickets

6. the ratio of the number of tickets marked "D2" to the total number of tickets

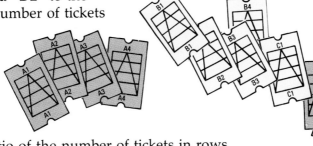

7. the ratio of the number of tickets in rows A through E to the total number of tickets

8. the ratio of tickets in Row F to the total number of tickets

Answer the following questions.

9. How would you write the ratio for Question 4 as a fraction in lowest terms?

10. What whole number is equal to the ratio you found for Question 7?

Probability also involves making choices.

11. How many choices do you have if you are asked to pick a letter of the alphabet?

12. How many choices do you have if you are asked to choose a whole number from 1 to 50?

13. How many choices do you have if you are asked to pick a whole number from 10 to 50?

14. How many choices do you have if you are asked to pick a month from the months in a year?

15. How many choices do you have if you are asked to pick a date in April?

13.2 Finding Simple Probability

One student from the list shown here will be selected to be an extra in a new film. If everyone has an equal chance, what is the probability that Alexis will be selected?

1. Virginia
2. Roberto
3. Sam
4. Alice
5. Kent
6. Marcus
7. Jamie
8. Evan
9. Anne
10. Laticia
11. Bruce
12. Carol
13. Estelle
14. Felice
15. Alfred
16. Eric
17. Marie
18. Brendon
19. Alexis
20. Debbie
21. Katherine
22. Erica
23. Juan
24. Ainslie
25. William
26. Liz
27. Vance
28. Jimmy
29. David
30. Renaldo

Examples

To find the probability of a given event, use the formula:

$$\text{Probability of an event} = \frac{\text{Number of favorable outcomes}}{\text{Total number of possible outcomes}}$$

Outcomes with equal chances of occurring are called **equally likely outcomes**.

A Find the probability that Alexis will be selected.

There are 30 possible outcomes; 1 outcome is favorable.

The probability that Alexis will be selected is $\frac{1}{30}$.

B Find the probability of picking a Friday from the dates shown. (All outcomes are equally likely.)

4 favorable outcomes: 7, 14, 21, 28; 28 possible outcomes

The probability of picking a Friday is $\frac{4}{28}$, or $\frac{1}{7}$.

▶ Think and Discuss

1. You are picking a whole number from 0 to 30 out of a jar. You want to know the probability that the number picked is divisible by 7. List the favorable outcomes.

2. Find the probability in Question 1.

3. List the favorable outcomes of rolling a number less than 3 with a number cube.

4. Refer to the introduction to this lesson. What is the probability that Alexis will be selected? Liz is a friend of Alexis's. Find the probability that Alexis or Liz will be chosen.

Exercises

Find the probability of choosing each of the following from the list shown at the right. (See Example A.)

5. a freshman

6. a name that begins with A

7. a girl

Find the probability of choosing each of the following from the dates shown. (All outcomes are equally likely.) (See Example B.)

8. a Thursday

9. a date that has a 5 in it

10. a Monday

11. a date that ends in 0

A P R I L						
S	M	T	W	T	F	S
		1	2	3	4	5
6	7 OPENING DAY	8	9	10	11	12
13	14	15	16	17	18	19
20	21	22	23	24	25	26
27	28	29	30			

▶ **Mixed Practice** (For more practice, see page 441.)

Find the probability of choosing each of the following from the zip code list shown here. (All outcomes are equally likely.)

60922
60901
60106
92753
93421

12. 60922

13. a zip code with a digit of 6

14. a zip code containing an even digit

15. a zip code with a last digit of 1

16. a zip code with a 9 in it

17. a zip code that is divisible by 3

▶ **Applications**

18. Mark has 6 pairs of white socks, 3 pairs of black socks, and 1 pair of red socks in his drawer. If he closes his eyes, what is the probability that Mark will pick the red pair?

19. The weather forecast predicts a forty percent chance of rain. What is the probability of rain?

▶ **Review** (Lesson 11.1)

Write each as a ratio in 3 different ways.

20. 1 prize: 10 people

21. 5 boys: 8 girls

22. 2 vases: 12 roses

13.3 Computing Odds

John felt pretty good after he learned about probability in mathematics class. "What are my chances of getting this job?" he asked on an interview at the department store that afternoon. "I'd say the odds are 1 to 3 that you will be hired," the manager replied.

John expected a probability and got odds instead. The two ideas are related.

Examples

To find odds, use the appropriate formula.

Odds of an event happening =

$$\frac{\text{Number of favorable outcomes}}{\text{Number of unfavorable outcomes}}$$

Odds against an event =

$$\frac{\text{Number of unfavorable outcomes}}{\text{Number of favorable outcomes}}$$

Odds can be stated in different ways: $\frac{1}{3}$, 1 : 3, and 1 to 3.

A **Find the odds of a license plate starting with a number rather than a letter. (Assume all outcomes are equally likely.)**

The number of favorable outcomes is 10 (digits 0 to 9).
The number of unfavorable outcomes is 26 (letters A to Z).
The odds of a license plate starting with a number are 10 to 26, or **5 to 13**.

B **Find the odds against being elected class president if 4 other people are running. (All outcomes are equally likely.)**

The number of unfavorable outcomes is 4.
The number of favorable outcomes is 1.
The odds against being elected president are **4 to 1**.

▶ Think and Discuss

1. What are two other ways of writing the odds $\frac{6}{7}$?

2. What are the odds against rolling a 2 on a number cube?

3. What are the odds of rolling a number less than 5 on a number cube?

4. Refer to the introduction to this lesson. What was the probability of John getting the job?

5. What do you get if you multiply the odds of an event occurring by the odds that the event will not occur?

Exercises

Sue Barr
Mark Foreman
Judy Steinberg
Bill Wilson
Amanda Myer
Melissa Jones
Jackson Smith

Find the odds of drawing each of the following from this list of names. (See Example A.)

6. a woman's name   7. a man's name   8. Sue Barr's name

9. a name with the initials JS   10. a name that begins with A

What are the odds against pulling each of the following from the pile shown here? (See Example B.)

11. a red sweat shirt

12. a number less than 20

13. a white sweat shirt

14. a number with a 5 in it

▶ Mixed Practice (For more practice, see page 441.)

A new company has telephone extensions 1 through 20. The numbers are assigned randomly. Find the following odds.

15. of having extension number 15

16. against having an even extension

17. of having an extension number greater than 10

18. against having an extension that is a multiple of 5

▶ Applications

19. Nine out of ten dentists recommend Nofuss floss. What are the odds that your dentist will recommend this floss?

20. The names of the 50 states are placed in a bag. What are the odds of choosing 1 of the 19 states that border the Pacific Ocean or the Atlantic Ocean?

▶ Review (Lesson 5.6)

Write in lowest terms.

21. $\frac{5}{15}$ 22. $\frac{16}{20}$ 23. $\frac{35}{60}$ 24. $\frac{12}{48}$ 25. $\frac{56}{90}$ 26. $\frac{78}{100}$ 27. $\frac{225}{375}$

13.4 Relating Odds to Probability

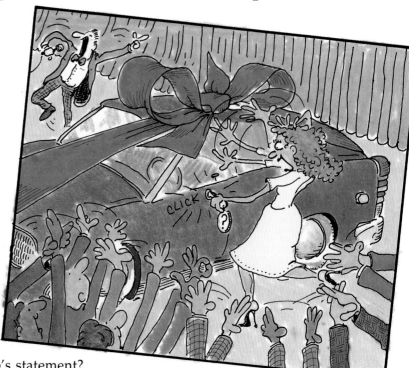

"*The key in your hand starts one of the five cars in front of you*," Skip Basil, the game show host, told the nervous contestant. "*That makes the odds 1 to 5 that you will be driving home in luxury tonight.*"

What is wrong with Skip's statement?

People often confuse probability and odds. Examine the situation of the contestant and you can see how odds and probability are related. On the game show, there are five cars in all. The key will start one car, but won't start the other four cars. So,

$$\text{Probability of starting} = \tfrac{1}{5}$$
$$\text{Odds of starting} = 1 \text{ to } 4$$
$$\text{Odds against starting} = 4 \text{ to } 1$$

1. Compare the answers given above for probability and odds. How are they alike? How are they different?

If you know the probability of an event occurring, you can find the odds of and against the event occurring. For example, the probability of the contestant starting the car is $\tfrac{1}{5}$.

Probability		**Odds Of**	**Odds Against**
$\dfrac{1}{5}$	Numerator remains the same. ⟶ Subtract numerator from denominator. ⟶	$\dfrac{1}{4}$	$\dfrac{4}{1}$

Reverse.

If you know the odds of (or against) an event occurring, you can find the probability of the event.

Odds Against   **Odds Of**   **Probability**

$\frac{4}{1}$ — Reverse. → $\frac{1}{4}$ — Numerator stays the same. / Add numerator and denominator. → $\frac{1}{5}$

2. Find the probability of the contestant starting the car if the key could start 2 cars out of 6.

3. Find the odds of the contestant starting the car if the key could start 2 cars out of 6.

4. Find the odds against the contestant starting the car if the key could start 2 cars out of 6.

5. According to a poll, Cleaver is considered a 3-to-2 favorite to win the election. Does the ratio $\frac{3}{2}$ represent the probability of Cleaver winning, the odds of Cleaver winning, or the odds against Cleaver winning?

6. A computer selects 10 names to receive prize notices out of 1000 different names on its list. Your name is on the list. What is the probability that you will receive a notice? What are the odds of receiving a notice? What are the odds against receiving a notice?

7. Tricia claimed, "The odds are 100 to 1 against me winning the contest." If her statement is correct, find the probability that she will win the contest.

8. The weather announcer states that there is a 40% chance of rain tomorrow. Find the probability that it will rain, the odds that it will rain, and the odds that it will not rain.

People use phrases involving probability loosely. Often they confuse probability and odds. State what is wrong with each statement below and rewrite each correctly.

9. Madelyn has a 50–50 chance.

10. The odds are 1 in a million.

11. The probability of winning is 2 to 1.

► **Review** (Lesson 6.5)

Estimate each product.

12. $\frac{8}{17} \times 20$ 13. $\frac{19}{21} \times 4\frac{1}{3}$ 14. $\frac{4}{15} \times 28$ 15. $1\frac{17}{19} \times 6\frac{2}{29}$

13.5 Using a Tree Diagram

Tyrone was planning his classes for the following school year. He could not decide between speech and literature for his English requirement, and among art, music, and typing for one elective. How many different pairs of classes could he choose?

A situation involving more than one decision or set of outcomes is called a **multiple event**.

Examples

To show all outcomes of a multiple event, draw a tree diagram.

A Draw a tree diagram that shows Tyrone's different choices.

Tree Diagram

English Requirement	Electives	Class Choices (Outcomes)
speech	art	speech and art
	music	speech and music
	typing	speech and typing
literature	art	literature and art
	music	literature and music
	typing	literature and typing

Tyrone has **6** possible pairs of classes.

B Find the probability of two heads, if a coin is tossed twice.

Draw a tree diagram.

1st Toss	2nd Toss	Outcomes	
heads	heads	heads	heads
	tails	heads	tails
tails	heads	tails	heads
	tails	tails	tails

There are 4 possible outcomes. One outcome is favorable.
The probability of tossing two heads is $\frac{1}{4}$.

▶ **Think and Discuss**

1. Find the probability of 6 and heads, if a number cube is rolled and a coin is tossed. Draw a tree diagram.

2. Refer to the introduction to this lesson. What are Tyrone's possible choices?

3. Where do you think the tree diagram got its name?

Exercises

Draw a tree diagram listing all possible outcomes for each of the following. (See Example A.)

4. rolling a number cube twice

5. tossing a coin three times

6. making a ham, cheese, or tuna sandwich with white, whole wheat, or rye bread

7. choosing a boy and a girl from Ann, Betty, Cathy, David, Ed, and Fred

If these two sets of letters are placed face down, what is the probability of drawing the following? (See Example B.)

8. a T and an A   9. a T and an M   10. an E and an A

▶ **Mixed Practice** (For more practice, see page 442.)

Draw a tree diagram and find the probability of the underlined event.

11. a coin is tossed and a number cube is rolled; <u>heads and 3</u>

12. a number cube is rolled twice; <u>1 and 5</u>, in that order

▶ **Applications**

13. Draw a tree diagram of all possible outcomes if you have a choice of ordering a pepperoni, sausage, or vegetarian pizza with a thin, thick, or deep-dish crust. How many possible pizzas are there?

14. In the 1976 World Series, the Reds beat the Yankees in four games. What are some of the other ways those four games could have ended? Draw a tree diagram to show how the four games could have ended.

▶ **Review** (Lessons 6.1, 6.2, 6.4)

Multiply. Write the answers in lowest terms.

15. $\frac{3}{5} \times \frac{1}{2}$ 16. $1\frac{1}{2} \times \frac{4}{9}$ 17. $\frac{7}{8} \times \frac{6}{7}$ 18. $2\frac{2}{3} \times 1\frac{1}{5}$ 19. $\frac{5}{15} \times \frac{18}{21}$

13.6 Using the Counting Principle

Patty and Paula are going to dinner and a movie. If they have 5 movies and 4 restaurants to choose from, how many different combinations are possible? You could use a tree diagram to find out, but there's a faster method.

Examples

To find the **number of possible outcomes of a multiple event,** use the Counting Principle, as shown below.

$$\text{Total number of possible outcomes} = \text{Number of outcomes for first event} \times \text{Number of outcomes for second event}$$

To find the **probability of a multiple event,** multiply the probabilities of each event together.

A How many possible outcomes exist for Patty and Paula's evening?

First event: 5 choices   Second event: 4 choices
There are $5 \times 4 = $ **20** possible outcomes.

B A number cube is rolled. Then a coin is tossed. Find the probability of getting a 5 and tails.

Probability of rolling a 5: $\frac{1}{6}$ Probability of tossing tails: $\frac{1}{2}$

The probability of getting a 5 and tails is $\frac{1}{6} \times \frac{1}{2} = \frac{1}{12}$.

▶ Think and Discuss

1. How many outcomes are possible if you spin a spinner with 4 different sections and roll a number cube?

2. Refer to the introduction to this lesson. How many different combinations of movies and restaurants are possible?

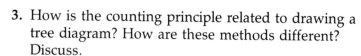

3. How is the counting principle related to drawing a tree diagram? How are these methods different? Discuss.

4. Which method do you prefer, the counting principle or a tree diagram? Why? Discuss.

Exercises

Find the number of possible outcomes for each of the following events. (See Example A.)

5. tossing a coin 3 times

6. rolling a number cube 4 times

7. rolling a number cube once and tossing a coin twice

8. tossing a coin twice and rolling a number cube 3 times

If a number cube is rolled twice, what is the probability of rolling each of the following? (See Example B.)

9. two 4s

10. two odd numbers

11. an even number and a 1, in that order

12. an even and an odd number, in that order

▶ Mixed Practice (For more practice, see page 442.)

Find the number of possible outcomes. What is the probability that a computer would randomly select each of the underlined events?

13. a city and mode of transportation are chosen from 25 cities and bus, plane, train, or car; New Orleans by plane

14. a blouse and skirt are chosen from 5 white blouses, 3 red blouses, 4 blue blouses, 2 black skirts, 3 gray skirts, and 2 white skirts; a red blouse and a white skirt

▶ Applications

15. Luke has five ties and three shirts. How many shirt and tie combinations can he wear?

16. Each day the cafeteria offers 2 soups, 4 kinds of sandwiches, 3 drinks, and 4 desserts. How many different meal combinations are possible?

▶ Review (Lessons 6.6, 6.7)

Divide. Write the answers in lowest terms.

17. $\frac{7}{8} \div \frac{3}{4}$

18. $2\frac{1}{5} \div \frac{2}{3}$

19. $\frac{3}{8} \div 12$

20. $4\frac{5}{6} \div 1\frac{5}{7}$

21. $\frac{5}{8} \div 2\frac{1}{3}$

13.7 Investigating Probability

At the start of each football game, the referee tosses a coin. The captain of the team that wins the toss can choose whether to kick or to receive.

The captain of Central High is unhappy. "I lost the toss again!" he complained. "That's 4 weeks in a row." "Relax," replied the water boy. "You're almost sure to guess right next week. The laws of probability are on your side. You can't possibly guess wrong 5 times in a row!"

1. Do you agree or disagree with the water boy's statement? Why?

2. What is the probability that the captain will win the toss next week?

3. Suppose the captain called "heads" 4 weeks in a row and lost each time. What should he call next week? Explain.

A coin cannot "remember" whether you won or lost last time. You can win (or lose!) 20 times in a row, and your chance of winning on the next toss is still $\frac{1}{2}$.

Sometimes, however, what happened earlier *can* affect probabilities.

At the freshman class party, the names of all the students are written on separate cards. The cards are then placed in a hat. A card is drawn to see who wins first prize. Then a card is drawn to see who wins the second prize.

4. Could the same student win both prizes? Discuss.

5. What happens to the card with the first winner's name after it is drawn? Does this matter?

Forty freshmen attend the party.

6. What is the probability that Carlos will win first prize?

7. Suppose Carlos wins first prize. His card is put back in the hat. What is the probability that he will win second prize?

8. Suppose Carlos wins first prize, and the card with his name on it is *not* put back in the hat. Can he win both prizes? What is the probability of Carlos winning second prize?

9. Explain why the following statement is true: Your chance of winning both prizes depends on the rules of the drawing.

We have discussed both *dependent* and *independent* events. If the first prize winner's card is not replaced, your chance of winning the second prize depends on whether or not you won the first prize. On the other hand, your chance of winning a coin toss is always $\frac{1}{2}$. It is *independent* of whether you won last time.

Sometimes it is not clear whether events are dependent or independent.

Complete each statement.
Explain your answer. Discuss.

10. Clevon Dribble sinks an average of 70% of his free throws. He has succeeded in his last 10 tries. If he makes the next shot, his team wins. His chance of making that shot is ___.

11. The Central High football team has won 9 games and lost 3 games. They have played two games in the rain this year, losing both times. If it rains this Saturday, their chance of winning is ___.

▶ **Review** (Lesson 5.7)

Write as a decimal.

12. $\frac{1}{9}$ **13.** $\frac{1}{11}$ **14.** $\frac{5}{6}$ **15.** $\frac{7}{9}$ **16.** $\frac{6}{11}$ **17.** $\frac{11}{12}$

18. $4\frac{3}{8}$ **19.** $7\frac{2}{3}$ **20.** $8\frac{3}{5}$ **21.** $9\frac{4}{11}$ **22.** $5\frac{1}{6}$ **23.** $2\frac{19}{100}$

Write as a fraction or mixed number in lowest terms.

24. 3.8   **25.** 9.04   **26.** 0.036   **27.** 8.125   **28.** 5.72   **29.** 1.555

Chapter 13 Review

A card is drawn from a set of 15 cards numbered 1–15. Find the probability of drawing each of the following. (Lesson 13.2)

1. 15   **2.** an even number   **3.** an odd number

A number cube is rolled. List the favorable outcomes for rolling each of the following. (Lesson 13.2)

4. an odd number   **5.** a number less than 3   **6.** a multiple of 4

A ticket is drawn from the set shown. Find the odds for and against drawing each of the following. (Lesson 13.3)

7. an orange ticket

8. a pink ticket

Draw a tree diagram of the following multiple event. (Lesson 13.5)

9. choosing a pair of shoes from 1 black pair, 1 brown pair, and 1 red pair, and a pair of socks from 1 white pair, 1 blue pair, and 1 green pair

Use the information from Question 9 to find the probability of each event. (Lesson 13.5)

10. red shoes with blue socks

11. white socks

Find the number of possible outcomes for each of the following events. (Lesson 13.6)

12. rolling a number cube and spinning a spinner with 4 sections

13. tossing a coin and drawing a card from a set of 12 cards numbered 1–12

A card is drawn from a set of 10 cards numbered 1–10 and a coin is tossed. Find the probability of each event. (Lesson 13.6)

14. 9 and heads   **15.** an odd number and tails   **16.** an even number and heads

Marbles like the ones shown are drawn from a container without replacement. Find the probability of drawing each of the following in the order listed. (Lesson 13.7)

17. red   red   **18.** green   blue   **19.** blue   blue   blue

A card is drawn from a set of 20 cards numbered 1–20. Find the odds for drawing each of the following.

1. 13   **2.** an even number   **3.** a multiple of 4

A number cube is rolled. Find the odds against rolling each of the following.

4. 3   **5.** an odd number   **6.** a number greater than 2

Draw a tree diagram of the following event.

7. choosing one boat from red, blue, green, and yellow boats, and one flag from white, purple, orange, and pink flags

Use the information above to find the probability of each event.

8. a red boat with an orange flag   **9.** a purple flag

Find the number of possible outcomes for each of the following.

10. tossing a coin and rolling a number cube

11. choosing an elective from 6 choices and a required course from 4 choices

A marble is picked from a set like the one shown. Find the probability of each event.

12. yellow marble   **13.** green marble   **14.** not a blue marble

A card is drawn from a set of 8 cards numbered 13–20. List the favorable outcomes for each event.

15. drawing a number less than 15

16. drawing an even number or a 17

A number cube is rolled and a letter is drawn from the letters P, E, S, T, and O. Find the probability of each of the following.

17. 6 and S   **18.** an odd number and E

Cards are drawn without replacement from a set of 10 cards numbered 1–10. Find the probability of drawing each of the following in the order listed.

19. 1   2   **20.** odd   10   **21.** even   odd   odd

▶ **Choose the letter that shows the correct answer.**

1. 8% of 175 is
   a. 140
   b. 14
   c. 1.4
   d. not given

2. 24 is what percent of 96?
   a. $33\frac{1}{3}\%$
   b. 300%
   c. 25%
   d. not given

3. 65% of 480 is
   a. 31,200
   b. 312
   c. 31.20
   d. not given

4. 348 is what percent of 116?
   a. 300%
   b. 30%
   c. $33\frac{1}{3}\%$
   d. not given

5. $\frac{5}{6} = \frac{30}{⬚}$
   a. 32
   b. 60
   c. 42
   d. not given

6. $\frac{⬚}{25} = \frac{10}{2}$
   a. 100
   b. 250
   c. 125
   d. not given

7. $\frac{15}{20} = \frac{⬚}{12}$
   a. 9
   b. 8
   c. 10
   d. not given

8. $\frac{8}{⬚} = \frac{56}{14}$
   a. 1
   b. 2
   c. 3
   d. not given

9. $37\frac{1}{2}\% = $ ⬚
   a. 0.375
   b. 3.75.
   c. 37.5
   d. not given

10. $0.005 = $ ⬚
   a. 50%
   b. 5%
   c. 500%
   d. not given

11. $0.08\overline{3} = $ ⬚
   a. $0.8\frac{1}{3}\%$
   b. 83%
   c. $8\frac{1}{3}\%$
   d. not given

12. $9.4 = $ ⬚
   a. 94%
   b. 0.094%
   c. 940%
   d. not given

▶ **Find the probability of choosing each of the following from a set of cards numbered 1–18.**

13. a 6
   a. $\frac{1}{18}$
   b. $\frac{1}{3}$
   c. $\frac{1}{6}$
   d. not given

14. an even number
   a. $\frac{1}{9}$
   b. $\frac{9}{1}$
   c. $\frac{1}{2}$
   d. not given

15. a multiple of 6
   a. $\frac{1}{3}$
   b. $\frac{1}{4}$
   c. $\frac{1}{9}$
   d. not given

16. a number less than 9
   a. $\frac{1}{6}$
   b. $\frac{1}{3}$
   c. $\frac{4}{9}$
   d. not given

17. What are the odds of rolling a 1, 3, or 7 on a 12-sided number cube?
   a. $\frac{1}{3}$
   b. $\frac{3}{1}$
   c. $\frac{1}{4}$
   d. not given

► **Use Jim's math scores for problems 18–21.**
83, 92, 91, 62, 83, 78, 83

18. The mode of the scores is
    a. 83   **b.** 78   **c.** 30   **d.** not given

19. The median score is
    a. 83   **b.** 78   **c.** 30   **d.** not given

20. The range of the scores is
    a. 83   **b.** 78   **c.** 30   **d.** not given

21. The mean score is
    a. 83   **b.** 78   **c.** 30   **d.** not given

► **Solve.**

22. A radio is on sale. The original price was $108. The new price is $75.60. What is the percent of decrease?
    a. 25%   **b.** 40%
    c. 30%   **d.** not given

23. The cost of dinner is $35.95. The sales tax is 6%. What is the total bill, to the nearest cent?
    a. $38.11   **b.** $57.52
    c. $33.79   **d.** not given

24. On a map scale 1 cm equals 65 km. The distance between two towns on the map is 5.5 cm. What is the actual distance?
    a. 11 km
    b. 325 km
    c. 357.5 km
    d. not given

25. Suki's room measures 15 feet by 13 feet. In a drawing she uses a scale of $\frac{1}{2}$ inch equals 1 foot. What is the size of her drawing?
    a. $7\frac{1}{2}$ inches by $6\frac{1}{2}$ inches
    b. 30 inches by 26 inches
    c. 10 inches by $4\frac{1}{3}$ inches
    d. not given

► **Compute.**

26. $3\frac{1}{5} \times \frac{5}{6}$
    a. $2\frac{1}{2}$
    b. $13\frac{1}{3}$
    c. $3\frac{21}{25}$
    d. not given

27. $6 - 2.094$
    a. 3.906
    b. 3.094
    c. 4.906
    d. not given

28. $8.4 \div 0.06$
    a. 1.4
    b. 14
    c. 140
    d. not given

29. $8\frac{3}{4} + 5\frac{2}{3}$
    a. $13\frac{5}{12}$
    b. $14\frac{5}{12}$
    c. $13\frac{2}{3}$
    d. not given

30. $1\frac{7}{8} \div 1\frac{3}{4}$

   a. $1\frac{1}{14}$

   b. $3\frac{9}{32}$

   c. $2\frac{1}{7}$

   d. not given

31. 8.14×2.6

   a. 211.64
   b. 2116.4
   c. 2.1164
   d. not given

32. $4587 \div 60$

   a. 760 R27
   b. 74 R47
   c. 76 R27
   d. not given

33. ___ are most effective for showing the parts of a whole.
   a. bar graphs   **b.** circle graphs   **c.** pictographs

34. Temperature trends from January through December are best displayed in a(n) ___ .
   a. line graph   **b.** pictograph   **c.** circle graph

35. The values of imports and exports for 7 countries are best displayed in a(n) ___ graph.
   a. line   **b.** double bar   **c.** circle

36. In a pictograph, a(n) ___ represents a quantity.
   a. symbol   **b.** vertical axis   **c.** tally

37. On a vertical bar graph showing the average price of gasoline in 5 cities, the label "Price in $" will be on the ___ .
   a. horizontal axis   **b.** vertical axis   **c.** not given

▶ **Solve.**

38. All ice skates are marked 25% off. The regular price is $85. What is the sale price?
   a. $21.25
   b. $68
   c. $63.75
   d. not given

39. In the last four games Dale scored 12, 19, 23, and 18 points. What was her average number of points per game?
   a. 18 points
   b. 72 points
   c. 13.5 points
   d. not given

40. Dionne works 15 hours a week and makes $4.85 an hour. How much does she make in 2 weeks?
   a. $72.75
   b. $145.50
   c. $68.55
   d. not given

41. Suki saved 40% on her new leotard, a savings of $9.80. What was the regular price?
   a. $24.50
   b. $14.70
   c. $34.30
   d. not given

TAKE 5

1 ▶ **Change For a Dollar**

How can you make $1.00 using 21 coins?

2 ▶ Letter Logic

Find the next letter in this pattern.

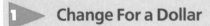

3 ▶ **An Improper Fraction**

Where do you often see $\frac{24}{31}$?

4 ▶ *Running Total*

Think of your favorite major league baseball team. What do you think will be greater next season, the sum of all the runs they score in their 162 games or the product?

5 ▶ Socks Match

A drawer contains 10 red socks and 10 blue socks. The room is dark and you can't see inside the drawer. How many socks must you choose before you are guaranteed to have a matching pair?

How Often?

• Once in a millennium
• Every other century
• At the sesquicentennial
• Semi-annually
• Twice a decade

"When we started our summer painting company, we had no idea that we needed to know so much about geometry. We were constantly figuring the area of different surfaces in order to estimate the amount of paint to buy."

Geometry: Perimeter and Area

San Francisco

Chicago

Los Angeles

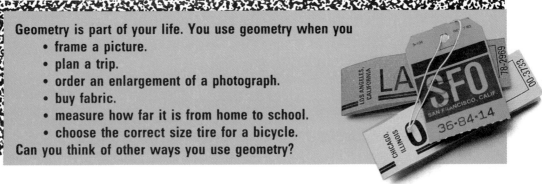

Geometry is part of your life. You use geometry when you
- frame a picture.
- plan a trip.
- order an enlargement of a photograph.
- buy fabric.
- measure how far it is from home to school.
- choose the correct size tire for a bicycle.

Can you think of other ways you use geometry?

Squaring Off

Materials: notebook paper, rulers, pencils

Place a blank sheet of notebook paper on your desk. Use a ruler to measure the length of each side of the paper in inches. Next, mark off one-inch intervals around the edges of the paper. Draw lines to connect the vertical and the horizontal markings on the paper to form a grid of squares, one inch on each side (**square inches**). Answer the following questions:

1. How many complete squares fit on your paper?
2. Suppose you could combine partial squares. Estimate how many squares (square inches) would be needed to cover your paper.
3. Place an object on your paper. Estimate how many squares (square inches) the object covers.

14.1 Introduction to 2-Dimensional Figures

Gary is studying for his driving test. He is learning what various road signs mean just by learning their shapes. Can you tell what some of these signs mean just by looking at their shapes?

These shapes and others like them are 2-dimensional figures. They have 2 dimensions—length and width. These figures are called **polygons**.

▶ Express Yourself

Here are some terms that will help you in this chapter.

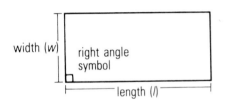 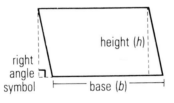 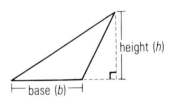

width (*w*) right angle symbol length (*l*)   right angle symbol height (*h*) base (*b*)   height (*h*) base (*b*)

To test for a right angle, place the corner of a card or book on the drawing.

Copy the figures below. Label the length and width and mark one angle as a right angle by using the symbol.

1.

2.

3.

4. Some polygons have names with a prefix that means the number of sides they have. For example, *tri-* in the word triangle means three. Look up the meanings of the following prefixes: *penta-*, *quad-*, *octa-*, *hexa-*, *deca-*.

Match each prefix in Question 4 to a figure below.

5. 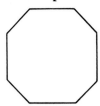   **6.**

7. 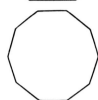   **8.**   **9.**

The sides of a **regular polygon** are all equal in length and its angles are all the same size.

Tell whether the figures below are regular polygons or not.

10.   **11.**

12. 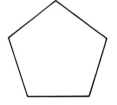   **13.**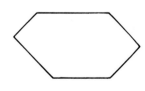

▶ Practice What You Know

You already know a lot about geometry because you know how to measure objects and how to add and multiply.

Add.

14. $15 + 29$ **15.** $20 + 30 + 50$ **16.** $2.3 + 3.1 + 8.5$

Multiply.

17. $21 \times 3 \times 9$ **18.** $16.3 \times 45.2 \times 3.1$ **19.** 125.8×333.3

In geometry you might see an expression such as $2 \times \pi \times r$. Another way to write this is $2\pi r$.

Rewrite each expression below.

20. $b \times h$ **21.** $l \times w$ **22.** $\pi \times d$ **23.** $2 \times l$

24. Would you use the shorthand described above to rewrite 2×3? Explain your answer.

14.2 Finding Perimeters of Polygons

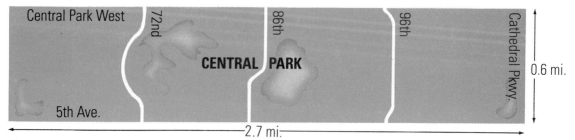

One Saturday Ed walked around the edge of Central Park. Ed could find how far he had walked by finding the distance around the park.

The distance around is called the **perimeter**.

Examples

To find the perimeter of a polygon, add the lengths of the sides.
To find the perimeter of a rectangle, use the formula
$P = 2l + 2w$.

P = perimeter, l = length, w = width

A **Find the perimeter of the triangle.**

Add the lengths.

$$\begin{array}{r} 4.5 \\ 3.5 \\ + \ 7.5 \\ \hline 15.5 \end{array}$$

The perimeter is **15.5 meters**.

4.5 m 7.5 m

3.5 m

B **Find the perimeter of the rectangle.**

12 ft.

8 ft. 8 ft.

12 ft.

$P = 2l + 2w$
$P = (2 \times l) + (2 \times w)$
$P = (2 \times 12) + (2 \times 8)$
$P = 24 + 16$
$P = 40$
The perimeter is **40 feet**.

▶ Think and Discuss

1. How many sides does a rectangle have?

2. What is the name of a polygon that has three sides?

3. What is the perimeter of a triangle with sides that measure 3 meters, 4 meters, and 5 meters?

4. Refer to the introduction to this lesson. How far did Ed walk?

5. How do you find the perimeter of a square? Write a formula for finding the perimeter of a square.

Exercises

Find the perimeter of each polygon. (See Example A.)

6. 4.6 cm, 5.5 cm, 8.7 cm, 8.1 cm

7.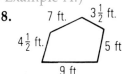
$4\frac{1}{2}$ in., $7\frac{1}{2}$ in., 6 in.

8. 7 ft., $3\frac{1}{2}$ ft., $4\frac{1}{2}$ ft., 5 ft., 9 ft.

Find the perimeter of each rectangle. (See Example B.)

9. $l = 5$ in.
$w = 12$ in.

10. $l = 175$ mm
$w = 225$ mm

11. $l = 22$ ft.
$w = 41$ ft.

12. $l = 3.6$ m
$w = 2.4$ m

13. $l = 10$ cm
$w = 15$ cm

14. $l = 5.5$ in.
$w = 20.5$ in.

15. $l = 12.1$ cm
$w = 16.9$ cm

16. $l = 33$ m
$w = 57$ m

▶ Mixed Practice (For more practice, see page 443.)

Find the perimeter of each polygon.

17. 2.8 cm, 6.2 cm

18. Rectangle
$l = 5.7$ m
$w = 3.4$ m

19. 2 m, 2 m, 5 m, 5 m, 5 m

20. 7 in., 7 in., 7 in., 7 in.

21. Rectangle
$l = 5.2$ cm
$w = 6$ cm

22. Rectangle
$l = 2\frac{1}{2}$ in.
$w = 1\frac{1}{4}$ in.

23. Rectangle
$l = 8.4$ m
$w = 9.5$ m

24. Rectangle
$l = 4\frac{2}{3}$ ft.
$w = 3\frac{1}{2}$ ft.

25. 6 ft., 6 ft., 6 ft., 6 ft., 6 ft., 6 ft.

26.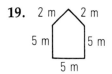
2.2 m, 5.9 m, 6.48 m

27. Rectangle
$l = 2$ cm
$w = 2.5$ cm

28. Rectangle
$l = 14$ in.
$w = 21$ in.

▶ Applications

29. A triangular-shaped garden has sides that measure $75\frac{3}{4}$ feet, $119\frac{1}{2}$ feet, and $98\frac{3}{4}$ feet. What is the perimeter of the garden?

30. A rectangular field is twenty feet wide and fifty-two feet long. How many times must you walk around the field to walk one mile?

▶ Review (Lesson 11.2)

Tell whether each statement is a proportion. Use = or ≠.

31. $\frac{6}{8}$ ⬚ $\frac{10}{12}$

32. $\frac{6}{20}$ ⬚ $\frac{3}{10}$

33. $\frac{4}{5}$ ⬚ $\frac{10}{8}$

34. $\frac{1}{5}$ ⬚ $\frac{8}{40}$

14.3 Estimating Lengths and Perimeters

In basketball, the free-throw line must be 15 feet from the basket. In baseball, the bases must be 90 feet apart. If you have played sports on an unmarked field, you know that being able to estimate lengths and perimeters is a useful skill.

1. Suppose you are playing each of the sports above. You have no ruler or other measuring tool with you. How would you decide where to place the free-throw line and the bases? Discuss and explain.

2. Suppose you have a 12-inch ruler. Now how would you decide where to place the free-throw line and the bases? Discuss and explain.

If you have a ruler, you can measure the length of your foot and the length of your stride.

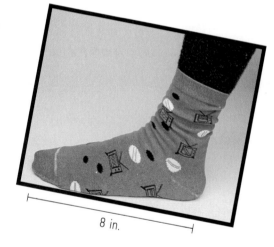

Then you can walk toe to heel to estimate short distances and walk in even strides to estimate long distances.

8 in.

24 in.

3. Bill's stride is 3 feet. How many strides from home plate will he place first base, if it should be 90 feet away?

4. Mary's stride is $2\frac{1}{2}$ feet. How many strides from home plate will she place third base?

5. Which method would you use to locate a free-throw line in basketball?

6. Where else might you use one of these methods to estimate lengths or perimeters? Discuss.

When a figure has an unusual shape, the problem of estimating its perimeter becomes more difficult. If a figure has curved sides, you cannot measure its perimeter with a ruler.

One way to estimate its perimeter is to first draw line
segments that approximate the figure. Then measure the
segments.

Draw line segments and measure
each. Add the measures.

5 cm + 3 cm + 7 cm = 15 cm

The perimeter of the curved figure is **about 15 centimeters**.

**Copy each figure below. Estimate the perimeters of each by
drawing and measuring line segments.**

7.   8.   9.   10.

Another method of estimating
perimeters is to use a piece of string
to trace the perimeter of the figure.
Then you can use a ruler to measure
the string.

Estimate the perimeter of each figure below using string.
(For more practice, see page 443.)

11.   12.   13.    14.

15. Estimate the perimeter of the top of an aluminum can, or
some other curved object in your classroom.

16. When would you use a tape measure?

▶ **Review** (Lessons 4.3, 4.6)

Choose the more reasonable measure.

17. The distance across the room is 9 ___. m   cm

18. Deanna and Crystal drank 1 ___ of orange juice.   mL L

19. Chloe's new earrings are 2 ___ long.   m cm

14.4 Finding Circumference

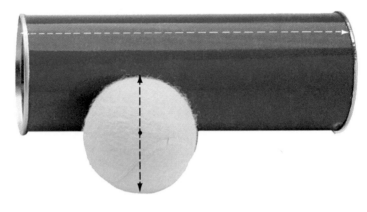

Which do you think is greater, the height of a can of three tennis balls or the distance around the can? You can answer this question without measuring if you understand circumference.

Circumference is the distance around a circle.

Examples

To find circumference, use the appropriate formula.

$C = \pi \times d$, or πd $C = 2 \times \pi \times r$, or $2\pi r$

C = circumference, d = diameter, r = radius

Use either 3.14 or $\frac{22}{7}$ for π,
or the π key on your calculator.

radius

diameter

A Find the circumference.
  $d = 6.4$ centimeters

  $C = \pi \times d$
  $C \approx 3.14 \times 6.4$
  $C \approx 20.096$
  $C \approx$ **20.1 centimeters**

B Find the circumference.
  $r = 9\frac{1}{4}$ millimeters

  $C = 2 \times \pi \times r$
  $C \approx 2 \times \frac{22}{7} \times 9\frac{1}{4}$
  $C \approx 58\frac{1}{7}$
  $C \approx 58\frac{1}{7}$ **millimeters**

▶ Think and Discuss

1. Write $\frac{22}{7}$ as a decimal. Which is greater, $\frac{22}{7}$ or 3.14?

2. Find the circumference of a circle with a radius of 4.5 m.

3. Refer to the introduction to this lesson. A tennis ball has circumference $\pi \times d$. The height of the can is about $3 \times d$. Which is greater? Are you surprised?

4. If you know the radius of a circle, how can you find its diameter?

5. If you know the circumference of a circle, how can you find its diameter?

6. Press the key marked π on a calculator. What does it display? When you use 3.14 for π, do you get an exact answer? Why?

7. When is it easier to use $\frac{22}{7}$ instead of 3.14 for π? When is it easier to use 3.14?

Exercises

Find each circumference. (See Example A.)

8. $d = 2.2$ cm   9. $d = 8$ in.   10. $d = 14$ m   11. $d = 4\frac{1}{2}$ ft.

Find each circumference. (See Example B.)

12. $r = 1.5$ m   13. $r = 5\frac{1}{4}$ in.   14. $r = 0.8$ km   15. $r = 2$ in.

▶ **Mixed Practice** (For more practice, see page 444.)

Find each circumference.

16.
3 m

17.
1.6 km

18.
$r = 7$ ft.

19.
$d = 10$ yd.

20. $r = 6$ in.   21. $d = 1$ m   22. $d = 9$ ft   23. $r = 1.9$ m

▶ **Applications**

24. The earth is about 93,000,000 miles from the sun and has a nearly circular orbit. Use a calculator to find how far the earth travels when it makes one orbit around the sun.

25. A bicycle wheel is 26 inches in diameter. How far forward will the bicycle move when the wheel makes one complete revolution?

▶ **Review** (Lessons 6.1, 6.2, 6.4)

Multiply. Write your answers in lowest terms.

26. $\frac{7}{11} \times \frac{6}{7}$ 27. $2\frac{1}{3} \times 5\frac{1}{4}$ 28. $\frac{1}{2} \times 6\frac{4}{5}$

14.5 Finding the Areas of Rectangles and Parallelograms

Kathi is a drummer for a band. She needs a place to practice where she won't disturb the neighbors, so she plans to cover an entire room with carpeting.

To find how much carpeting to buy, Kathi has to find areas.

Examples

To find the area of a parallelogram or a rectangle, use the appropriate formula.

parallel sides two sides that are an equal distance apart at all points

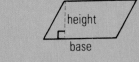

parallelogram a four-sided figure whose opposite sides are equal and parallel

$A = b \times h$, or bh
A = area, b = base, h = height

rectangle a parallelogram with four right angles

$A = l \times w$, or lw
A = area, l = length, w = width

A Find the area.

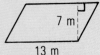

7 m

13 m

$A = b \times h$
$A = 13 \times 7$
$A = 91$

The area is **91 square meters.**

B Find the area.

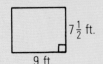

$7\frac{1}{2}$ ft.

9 ft.

$A = l \times w$
$A = 9 \times 7\frac{1}{2}$
$A = 67\frac{1}{2}$

The area is **$67\frac{1}{2}$ square feet.**

▶ Think and Discuss

1. Find the area of the parallelogram. $b = 15$ ft., $h = 5$ ft.

2. Refer to the introduction to this lesson. Kathi's drum room is 10 feet high, 10 feet wide, and 10 feet long. How much carpeting does she need to buy?

3. Each side of a square measures $5\frac{1}{2}$ yards. Find its area.

4. If a rectangle has an area of 24 square meters, can you find the lengths of the sides? Discuss.

Exercises

Find the area of each parallelogram. (See Example A.)

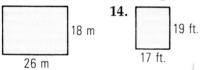

5. 11 in. — 24 in.

6. 9 ft. — 18 ft.

7. 2 m — 2.5 m

8. 12 cm — 4.5 cm

9. $b = 130$ m
   $h = 50$ m

10. $b = 25$ in.
    $h = 15$ in.

11. $b = 18$ yd.
    $h = 18$ yd.

12. $b = 0.13$ km
    $h = 0.28$ km

Find the area of each rectangle. (See Example B.)

13. 18 m — 26 m

14. 19 ft. — 17 ft.

15. 20 yd. — 125 yd.

16. 32 km — 25 km

17. $l = 12$ m
    $w = 7$ m

18. $l = 29$ in.
    $w = 36$ in.

19. $l = 20$ cm
    $w = 9$ cm

20. $l = 7.2$ km
    $w = 1.9$ km

▶ Mixed Practice (For more practice, see page 444.)

Find the area of each parallelogram or rectangle.

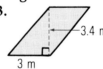
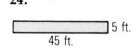

21. 5 m — 5 m

22. 20 in. — 15 in.

23. 3.4 m — 3 m

24. 5 ft. — 45 ft.

25. $b = 2.0$ cm
    $h = 1.6$ cm

26. $l = 1.5$ m
    $w = 3.2$ m

27. $b = 28$ in.
    $h = 15$ in.

28. $l = 62$ yd.
    $w = 31$ yd.

▶ Applications

29. A sheet of looseleaf paper is 11 inches long and $8\frac{1}{2}$ inches wide. Find its area.

30. A mile is 1760 yards long. An acre is 4840 square yards. How many acres are in a square mile?

▶ Review (Lessons 12.7, 12.8, 12.9)

Solve each percent problem.

31. What is 2% of 3150?

32. What percent of 60 is 3?

33. 45 is 15% of what number?

34. What is 87% of 500?

14.6 Finding the Area of a Triangle

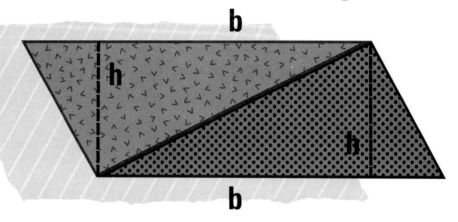

b

h

h

b

Draw a parallelogram like the one shown. Then cut the parallelogram so that two triangles are formed. By comparing the triangles, you can see that each has area half that of the original parallelogram.

Examples

To find the area of a triangle, use the formula
$A = \frac{1}{2} \times b \times h$, or $\frac{1}{2}bh$.
A = area, b = base, h = height

height

base

A Find the area of the triangle.

10 cm

⊢8 cm⊣

$A = \frac{1}{2} \times b \times h$

$A = \frac{1}{2} \times 8 \times 10$

$A = \frac{1}{2} \times 80 = 40$

The area is **40 square centimeters**.

B Find the area of the triangle.

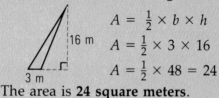

16 m

3 m

$A = \frac{1}{2} \times b \times h$

$A = \frac{1}{2} \times 3 \times 16$

$A = \frac{1}{2} \times 48 = 24$

The area is **24 square meters**.

▶ Think and Discuss

1. Write the formulas for the areas of a parallelogram and a triangle. How are they related?

2. The area of a triangle is 24 square inches. The base is 6 inches. What is its height?

3. How many bases does a triangle have? Explain.

4. What happens to the area of a triangle if both its base and its height are doubled?

Exercises

Find the area of each triangle. (See Example A.)

5. $b = 8.6$ cm
$h = 5.4$ cm

6. $b = 16$ ft.
$h = 12$ ft.

7. $b = 19$ m
$h = 9.5$ m

8. $b = 6$ in.
$h = 22$ in.

Find the area of each triangle. (See Example B.)

9.
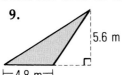
5.6 m
⊢4.8 m⊣

10.
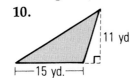
11 yd.
⊢15 yd.⊣

11.
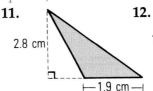
2.8 cm
⊢1.9 cm⊣

12.
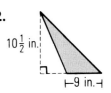
$10\frac{1}{2}$ in.
⊢9 in.⊣

▶ **Mixed Practice** (For more practice, see page 445.)

Find the area of each triangle.

13.
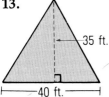
35 ft.
⊢40 ft.⊣

14.
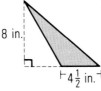
8 in.
⊢$4\frac{1}{2}$ in.⊣

15.
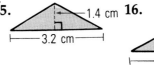
1.4 cm
⊢3.2 cm⊣

16.
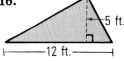
5 ft.
⊢12 ft.⊣

17. $b = 31$ m
$h = 25$ m

18. $b = 22$ in.
$h = 19$ in.

19. $b = 15$ ft.
$h = 24$ ft.

20. $b = 11$ mm
$h = 25$ mm

21. $b = 9.8$ cm
$h = 4$ cm

22. $b = 5\frac{1}{2}$ in.
$h = 7$ in.

23. $b = 7.2$ m
$h = 3.5$ m

24. $b = 6\frac{1}{2}$ in.
$h = 3\frac{3}{4}$ in.

▶ **Applications**

25. How many square inches of material are in the pennant below?

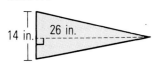
14 in. 26 in.

26. What is the area of the house wall shown below?

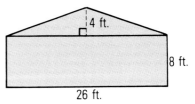
4 ft.
8 ft.
26 ft.

▶ **Review** (Lessons 1.9, 1.10, 1.11)

Add or subtract.

27. $\begin{array}{r} \$89.65 \\ + 37.87 \end{array}$

28. $\begin{array}{r} 5007 \\ - 2348 \end{array}$

29. $\begin{array}{r} 13 \\ - 6.91 \end{array}$

30. $\begin{array}{r} 26.75 \\ + 9.856 \end{array}$

14.7 Finding the Area of a Trapezoid

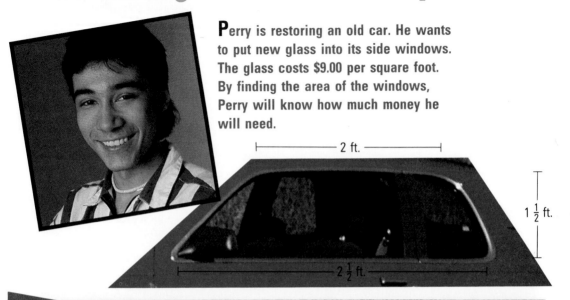

Perry is restoring an old car. He wants to put new glass into its side windows. The glass costs $9.00 per square foot. By finding the area of the windows, Perry will know how much money he will need.

Examples

To find the area of a trapezoid, use the formula.

trapezoid a four-sided figure that has exactly one pair of parallel sides

$A = \frac{1}{2} \times (a + b) \times h,$

or $\frac{1}{2}(a + b)h$

A = area, h = height, a = top base, b = bottom base

Find the area of the trapezoid.

$A = \frac{1}{2} \times (2 + 2\frac{1}{2}) \times 1\frac{1}{2}$

$= \frac{1}{2} \times (4\frac{1}{2} \times 1\frac{1}{2})$

$= \frac{1}{2} \times 6\frac{3}{4}$

$= 3\frac{3}{8}$

The area is $3\frac{3}{8}$ square feet.

▶ **Think and Discuss**

1. Solve. $\frac{1}{2} \times (31 + 47)$

2. Divide the trapezoid pictured to the right into 2 triangles. Can you think of another way to find its area?

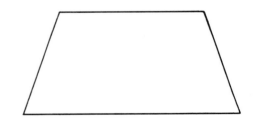

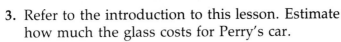

3. Refer to the introduction to this lesson. Estimate how much the glass costs for Perry's car.

4. Explain how the area of a trapezoid can be found using the average of its two bases.

5. How do you identify the bases of a trapezoid?

Exercises

Find the area of each trapezoid. (For more practice, see page 445.)

6.

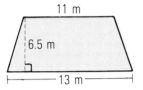

7.

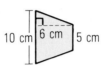

8.

9. $a = 1.6$ m
$b = 2.8$ m
$h = 1.4$ m

10. $a = 34$ in.
$b = 28$ in.
$h = 15$ in.

11. $a = 5$ cm
$b = 12$ cm
$h = 18$ cm

12.

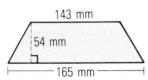

13.

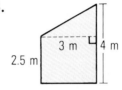

14.

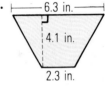

15. $a = 8$ in.
$b = 19$ in.
$h = 6$ in.

16. $a = 1.5$ m
$b = 2.5$ m
$h = 2.0$ m

17. $a = 3.0$ cm
$b = 7.1$ cm
$h = 4.2$ cm

▶ Applications

18. A trapezoid-shaped field measures 100 yards and 150 yards on its parallel sides and 200 yards across. What is the area of the field?

19. One quart of paint covers 60 square feet. How many gallons should a painter buy to cover the wall shown here?

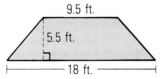

▶ Review (Lessons 6.1, 6.4, 6.6, 6.7)

Multiply or divide.

20. $4\frac{1}{2} \times 3\frac{5}{8}$

21. $8\frac{5}{12} \div 7\frac{2}{3}$

22. $\frac{1}{2} \times \frac{3}{8}$

23. $\frac{6}{3} \div \frac{2}{5}$

14.8 Finding the Area of a Circle

Theora Baines is submitting a bid to paint the landing circle for the Centerville Hospital helicopter port. The diameter of the circle is 30 feet. Before she can figure out how much paint she will need, Theora must find the area of the circle.

Examples

To find the area of a circle, use the formula $A = \pi \times r \times r$.
A = area, r = radius
Use either 3.14 or $\frac{22}{7}$ for π.

A Find the area of the circle.

$A = \pi \times r \times r$

$A \approx \frac{22}{7} \times 7\frac{1}{2} \times 7\frac{1}{2}$

$A \approx \frac{22}{7} \times \frac{225}{4}$

$A \approx 176\frac{11}{14}$

$A \approx 176\frac{11}{14}$ square inches

(circle labeled $7\frac{1}{2}$ in.)

B Find the area of the circle.

First find the radius. $r = \frac{1}{2} \times d$

$r = \frac{1}{2} \times 30 = 15$

$A = \pi \times r \times r$

$A \approx 3.14 \times 15 \times 15$

$A \approx 3.14 \times 225$

$A \approx 706.5$

$A \approx 706.5$ square feet

(circle labeled 30 ft)

▶ Think and Discuss

1. What is the area of a circle that has a radius of 10 inches?

2. What is the area of a circle that has a diameter of 14 meters?

3. If the radius of a circle is doubled, what happens to the area?

4. Refer to the introduction to this lesson. What is the area of the landing circle?

5. The circumference of a circle is 62.8 centimeters. Discuss ways to find its area. Then find it.

Exercises

Find the area. (See Example A.)

6. $r = 7$ m   7. $r = 12$ ft.   8. $r = 4.2$ cm   9. $r = 8\frac{1}{4}$ in.   10. $r = 5$ mi.

Find the area. Round answers to the nearest tenth.
(See Example B.)

11. $d = 13$ m   12. $d = 8$ mi.   13. $d = 42$ m   14. $d = 3$ km   15. $d = 5$ yd.

▶ Mixed Practice (For more practice, see page 446.)

Find the area. Round answers to the nearest tenth as needed.

16. $r = 3$ ft.   17. $r = 6.1$ m   18. $d = 36$ in.   19. $r = 10$ yd.   20. $r = 4\frac{1}{4}$ in.

21.
30 in.

22.
400 yd.

23.
$62\frac{1}{4}$ in.

24.
100 cm

25.
1 mi.

26.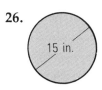
15 in.

27.
9.5 mi.

28.
5.2 m

29.
7.3 mm

30.
2 ft.

▶ Applications

31. A radio station has a broadcast radius of 55 miles. How large is the area that the station serves?

32. The inside of a circular sidewalk is 20 feet from the center of its circle. The outside is 24 feet away. Find the area of the sidewalk.

24 ft.
20 ft.

▶ Review (Lesson 3.9)

Simplify.

33. $3 \times 2 \times 7 + 15$ 34. $19 - 16 \div 4 \times 3$ 35. $6 \times 5 + 8 \times 8$

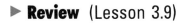

14.9 Analyzing Diagrams

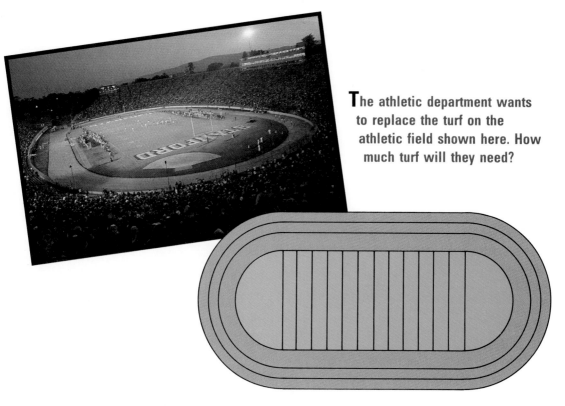

The athletic department wants to replace the turf on the athletic field shown here. How much turf will they need?

You can find the area of the entire field by adding the areas of each part. First, break the field into three parts: a rectangle and two semicircles.

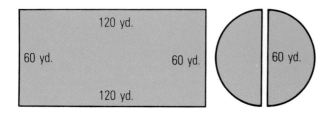

1. The two semicircles can be thought of as one circle, split into two pieces. What is the radius of the circle?

2. Find the area of the circle. Let $\pi = 3.14$.

3. What is the area of the rectangle?

4. What is the total area of the athletic field?

To solve area and perimeter problems based on diagrams, you may need to use several strategies, such as finding missing information and breaking the diagram into parts.

Here is the floor plan of a kitchen with parts cut out where built-in cabinets and appliances are located. Suppose you wanted to find the area needed for floor covering.

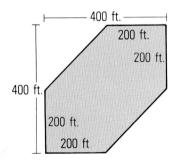

5. What is the total distance from the west (left) wall to the east (right) wall?

6. What is the total distance from the north (top) wall to the south (bottom) wall?

7. Find *a*.

8. Find *b*.

9. What is the amount of floor covering needed?

10. What is the area of the room's ceiling?

11. The figure to the right shows the floor of a living room with a semicircular window. What is the area of the floor?

12. At $0.30 a foot, what would it cost to buy molding to go around the entire room?

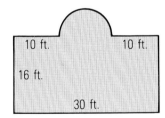

13. The figure to the left shows the dimensions of a plot of land. What is the total area of the plot of land?

14. If the land in Question 13 sells for $5000 an acre, about how much would this plot cost?
1 acre = 43,560 square feet

▶ **Review** (Lesson 11.5)

On a scale drawing, 1 inch = 12 feet. Find the following scale measurements.

15. $2\frac{1}{2}$ inches

16. $\frac{3}{4}$ inch

17. 1.25 inches

Complete each statement. (Lesson 14.1)

1. All sides of a(n) ___ polygon have the same length.

2. Perimeter means ___ .

Find the perimeters of each polygon. (Lesson 14.2)

3.

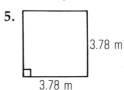

9 in. 9 in.
9 in. 9 in.
9 in. 9 in.

4. Triangle with sides 9.3 cm, 8.5 cm, 10.8 cm

5.

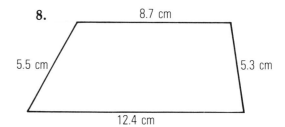

3.78 m
3.78 m

6. Rectangle with sides 6 ft. 5 in., and 9 ft. 8 in.

7.

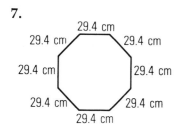

29.4 cm
29.4 cm   29.4 cm
29.4 cm   29.4 cm
29.4 cm   29.4 cm
29.4 cm
29.4 cm

8.

8.7 cm
5.5 cm   5.3 cm
12.4 cm

Find the circumference and area of each circle. (Lessons 14.4, 14.8)

9.

8 m

10.

6.4 cm

11. $r = 13$ m

Find the area of each figure. (Lessons 14.5, 14.6, 14.7)

12.

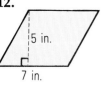

5 in.
7 in.

13.

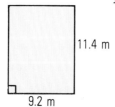

11.4 m
9.2 m

14.

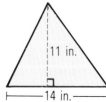

11 in.
14 in.

15.

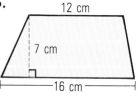

12 cm
7 cm
16 cm

Chapter 14 Test

Find the circumference and area of each circle. Round answers to the nearest hundredth.

1.

6.1 cm

2.

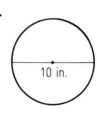

7 m

3.

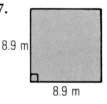

10 in.

Find the area of each figure.

4.

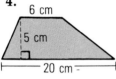

6 cm
5 cm
20 cm

5.

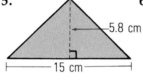

5.8 cm
15 cm

6.

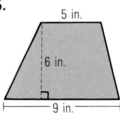

5 in.
6 in.
9 in.

7.

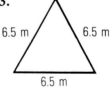

8.9 m
8.9 m

Complete each statement.

8. The formula for the _____ of a rectangle is $P = 2l + 2w$.

9. The distance around a circle is called the _____ .

10. The letter _____ represents the height in formulas for triangles, trapezoids, and parallelograms.

Find the perimeter of each polygon.

11.

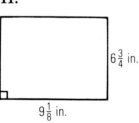

$6\frac{3}{4}$ in.
$9\frac{1}{8}$ in.

12.

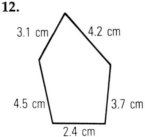

3.1 cm
4.2 cm
4.5 cm
3.7 cm
2.4 cm

13.

6.5 m
6.5 m
6.5 m

"When I started working at Camping Specialists, I wasn't really sure what it meant that a backpack had a capacity of 2270 cubic inches. Then I realized I could apply what I had learned in math class to things like camping."

Geometry: Surface Area and Volume

Many everyday tasks involve the use of geometry. Understanding surface area and volume is helpful when a person
- paints a house.
- fills a swimming pool.
- wallpapers a room.
- wraps a present.
- delivers heating oil to homes.
- tiles a bathroom.

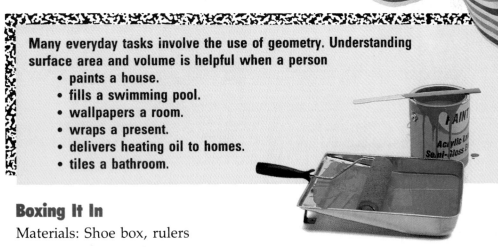

Boxing It In

Materials: Shoe box, rulers

Work with 2 or 3 classmates. Find the area of each side of a shoe box by measuring its dimensions.

1. What formula did you use to find each area?
2. Which areas are the same?
3. How many measurements must you find? 24? 12? 3? 1?
4. Find the sum of the areas. What unit is your answer in?

15.1 Introduction to 3-Dimensional Figures

Look at the objects on this page. How would you determine the amount of paper needed to cover each object or the amount of water each object could hold? You already know a lot about this question because you know how to work with polygons and circles, how to measure, and how to multiply.

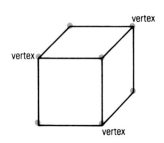

Figures like the ones above are called 3-dimensional figures. The 3 dimensions are length, width, and height.

1. Where do you see polygons and circles in the figures above?

2. Identify 2 objects that resemble each of the figures above.

▶ Express Yourself

Here are some terms that will help you in this chapter.

polyhedron   a 3-dimensional figure whose faces are polygons

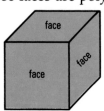

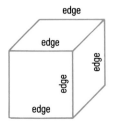

The polyhedrons shown above each have 6 **faces**, 12 **edges**, and 8 **vertices** (plural of vertex).

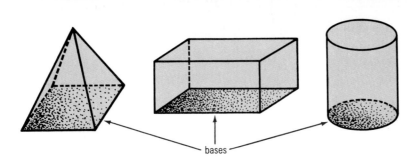

bases

surface area   the total area of the outside surface of a 3-dimensional figure

volume   a measure of the amount of space inside a 3-dimensional figure

cubic inches, cubic centimeters

units of volume

1 cm
1 cm
1 cm

▶ Practice What You Know

3. Look at the faces of the polyhedron shown to the right. What types of polygons are they?

4. Find the area of a rectangle 4 feet long and $3\frac{1}{4}$ feet wide. What formula did you use?

5. Find the area of a square with sides 5.6 centimeters long.

6. Find the area of a triangle whose base is 18 inches and whose height is $10\frac{1}{2}$ inches. What formula did you use?

7. Use the following steps to draw a cube.

| **a.** Draw a square. | **b.** Draw another square the same size. | **c.** Connect the corners of each square. | **d.** Erase lines that would not be seen. |

8. How many faces does a cube have? Edges? Vertices?

9. Refer to the first polyhedron on this page. Find the number of faces, edges, and vertices it has.

15.2 Constructing Polyhedron Models

Model building is an enjoyable way to learn more about polyhedrons.

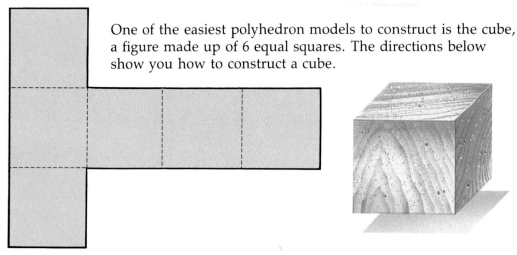

One of the easiest polyhedron models to construct is the cube, a figure made up of 6 equal squares. The directions below show you how to construct a cube.

Copy the figure above and cut out your drawing. Fold along dotted lines. Use tape to keep the edges together.

1. How many faces does a cube have? How many edges? How many vertices?

2. What shape is each face of a cube?

A T-shape is not the only shape you can use to construct a cube.

Copy and cut out each figure. Try to fold each into a cube. Which ones form a cube?

3.   **4.**   **5.**

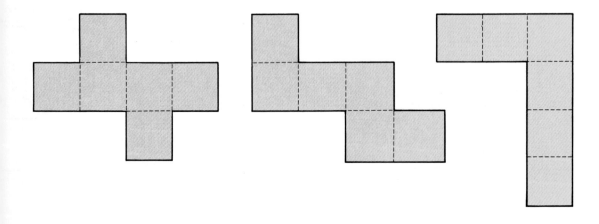

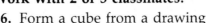

PROBLEM SOLVING

Work with 2 or 3 classmates.

6. Form a cube from a drawing like the one shown at the left.

7. Make three drawings of your own that can be folded into cubes.

8. Compare your drawings with those of your classmates. How many different drawings did your class come up with?

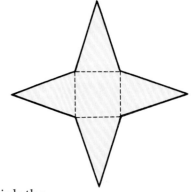

A different polyhedron can be built using the figure at the right as a model. Before constructing the polyhedron, answer the following questions.

9. What 2-dimensional figure makes up the base?

10. What will the polyhedron look like?

11. How many faces, edges, and vertices do you think the polyhedron will have?

12. Copy and cut out the figure above. Fold the figure along dotted lines. Use tape to join edges.

13. How many faces, edges, and vertices does your figure have?

▶ **Review** (Lesson 14.5)

Find the area of each figure.

14. Rectangle
   $l = 4.6$ cm
   $w = 3.3$ cm

15. Parallelogram
   $b = 5\frac{1}{4}$ in.
   $h = 2\frac{1}{2}$ in.

16. Square
   $s = 9.8$ cm

17. Rectangle
   $l = 6\frac{3}{8}$ in.
   $w = 5\frac{1}{2}$ in.

18.
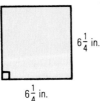
$6\frac{1}{4}$ in.

$6\frac{1}{4}$ in.

19.
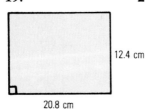
12.4 cm

20.8 cm

20.
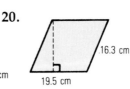
16.3 cm

19.5 cm

21.
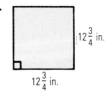
$12\frac{3}{4}$ in.

$12\frac{3}{4}$ in.

Geometry: Surface Area and Volume   **335**

15.3 Finding the Surface Areas of Cubes and Prisms

Meghan wants to make a terrarium like the one shown here. Which sheet of plexiglass will she need to buy? Meghan should find the surface area of the terrarium before making her decision.

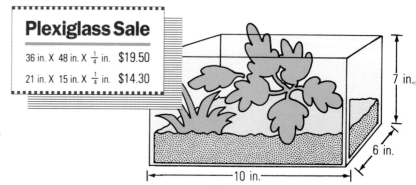

Plexiglass Sale

36 in. X 48 in. X $\frac{1}{4}$ in. $19.50

21 in. X 15 in. X $\frac{1}{4}$ in. $14.30

7 in.
6 in.
10 in.

Examples

To find the surface area of a cube or a prism, add the areas of its faces.

cube   a polyhedron with squares as faces
$A = 6(s \times s)$
A = surface area, s = length of edge

s = 9 cm

rectangular prism   a polyhedron with rectangles as faces
$A = 2(lw) + 2(wh) + 2(lh)$
A = surface area, l = length,
w = width, h = height

Note: A prism is named for its base. There are other types of prisms, but we will study only rectangular prisms.

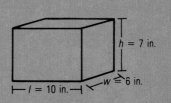

h = 7 in.
w = 6 in.
l = 10 in.

A Find the surface area of the cube above.

$A = 6(s \times s)$
$A = 6(9 \times 9)$ Substitute 9 cm for s.
$A = 6(81)$ Multiply.
$A = 486$
$A =$ **486 square centimeters**

B Find the surface area of the prism above.

$A = 2(lw) + 2(wh) + 2(lh)$
$A = 2(10 \times 6) + 2(6 \times 7)$
$\quad + 2(10 \times 7)$
$A = 120 + 84 + 140$
$A = 344$
$A =$ **344 square inches**

▶ Think and Discuss

1. Find the surface area of a cube with edges 6 feet long.

2. What is the formula for the area of each face of a cube?

3. If you know the surface area of a cube, can you find the area of each face? Explain.

4. Refer to the introduction to this lesson. Which sheet of plexiglass should Meghan buy?

Exercises

Find the surface area of each cube. (See Example A.)

5. $s = 12$ cm

6. $s = 7\frac{1}{2}$ in.

7. $s = 2.6$ cm

8. $s = 6\frac{1}{4}$ in.

Find the surface area of each prism. (See Example B.)

9. $l = 15$ cm
$w = 20$ cm
$h = 8$ cm

10. $l = 18$ cm
$w = 12$ cm
$h = 20$ cm

11. $l = 4.1$ cm
$w = 6.2$ cm
$h = 8.8$ cm

12. $l = 11$ in.
$w = 6$ in.
$h = 7$ in.

▶ Mixed Practice (For more practice, see page 446.)

Find the surface area of each figure.

13.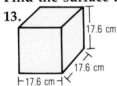
17.6 cm
17.6 cm
17.6 cm

14.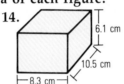
6.1 cm
10.5 cm
8.3 cm

15.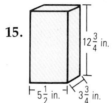
$12\frac{3}{4}$ in.
$5\frac{1}{2}$ in.
$3\frac{3}{4}$ in.

16.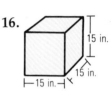
15 in.
15 in.
15 in.

17. Cube
$4\frac{1}{2}$-inch edges

18. Prism
$l = 9$ cm
$w = 7$ cm
$h = 3.15$ cm

19. Prism
$l = 3\frac{1}{2}$ in.
$w = 6$ in.
$h = 8\frac{1}{4}$ in.

20. Cube
$s = 5.7$ cm

▶ Applications

21. Each edge of Cube 1 is two inches long. Each edge of Cube 2 is twice as long. Is the surface area of Cube 2 twice that of Cube 1? Compute and show.

22. A company sells thinner in cans like the one shown. How many square inches of metal are needed to make each can (excluding the cap and spout)?

$9\frac{1}{2}$ in.
$6\frac{1}{2}$ in.
4 in.

▶ Review (Lessons 14.4, 14.6, 14.8)

Find the area of each triangle. Find the circumference and area of each circle.

23. Triangle
$b = 12$ in.
$h = 18$ in.

24. Circle
$r = 40$ cm

25. Triangle
$b = 13$ in.
$h = 7$ in.

26. Circle
$d = 6$ m

15.4 Finding the Surface Areas of Pyramids and Cylinders

How much cardboard is used to make each of the mailing tubes shown here?

$h = 18$ in.

$d = 3\frac{1}{2}$ in.

Examples

To find the surface area of a pyramid, add the areas of its faces. To find the surface area of a cylinder, add the areas of the side and bases.

square pyramid a polyhedron with a square base whose other faces are triangles. All the triangular faces meet at a vertex.

$A = (s \times s) + 4(\frac{1}{2}sh)$

A = surface area, s = side, h = height

$h = 5$ cm

$s = 6$ cm

$s = 6$ cm

cylinder a 3-dimensional figure with two bases that are equal circles

$A = 2(\pi \times r \times r) + (Ch)$

A = surface area, r = radius, C = circumference, h = height

Use 3.14 or $\frac{22}{7}$ for π.

$r = 1\frac{3}{4}$ in.

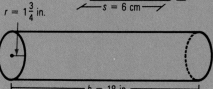

$h = 18$ in.

Note: A pyramid is named for its base. There are other types of pyramids, but we will study only square pyramids.

A Find the surface area of the pyramid above.

$A = (s \times s) + 4(\frac{1}{2}sh)$

$A = (6 \times 6) + 4(\frac{1}{2} \times 6 \times 5)$ Substitute values for s and h.

$A = 36 + 60 = 96$ Simplify.

$A = $ **96 square centimeters**

B Find the surface area of the cylinder above.

$A = 2(\pi \times r \times r) + (Ch)$ Use $C = 2\pi r$.

$A \approx 2(\frac{22}{7} \times \frac{7}{4} \times \frac{7}{4}) + (2 \times \frac{22}{7} \times \frac{7}{4} \times 18)$ Substitute values for r and h. Use $\frac{22}{7}$ for π.

$A \approx 19\frac{1}{4} + 198 = 217\frac{1}{4}$ Simplify.

$A \approx $ **$217\frac{1}{4}$ square inches**.

▶ Think and Discuss

1. Refer to the introduction to this lesson. How much cardboard was needed to make the first cylinder?

2. How many triangular faces does a square pyramid have? What is the formula for the area of each triangular face?

3. Explain where the formula $A = (s \times s) + 4(\frac{1}{2}sh)$ comes from.

4. A printed label covers all but the ends of a mailing tube. What shape is the label?

Exercises

Find the surface area of each pyramid. (See Example A.)

5. $s = 4$ in.
$h = 5$ in.

6. $s = 3$ cm
$h = 6$ cm

7. $s = 10$ in.
$h = 9$ in.

8. $s = 10.4$ cm
$h = 12.6$ cm

Find the surface area of each cylinder. (See Example B.)

9. $h = 5$ in.
$r = 5$ in.

10. $h = 15$ in.
$r = 10$ in.

11. $h = 25$ cm
$r = 7$ cm

12. $h = 17.8$ cm
$r = 6.3$ cm

▶ Mixed Practice (For more practice, see page 447.)

Find the surface area of each figure.

13. Cylinder
$h = 25$ in.
$r = 14$ in.

14. Pyramid
$s = 10$ cm
$h = 12$ cm

15. Cylinder
$r = 4$ in.
$h = 12\frac{1}{2}$ in.

16. Pyramid
$h = 6.6$ cm
$s = 4.2$ cm

▶ Applications

17. How many square feet of canvas are needed to make the tent shown here?

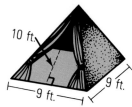

10 ft.

9 ft.

9 ft.

18. A can of tomatoes is 5 inches tall and 3 inches in diameter. How many square inches of paper (disregarding overlap) make up the label that covers the can?

▶ Review (Lesson 14.5)

Find the area of each figure.

19. Parallelogram
$b = 3$ m
$h = 1.6$ m

20. Rectangle
$l = 4.5$ cm
$w = 2.8$ cm

21. Rectangle
$l = 9\frac{1}{2}$ ft.
$w = 7$ ft.

22. Parallelogram
$b = 13$ cm
$h = 45$ cm

15.5 Finding the Volumes of Cubes, Prisms, and Pyramids

Do you think the larger cube shown has twice as much space (volume) inside as the smaller cube? Three times as much? More? The answer might surprise you.

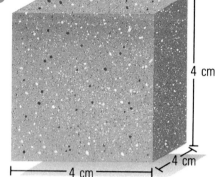

2 cm

2 cm

2 cm

4 cm

4 cm

4 cm

Examples

To find the volume of cubes, prisms, and pyramids, use the appropriate formula.

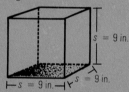

$s = 9$ in.

$s = 9$ in.

$s = 9$ in.

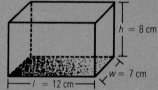

$h = 8$ cm

$w = 7$ cm

$l = 12$ cm

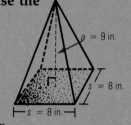

$a = 9$ in.

$s = 8$ in.

$s = 8$ in.

$V = s \times s \times s$
V = volume,
s = side

$V = Bh$
V = volume, h = height,
B = area of base (lw)

$V = \frac{1}{3} Ba$
V = volume, a = altitude,
B = area of base ($s \times s$)

A Find the volume of the cube above.

$V = s \times s \times s$
$V = 9 \times 9 \times 9$
$V = 729$
$V =$ **729 cubic inches**

B Find the volume of the prism above.

$V = Bh$
$V = lwh$
$V = 12 \times 7 \times 8$
$V = 672$
$V =$ **672 cubic centimeters**

C Find the volume of the pyramid above.

$V = \frac{1}{3} Ba$
$V = \frac{1}{3} \times s \times s \times a$
$V = \frac{1}{3}(8 \times 8 \times 9)$
$V = 192$
$V =$ **192 cubic inches**

▶ Think and Discuss

1. Refer to the introduction to this lesson. Find the volume of the two cubes. How many times bigger is the second cube?

2. Find the volume of a prism with $l = 6$ in., $w = 6$ in., and $h = 3$ in. Find the volume of a pyramid with the same dimensions.

---

3. Suppose you filled the prism in Question 2 with sand. How many pyramids from Question 2 could you fill with that sand? How does this relate to the formulas?

4. What is the formula for the volume of a cube? What is the formula for the area of a square? How are they related?

Exercises

Find the volume of each cube. (See Example A.)

5. $s = 10$ cm   6. $s = 9$ in.   7. $s = 16$ cm   8. $s = 5.6$ cm

Find the volume of each prism. (See Example B.)

9. $l = 7$ m
$w = 15$ m
$h = 6$ m

10. $l = 8$ cm
$w = 4$ cm
$h = 2$ cm

11. $l = 10$ ft.
$w = 16$ ft.
$h = 19$ ft.

12. $l = \frac{1}{4}$ in.
$w = \frac{3}{4}$ in.
$h = \frac{5}{8}$ in.

Find the volume of each pyramid. (See Example C.)

13. $s = 6$ in.
$a = 5$ in.

14. $s = 18$ cm
$a = 9$ cm

15. $s = 12$ in.
$a = 6$ in.

16. $s = 5.2$ m
$a = 3$ m

▶ **Mixed Practice** (For more practice, see page 447.)

Find the volume of each figure.

17.
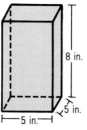
8 in.
5 in.
5 in.

18.
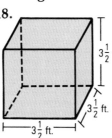
$3\frac{1}{2}$ ft.
$3\frac{1}{2}$ ft.
$3\frac{1}{2}$ ft.

19.
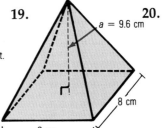
$a = 9.6$ cm
8 cm
8 cm

20.
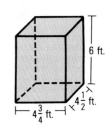
6 ft.
$4\frac{3}{4}$ ft.
$4\frac{1}{2}$ ft.

▶ **Applications**

21. The Great Pyramid of Egypt was originally 481 feet high. Its base was 756 feet on each side. Find the volume of the original pyramid. Use a calculator.

22. A driveway is ninety-six feet long and twelve feet wide. It is paved with concrete six inches thick. At a cost of $3.50 per cubic foot, what did the concrete cost?

▶ **Review** (Lesson 11.2)

Tell whether each statement is a proportion. Use = or ≠.

23. $\frac{1}{3}$ ※ $\frac{15}{45}$ 24. $\frac{4}{70}$ ※ $\frac{12}{15}$ 25. $\frac{12}{20}$ ※ $\frac{30}{50}$ 26. $\frac{8}{9}$ ※ $\frac{72}{80}$

15.6 Breaking a Problem into Parts

A pet store is holding a contest. A fish tank is filled with marbles. The person who gives the closest guess of how many marbles are in the tank wins a platypus named Mark.

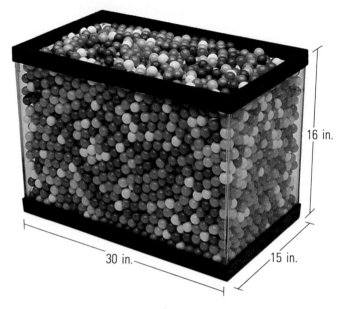

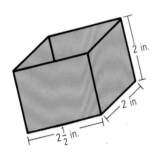

Sam and Brian took a very organized approach to solving this problem and solved the problem in two parts.

First Part   They took a smaller box, and they were able to fit 146 marbles in the smaller box.

Second Part   They determined the volume of the small box and the volume of the fish tank.

1. What is the volume of the small box in cubic inches?

2. What is the volume of the fish tank in cubic inches?

3. About how many of the small boxes would fill the fish tank?

By putting the parts together, Sam and Brian estimated the number of marbles that would fit in the fish tank.

4. How did Sam and Brian use their answers from Part 1 to estimate the number of marbles in the fish tank?

5. How many marbles did they estimate would fit in the fish tank?

6. Did Sam and Brian's method give an exact answer or just a good guess? Explain.

The fish-tank problem was solved by breaking the problem into parts. The next problem shows other methods of breaking problems into parts.

The swimming pool in Stockton is shown here. Notice the irregular shape and sloping bottom of the pool. The village officials need to find out the volume of the pool to determine a water tax.

An official spoke to the man responsible for filling the pool each spring. He reported that the pump supplies water at 250 gallons per minute. It takes about 90 minutes to fill the pool.

7. Explain how you can estimate how many gallons of water the pool holds.

8. About how many gallons of water does the pool hold?

9. One gallon of water has a volume of 0.13 cubic feet. Estimate the volume of the pool.

10. The village collects a 10-cent water tax for each cubic foot of water used in pools. What is the tax for the water used in this pool?

Describe how you would break the following problems into parts to solve them.

11. Calculate the weight of 2000 nickels.

12. Find how much 500 pounds of pennies is worth.

13. Decide how many gallons of punch to have at a school dance.

14. Estimate how long it will take to drive 1000 miles.

15. Estimate how many gallons of gas a car would use in a year.

▶ **Review** (Lesson 3.8)

Write which operation you would use to solve each problem. Then solve.

16. A drawing of Laura's bedroom is shown at the right. How many square feet of carpeting does she need to buy?

17. One type of carpet sells for $18.69 a square yard. Another type is on sale for one-third that price. How much does the sale carpet cost per square yard?

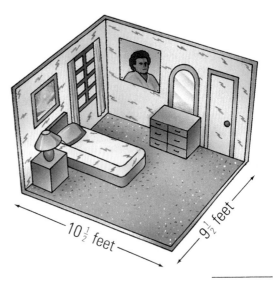

$10\frac{1}{2}$ feet

$9\frac{1}{2}$ feet

15.7 Finding the Volumes of Cylinders and Cones

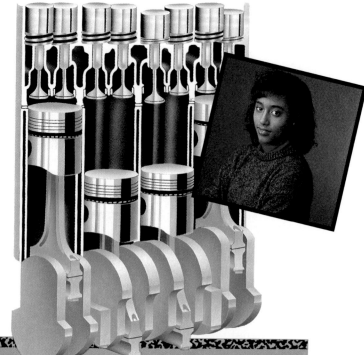

Tanika's car has a 4-cylinder, 1600-cubic-centimeter engine. That means that each cylinder has a displacement of 400 cubic centimeters (for a total of 1600 cubic centimeters). Displacement is the term used for the volume of a cylinder in an automobile.

Examples

To find the volumes of cylinders and cones, use the appropriate formula.

$V = Bh$
V = volume
B = area of the base
$(\pi \times r \times r)$
h = height

$h = 9$ cm

$r = 4$ cm

cone   a 3-dimensional figure with a circular base connected to a vertex

$V = \frac{1}{3}Ba$
$B = \pi \times r \times r$
a = altitude

$a = 21$ cm

$r = 10$ cm

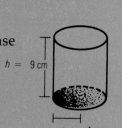

A Find the volume of the cylinder.

$V = Bh$
$V = \pi \times r \times r \times h$
$V \approx (3.14 \times 4 \times 4) \times 9$
$V \approx 50.24 \times 9$
$V \approx$ **452.16 cubic centimeters**

B Find the volume of the cone.

$V = \frac{1}{3} \times Ba$
$V = \frac{1}{3} \times \pi \times r \times r \times a$
$V \approx \frac{1}{3} \times (\frac{22}{7} \times 10 \times 10) \times 21$
$V \approx \frac{1}{3} \times \frac{2200}{7} \times 21$
$V \approx$ **2200 cubic centimeters**

▶ Think and Discuss

1. Find the volume of a cylinder and the volume of a cone with the following dimensions. h (or a) = 8 ft. r = 6 ft.

2. How do the volumes compare in Question 1?

3. In talking about volume, someone said, "Cones are to cylinders as pyramids are to prisms." What do you think he was talking about? Discuss.

Exercises

Find the volume of each cylinder. (See Example A.)

4. r = 10 ft.
   h = 18 ft.

5. h = 8 in.
   r = 3 in.

6. r = 3 cm
   h = 16 cm

7. h = 7.4 m
   r = 6 m

Find the volume of each cone. (See Example B.)

8. r = 8 ft.
   a = 21 ft.

9. a = 12 cm
   d = 6 cm

10. r = 9 in.
    a = 18 in.

11. a = 16.8 m
    d = 6.4 m

▶ Mixed Practice (For more practice, see page 448.)

Find the volume of each figure.

12. Cone
    d = 6 cm
    a = 9 cm

13. Cylinder
    h = 18 ft.
    r = 10 ft.

14. Cylinder
    r = 6 cm
    h = 14 cm

15. Cone
    a = 5.5 m
    d = 3 m

16. Cylinder
    h = 6 cm
    r = 14 cm

17. Cone
    d = 7 in.
    a = 15 in.

18. Cylinder
    r = $8\frac{3}{4}$ in.
    h = 16 in.

19. Cone
    r = $4\frac{1}{5}$ ft.
    a = 10 ft.

▶ Applications

20. How much water can the drinking cup hold? Round your answer to the nearest cubic inch.

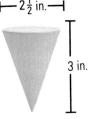
$\vdash$ $2\frac{1}{2}$ in. $\dashv$

3 in.

21. Each cylinder in an 8-cylinder engine has a 3-inch diameter and covers 4 inches (height). Find the engine's total displacement (volume). Round your final answer to the nearest cubic inch.

▶ Review (Lesson 3.9)

Simplify.

22. $6(8 + 3) + 24$

23. $9 + 20 \times 4 + 3 \times 3$

24. $5 \times 9 - 8 \times 3$

25. $15 - 6 \div 2 + 8 \times 5$

26. $49 \div 7 + 8 \div 4$

27. $9(6) - 3(5)$

15.8 Geometry: A Career Application

Architects are among the many people involved in the planning and building of skyscrapers. They often need to compute surface areas and volumes. Let's look at some of the factors an architect considers while working on the plans for a skyscraper.

One of the concerns of an architect is to assure that the building has a solid and safe foundation. The dimensions of pillars (and their volume) are an important factor when estimating costs of a foundation.

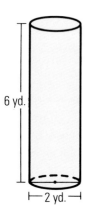

6 yd.

├─ 2 yd. ─┤

1. Find the number of cubic yards of concrete needed for each pillar shown.

2. At $50 per cubic yard of concrete, how much would 28 pillars cost?

If the building is made of brick, an architect works with a bricklayer.

3. Determine the volume and surface area of the brick shown.

4. One story of a building is 9 feet high and the bricks will be separated by $\frac{1}{2}$ inch of mortar. Determine how many rows of bricks are needed for one story.

$2\frac{1}{4}$ in.

$3\frac{3}{4}$ in.

8 in.

Another concern of architects is window space and location. Architects want to assure proper energy conservation.

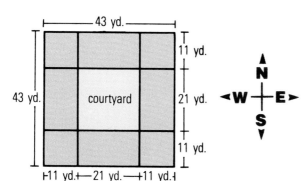

43 yd.

11 yd.

43 yd.

courtyard

21 yd.

11 yd.

11 yd. — 21 yd. — 11 yd.

N

W   E

S

5. A skyscraper is ten stories tall. Each story is 9 feet high. All outside walls are glass. Find the surface area of the glass walls for one story. What is the surface area of the glass walls for the entire skyscraper?

6. If glass is $7.40 a square yard, what would one story of glass walls cost?

7. The west, north, and east windows that do not face the courtyard need special glass for extra protection against the wind and extreme temperatures. How many square yards of this special glass must be ordered?

A building must also be properly insulated.

8. Fiberglass insulation is going to be used for a wall 90 feet wide and 9 feet high. If a 75-square-foot roll of insulation sells for $17.94, how much will it cost to insulate the wall?

▶ **Review (Lesson 2.11)**

Solve. Tell whether you used mental math, paper-and-pencil, or a calculator.

9. the cost of 5 dozen eggs at $0.79 a dozen

10. the total number of legs on 25 dogs

11. the average miles per gallon of gasoline used if you traveled 257.8 miles on 7 gallons of gas

Chapter 15 Review

Complete each statement. (Lessons 15.1, 15.2, 15.3, 15.4, 15.7)

1. A soda can is an example of a(n) ___ .

2. A cereal box is an example of a(n) ___ .

3. The ___ in Egypt are examples of 3-dimensional figures with a square bottom and triangular faces.

Find the surface area of each figure. (Lessons 15.3, 15.4)

4. Cube
   $s = 4$ in.

5. Prism
   $l = 3$ m
   $w = 2$ m
   $h = 7$ m

6.
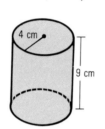

7.

8. Cylinder
   $d = 10$ in.
   $h = 15$ in.

9.
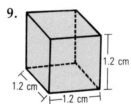

10. Pyramid
    $s = 20$ cm
    $h = 14$ cm

11.
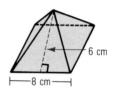

Find the volume of each figure. (Lessons 15.5, 15.7)

12.

13. Prism
    $l = 8$ ft.
    $w = 5$ ft.
    $h = 7$ ft.

14. Pyramid
    $s = 12$ in.
    $a = 15$ in.

15.
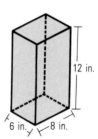

16.
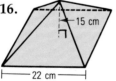

17. Cone
    $r = 7$ in.
    $a = 18$ in.

18.
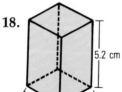

19. Cylinder
    $h = 22$ cm
    $d = 12$ cm

Chapter 15 Test

Find the volume of each figure.

1. Cylinder
  $r = 1.5$ m
  $h = 4$ m

2. Cube
  $s = 7$ in.

3.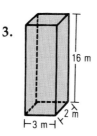
16 m
2 m
3 m

4.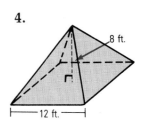
8 ft.
12 ft.

5.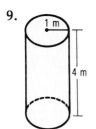
5.2 cm
5.2 cm
5.2 cm

6. Prism
  $l = 12$ in.
  $w = 10$ in.
  $h = 16$ in.

7. $r = 7$ cm
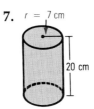
20 cm

8. Pyramid
  $s = 40$ in.
  $a = 52$ in.

Find the surface area of each figure.

9.
1 m
4 m

10. Cube
  $s = 9$ in.

11.
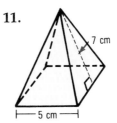
7 cm
5 cm

12. Prism
  $l = 6$ cm
  $w = 4$ cm
  $h = 2$ cm

13.
2
$2\frac{1}{2}$ in.
$2\frac{1}{2}$ in.

14. Cylinder
  $h = 22$ cm
  $r = 7$ cm

15.
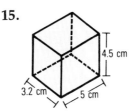
4.5 cm
3.2 cm
5 cm

16. Pyramid
  $s = 10$ in.
  $h = 12$ in.

Complete each statement.

17. A figure with 2 circular bases is called a(n) ____ .

18. A figure with 6 square faces is called a(n) ____ .

19. A figure with 4 triangular faces and 1 square base is called
  a(n) ____ .

20. A figure with 6 rectangular faces is called a(n) ____ .

"When my buddies and I go through our paces, we really get going fast. I never would have thought that math could help us figure out how our speed changes on different courses. Using formulas has given us a competitive edge."

Pre-Algebra: Expressions and Sentences

Algebra helps you see how information fits together through formulas. You can use algebra when you
- determine speeds.
- compute compound interest.
- convert currency.
- compute batting averages.

In all of these situations you could substitute numbers in algebraic phrases and sentences to see the whole picture.

Missing Numbers

Work with 2 or 3 classmates. Look at the following list of numbers: 15, 20, 8, 4, 10. Then follow the rules below.

1. Let a small square represent any one-digit number. Use this symbol to represent the same number in each expression.
2. Write each number from the list in four different ways.
3. Use a different operation for each expression.

For example, 15 can be written as

$$3 \times \square,$$
$$\square + \square + \square,$$
$$75 \div \square,$$
$$\text{and } 20 - \square,$$

where $\square$ stands for 5.

20 can be written as

$$5 \times \square,$$
$$\frac{80}{\square},$$
$$16 + \square,$$
$$\text{and } 24 - \square,$$

where $\square$ stands for 4.

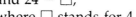

16.1 Introduction to Algebra

In the last two chapters, you used formulas like $C = 2\pi r$ and $A = lw$. These formulas are examples of the use of algebra in mathematics. Algebra takes words and sentences and replaces them with numbers and symbols. After that, you just apply the mathematics you already know.

▶ Express Yourself

Here are some of the terms used in algebra:

unknown or **variable**   any symbol, such as b or n, that may be replaced by a number or numbers

expression   a mathematical phrase, such as 6, $n + 2$, or $6b$

equation   a mathematical sentence stating that two expressions are equal, such as $6 = n + 2$

Choose one of the above words to complete each sentence.

1. The letter b found in $9b + 7$ is called a(n) ___※___ or a(n) ___※___ .

2. The phrase $b - 4$ is called a(n) ___※___ .

3. The sentence $14 + n = 6$ is called a(n) ___※___ .

▶ Practice What You Know

Throughout this book you have been answering questions based on situations that involve mathematics. Now you will ask mathematical questions based on information supplied. Take, for example, the statement "Sheila read a 485-page book in five days." A question might be "What was the average number of pages Sheila read each day?"

For each situation below, ask a mathematical question.

4. Patrick went out to lunch four times last week. The average price of the four lunches was $3.52.

5. Nora earns $75 a week. Her employer takes out $12.50 a week for taxes.

6. Tickets to a play cost $15 and $12. Miss Wu has $300 to take 24 students to the play.

7. The Morgans just sold their house for $184,000. When they bought the house in 1987, they paid $137,000.

8. Rob is a waiter. He earns $3.75 an hour and works 6 hours a day, 5 days a week. This week he made $175 in tips.

9. Marcus is making a costume for the school play. He needs $2\frac{1}{4}$ yards of velvet at $8.99 a yard and $1\frac{1}{2}$ yards of satin at $12.59 a yard.

10. You will need to use the words *increase* and *decrease* in this chapter. Use *increase* and *decrease* in a sentence. Write $m + 5$ in words, using *increase*. Write $m - 3$ in words, using *decrease*.

More than one mathematical operation may occur in an expression or equation, for example: $6 + 8 \times 2 = a$. To solve problems with more than one operation, recall the order of operations:

1. Do all operations within parentheses.
2. Do all multiplications and divisions from left to right.
3. Do all additions and subtractions from left to right.

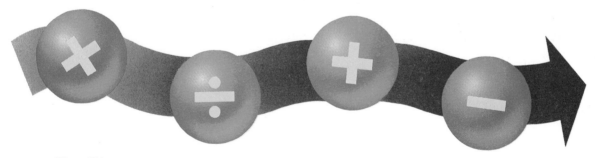

Simplify.

11. $3 + (6 \times 4)$ **12.** $15 \div 3 + 2$ **13.** $18 - 9 + 4$

14. $18 \div (6 + 3)$ **15.** $(5 + 4) + 6$ **16.** $12 \times 2 + 15$

17. $42 \div 3 + 4$ **18.** $14 + 6 - 3$ **19.** $4 \times (7 + 8) \div 2$

20. three plus seven times eight

21. eighteen divided by six, plus three times four

16.2 Writing Expressions

"**H**ow did you do selling ceramics last weekend?" James asked Rochelle. Rochelle smiled and replied, "My best day was Saturday, when I made fifteen dollars less than three times what I made on Sunday." "Hold on," said James, annoyed. "I've got to write an algebraic expression to understand what that means!"

Examples

To write an algebraic expression, first choose a variable if needed. Then decide what operations are involved.

A Write as an algebraic expression: forty-nine divided by seven.

$$49 \div 7 \text{ or } \frac{49}{7}$$

B Write as an algebraic expression: six less than some number. Let b be the variable.

some number $\rightarrow$ $b - 6$ $\leftarrow$ six less

C Write as an algebraic expression: fifteen more than three times some number. Let n be the variable.

three times some number $\rightarrow$ $3n + 15$ $\leftarrow$ fifteen more

D Write in words: $2n + 12$.

Twelve more than twice some number

▶ Think and Discuss

1. Write as an algebraic expression: twelve times twenty.

2. Refer to the introduction to this lesson. Write the amount that Rochelle made on Saturday as an algebraic expression.

3. Write $12b + 13$ in words.

4. What mathematical symbol is used for the term *times*?

5. What mathematical symbol is used for the term *less than*?

Exercises

Write as an algebraic expression. (See Example A.)

6. fifty-six divided by eight   7. seventy-two minus twelve

8. twenty-nine times five   9. forty plus eighteen

Write as an algebraic expression. (See Example B.)

10. seventeen minus some number   11. ninety-nine times some number

12. some number divided by six   13. the sum of five and some number

Write as an algebraic expression. (See Example B.)

14. four times some number, plus five   15. sixteen minus twice some number

16. forty-two divided by the product of three and some number

17. five less than five times some number

Write in words. (See Example D.)

18. $4 + 7b$ 19. $m \div 3$ 20. $17 - b$ 21. $36a + 7$ 22. $14t - 7$

▶ **Mixed Practice** (For more practice, see page 448.)

Write as an algebraic expression or in words.

23. thirty-six divided by three   24. twelve times twelve

25. seven decreased by some number   26. the sum of two and some number

27. eleven times some number   28. $4y - 7$

▶ **Applications**

29. The New York Yankees have played in 33 World Series. They have won 22. Write the number of times they have lost as an algebraic expression.

30. The coastline of Alaska is five hundred miles plus nine times the length of the California coastline. Let c equal the California coastline. Write the length of the Alaskan coast as an algebraic expression.

▶ **Review** (Lesson 5.2)

Rewrite as a whole number or mixed number.

31. $\frac{7}{7}$ 32. $\frac{21}{6}$ 33. $\frac{24}{4}$ 34. $\frac{29}{8}$ 35. $\frac{57}{9}$ 36. $\frac{24}{12}$

37. $\frac{16}{4}$ 38. $\frac{10}{10}$ 39. $\frac{18}{6}$ 40. $\frac{60}{12}$ 41. $\frac{81}{9}$ 42. $\frac{41}{5}$

16.3 Finding the Value of an Expression

Rebecca's teacher said that in Italy you get 1260 lire for every American dollar. Rebecca wrote the expression 1260d to remember this. Rebecca wondered what a $5 movie ticket would cost in lire.

Examples

To find the value of an expression, first substitute the value given for the variable, and then compute.

A Find the value of $a + 6$ when $a = 20$.

$a + 6 = 20 + 6$ ← Substitute the value
$a + 6 = \textbf{26}$ of a.

B Find the value of $3d - 5$ when $d = 4$.

$3d - 5 = (3 \times d) - 5$
$3d - 5 = (3 \times 4) - 5$
$3d - 5 = 12 - 5$
$3d - 5 = \textbf{7}$

▶ Think and Discuss

1. Find the value of $5c$ when $c = 6$.

2. Find the value of $\frac{36}{e} - e$ when $e = 6$.

3. Refer to the introduction to this lesson. What would a $5 movie ticket cost in lire?

4. Are there any values p cannot have in the expression $2p$? Explain.

Exercises

Find the value of each expression. (See Example A.)

5. $n + 9$ when $n = 7$

6. $23 - m$ when $m = 9$

7. $\frac{d}{7}$ when $d = 42$

8. $15 + k$ when $k = 19$

9. $17 - r$ when $r = 9$

10. $\frac{16}{b}$ when $b = 2$

Find the value of each expression. (See Example B.)

11. $12a - 6$ when $a = 11$

12. $9d + 22$ when $d = 6$

13. $4n - 7$ when $n = 5$

14. $11h \div 4$ when $h = 12$

15. $6b + 32$ when $b = 15$

16. $\frac{17n}{5}$ when $n = 10$

17. $18 + n - n$ when $n = 10$

18. $\frac{21}{n} - n$ when $n = 3$

▶ **Mixed Practice** (For more practice, see page 449.)

Find the value of each expression.

19. $\frac{f}{8}$ when $f = 16$

20. $d + \frac{15}{d}$ when $d = 5$

21. $32 - 2h$ when $h = 9$

22. $6f - 9$ when $f = 4$

23. $m + 4m$ when $m = 2$

24. $a + 19$ when $a = 1$

▶ **Applications**

25. Alice played in five more soccer games this month than last month. Write an algebraic expression to represent the number of games Alice played in this month. Find the value of the expression if Alice played in nine games last month.

26. Al sells shoes. He earns $45 a day plus $3 for each pair he sells. On Monday he sold the shoes shown here. Write an algebraic expression to represent his daily earnings. Then find its value for Monday.

▶ **Review** (Lesson 5.3)

Find 3 fractions equivalent to each fraction.

27. $\frac{3}{4}$

28. $\frac{2}{3}$

29. $\frac{1}{2}$

30. $\frac{4}{5}$

31. $\frac{1}{12}$

32. $\frac{7}{10}$

Use division to find a fraction equivalent to each fraction.

33. $\frac{12}{15}$

34. $\frac{15}{30}$

35. $\frac{30}{45}$

36. $\frac{65}{100}$

37. $\frac{21}{28}$

38. $\frac{33}{36}$

16.4 Writing Equations

A complicated sentence like "The distance traveled by a moving object is equal to its speed multiplied by its time in motion" is easier to understand when it is written as the equation $d = rt$. You might use this equation to figure out how far you can go in 2 hours at 35 miles per hour: $d = 35 \times 2$. In this lesson you'll be writing your own equations.

Examples

To write equations, first choose a letter for your variable. Then translate words into symbols.

A Write as an equation: Some number minus sixteen is forty-three.

$$e - 16 = 43$$

some number — minus sixteen — is forty-three

B Write as an equation: Nine plus the product of two and some number is fifty-nine.

$$9 + 2f = 59$$

nine plus — the product of two and some number — is fifty-nine

▶ Think and Discuss

1. Write as an equation: Some number plus twelve is eighteen.

2. Write as an equation: The product of four and some number is thirty-two.

3. Refer to the introduction to this lesson. If you were in a car traveling 55 miles per hour and you went 220 miles, what equation could you write to find how many hours you had traveled? What letter did you choose for your variable? Why?

4. What is the difference between an expression and an equation? Give an example of each.

Exercises

Write as an equation. (See Example A.)

5. Some number plus eight is thirteen.

6. Two times some number is fifteen.

7. Twelve divided by some number is four.

8. The difference between some number and one is eight.

Write as an equation. (See Example B.)

9. Two times some number, minus one, is thirty-seven.

10. Some number divided by five, plus nine, is sixteen.

11. Four times some number divided by two, plus two, is ten.

12. The sum of eight and some number times three is eleven.

13. Twelve divided by some number, plus two, is three.

▶ Mixed Practice (For more practice, see page 449.)

Write as an equation.

14. Some number divided by five is fifteen.

15. Three times some number, plus eight, is seventeen.

16. Eight less than some number is zero.

17. Some number divided by two is twenty-eight.

▶ Applications

18. In 1980, the population of Washington, D.C., was 638,432, or 118,236 less than its population in 1970. Let p represent the population in 1970. Write an equation using p to describe the population in 1980.

19. In 1988, Chicago Cub Andre Dawson was awarded a salary of $1.85 million. That's $450,000 more than twice his 1987 salary. Let s represent his 1987 salary. Write an equation using s to describe Dawson's 1988 salary.

▶ Review (Lessons 1.9, 1.10, 1.11)

Add or subtract.

20. $679 + 847$

21. $12.14 + 11.0$

22. $99.8 - 89.9$

23. $1381 - 1270$

24. $\begin{array}{r} 95.7 \\ -\ 50.83 \end{array}$

25. $\begin{array}{r} 6874 \\ +\ 5199 \end{array}$

26. $\begin{array}{r} 7.965 \\ +\ 6.037 \end{array}$

27. $\begin{array}{r} 5000 \\ -\ 2857 \end{array}$

16.5 Using Guess and Check to Solve Equations

1. $2 + a = 3$
2. $b + 5 = 10$
3. $4x = 24$
4. $y - 6 = 11$
5. $2c + 1 = 15$
6. $2x = 7$

On the quiz program "Name That Number," the contestants had one minute to solve the six equations shown above. To solve an equation means to find a number that can replace the variable and make the statement true.

1. Give yourself one minute. Do as many problems as you can.

2. What methods did you use?

Now you will solve equations like the ones shown above using Guess and Check. For each problem, follow the steps in the chart.

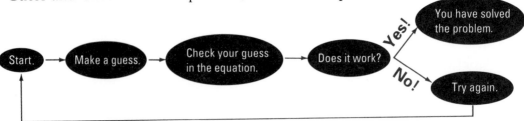

Solve the equation $c + 25 = 65$.

3. Guess 20 for c. Does 20 plus 25 equal 65?

4. Should your second guess be greater or less than 20? Why?

5. Guess another number and substitute it for c. Does your guess solve the equation? Continue until you have solved the equation.

6. What value of c solves the equation?

Exercises

Use Guess and Check to solve each equation.

7. $s + 25 = 51$ **8.** $42 - c = 33$ **9.** $v - 13 = 17$

10. $32 + m = 99$ **11.** $3y = 102$ **12.** $5d = 65$

13. $\frac{b}{10} = 10$ **14.** $\frac{b}{9} = 10$ **15.** $2d + 2 = 6$

16. $5c - 1 = 99$ **17.** $2a = 5$ **18.** $5 = \frac{25}{c}$

19. $17 = t + 4$ **20.** $22 = 44 - h$ **21.** $3z - 1 = 29$

22. $4 + c = 7$ **23.** $\frac{99}{z} = 3$ **24.** $7b + 1 = 50$

For the final round of the quiz program, the contestants were asked to write the following problems as equations and then solve each equation. Do the same.

25. Two added to a number equals twenty-two.

26. Three times a number equals ninety-six.

27. If you take 5 away from a number, the result is 13.

28. If you divide a number by six, the result is one.

29. If you multiply a number by 5 and then add 6 to the product, the result is 31.

30. If you multiply a number by 3 and then subtract 1 from the product, the result is 17.

31. Seven subtracted from a number is equal to forty-one.

32. If you divide a number by 3 and then subtract 1 from the quotient, the result is 2.

33. Now you have used Guess and Check to solve equations. Can you suggest any shortcuts to solve equations such as $2x = 16$ or $x + 8 = 36$? Explain.

▶ Review (Lesson 2.11)

Solve. Tell whether you used mental math, paper-and-pencil, or a calculator.

34. Four items cost $16.50 each and three items cost $5.85 each. There is a shipping charge of 5% of the total cost. Find the final cost of the order.

35. What is the change from $20 for 2 items at $3 each and 4 items at $1 each?

16.6 Solving Equations Using Addition or Subtraction

What number should be placed on the empty tray to balance the scale shown above? A balance scale is like an equation. If you add to or subtract from one side, you must do the same on the other side to keep both sides in balance.

Examples

To solve an equation using addition or subtraction, isolate the variable on one side.

A Solve. $m + 14 = 20$

$$m + 14 = 20$$
$$m + 14 - 14 = 20 - 14 \qquad \text{Subtract 14 from both sides.}$$
$$m + 0 = 6 \qquad\qquad\quad 14 - 14 = 0$$
$$m = 6$$
$$\text{Check: } 6 + 14 \stackrel{?}{=} 20 \qquad \text{Substitute 6 for } m \text{ in the original equation.}$$
$$20 = 20$$

B Solve. $p - 7 = 11$

$$p - 7 = 11$$
$$p - 7 + 7 = 11 + 7 \qquad \text{Add 7 to both sides.}$$
$$p + 0 = 18 \qquad\qquad\quad \text{Simplify.}$$
$$p = 18$$
$$\text{Check: } 18 - 7 \stackrel{?}{=} 11 \qquad \text{Substitute 18 for } p \text{ in the original equation.}$$
$$11 = 11$$

▶ Think and Discuss

1. Solve. $a + 11 = 21$

2. Solve. $d - 4 = 13$

3. One student solved the equation $g - 4 = 8$ and got the answer $g = 4$. Show that this answer is incorrect. How do you think the error was made? Explain.

4. Is it easier to solve $n + 103 = 267$ by the method in this lesson or by Guess and Check? Explain.

Exercises

Solve. (See Example A.)

5. $n + 8 = 14$ 6. $a + 15 = 24$ 7. $d + 19 = 36$

8. $b + 5 = 31$ 9. $c + 12 = 19$ 10. $f + 21 = 30$

11. $a + 7 = 34$ 12. $d + 25 = 26$ 13. $n + 3 = 26$

Solve. (See Example B.)

14. $p - 9 = 8$ 15. $f - 12 = 3$ 16. $b - 19 = 24$

17. $n - 2 = 21$ 18. $f - 27 = 48$ 19. $c - 7 = 0$

20. $d - 15 = 15$ 21. $a - 18 = 8$ 22. $n - 5 = 21$

▶ **Mixed Practice** (For more practice, see page 450.)

Solve.

23. $b - 21 = 19$ 24. $f + 9 = 11$ 25. $d + 16 = 23$

26. $f - 11 = 12$ 27. $p + 11 = 12$ 28. $c - 7 = 14$

29. $n + 15 = 22$ 30. $d - 8 = 29$ 31. $f - 19 = 1$

▶ **Applications**

32. After spending $9 for two movie tickets, Michael had $34 left. How much did he have before buying the tickets? Write an equation and then solve.

33. The diameter of Jupiter is 80,773 miles greater than the diameter of the Earth. Use the information in the photo to determine the diameter of the Earth. Write an equation and then solve.

88,700 miles

▶ **Review** (Lessons 2.4, 2.8, 3.5, 3.12)

Multiply or divide.

34. 49×72 35. $459 \div 27$ 36. 5×0.49 37. $391.4 \div 4.12$

38. $462 \div 17$ 39. 98×44.5 40. $682 \div 22$ 41. $1560 \div 24$

42. 1240×60 43. $40.2 \div 8.04$ 44. 76.12×11 45. $60.8 \div 3.8$

16.7 Solving Equations Using Multiplication or Division

Tenesha works for a veterinary hospital. She is in charge of feeding the animals. One Great Dane named Soren gets 24 ounces of food each day. If Tenesha feeds Soren twice a day, how many ounces must she weigh out for each meal?

You could write an equation to describe this problem.

Examples

To solve equations using multiplication or division, isolate the variable on one side of the equation.

A Solve. $2y = 24$

$2y = 24$

$\frac{2y}{2} = \frac{24}{2}$ Divide both sides of the equation by 2.

$1y = 12$ $\frac{2}{2} = 1$

$y = \mathbf{12}$

Check: $2 \times 12 \overset{?}{=} 24$ Substitute 12 for y in the original equation.

$ 24 = 24$

B Solve. $\frac{z}{3} = 13$

$\frac{z}{3} = 13$

$3 \times \frac{z}{3} = 3 \times 13$ Multiply both sides of the equation by 3.

$1z = 39$ $\frac{3}{3} = 1$

$z = \mathbf{39}$

Check: $\frac{39}{3} \overset{?}{=} 13$ Substitute 39 for z in the original equation.

$ 13 = 13$

▶ Think and Discuss

1. Solve. $7t = 21$

2. Solve. $\frac{a}{9} = 5$

3. Refer to the introduction to this lesson. How much food should Tenesha feed Soren at each meal?

4. Can you tell, without solving them, whether the equations $5 \times 5y = 50$ and $25y = 50$ have different solutions? Explain.

Exercises

Solve. (See Example A.)

5. $6d = 42$ 6. $9n = 81$ 7. $5b = 55$ 8. $16y = 64$

9. $13r = 91$ 10. $17m = 85$ 11. $4h = 52$ 12. $7t = 7$

13. $15k = 90$ 14. $22b = 88$ 15. $13c = 169$ 16. $25b = 2500$

Solve. (See Example B.)

17. $\frac{s}{7} = 8$ 18. $\frac{m}{16} = 7$ 19. $\frac{n}{10} = 4$ 20. $\frac{d}{3} = 17$

21. $\frac{t}{5} = 5$ 22. $\frac{d}{12} = 9$ 23. $\frac{r}{8} = 11$ 24. $\frac{s}{9} = 3$

25. $\frac{k}{14} = 14$ 26. $\frac{k}{7} = 21$ 27. $\frac{y}{21} = 5$ 28. $\frac{z}{15} = 4$

▶ Mixed Practice (For more practice, see page 450.)

Solve.

29. $8k = 96$ 30. $\frac{t}{4} = 17$ 31. $\frac{y}{11} = 12$ 32. $6d = 54$

33. $13f = 78$ 34. $\frac{v}{8} = 15$ 35. $\frac{m}{17} = 5$ 36. $7n = 84$

37. $\frac{k}{25} = 4$ 38. $9b = 81$ 39. $\frac{w}{6} = 18$ 40. $18c = 144$

▶ Applications

41. James prints T-shirts. He earns $4 an hour. How many hours must he work to earn $72? Write an equation and solve.

42. Al bowled three games. His average score was 167. What was his total for the three games? Write an equation and solve.

▶ Review (Lessons 6.1, 6.2, 6.4)

Multiply. Write the answers in lowest terms.

43. $\frac{3}{8} \times \frac{3}{4}$ 44. $7 \times 5\frac{1}{7}$ 45. $4\frac{1}{2} \times \frac{2}{15}$ 46. $1\frac{1}{3} \times 1\frac{7}{8}$ 47. $\frac{7}{15} \times 5$

16.8 Solving 2-Step Equations

Max was saving up to buy a new pair of hockey skates. Then disaster struck. In one day he had to spend half his savings on new shoes for track and pay back \$25 he borrowed from his sister. That left him only \$15.

To find out how much Max had saved, you can solve an equation.

Examples

To solve 2-step equations, first add or subtract. Then multiply or divide.

A Solve. $5r + 6 = 41$

$5r + 6 = 41$

$5r + 6 - 6 = 41 - 6$ — Subtract 6 from both sides.

$5r = 35$

$\frac{5r}{5} = \frac{35}{5}$ — Divide both sides by 5.

$1r = 7$

$r = \mathbf{7}$

Check: $(5 \times 7) + 6 \stackrel{?}{=} 41$ — Substitute 7 for *r*.

$35 + 6 \stackrel{?}{=} 41$

$41 = 41$

B Solve. $\frac{q}{2} - 25 = 15$

$\frac{q}{2} - 25 = 15$

$\frac{q}{2} - 25 + 25 = 15 + 25$ — Add 25 to both sides.

$\frac{q}{2} = 40$

$2 \times \frac{q}{2} = 2 \times 40$ — Multiply both sides by 2.

$1q = 80$

$q = \mathbf{80}$

Check: $\frac{80}{2} - 5 \stackrel{?}{=} 35$ — Substitute 80 for *q*.

$40 - 5 \stackrel{?}{=} 35$

$35 = 35$

► Think and Discuss

1. Solve. $7p + 6 = 55$

2. Solve. $\frac{c}{10} - 5 = 5$

3. Refer to the introduction to this lesson. How much money had Max saved before his disaster?

Exercises

Solve. (See Example A.)

4. $4t - 9 = 15$ 5. $7y + 8 = 36$ 6. $12b + 3 = 39$ 7. $9f - 12 = 51$

8. $15k - 9 = 66$ 9. $3n + 15 = 21$ 10. $4a - 10 = 38$ 11. $7q + 4 = 60$

Solve. (See Example B.)

12. $\frac{m}{8} + 5 = 7$ 13. $\frac{r}{10} - 13 = 6$ 14. $\frac{s}{7} - 24 = 1$ 15. $\frac{b}{17} + 16 = 18$

16. $\frac{a}{5} + 19 = 20$ 17. $\frac{d}{16} - 5 = 11$ 18. $\frac{z}{6} + 12 = 15$ 19. $\frac{c}{12} - 5 = 0$

► Mixed Practice (For more practice, see page 451.)

Solve.

20. $8r + 3 = 67$ 21. $\frac{h}{9} - 7 = 4$ 22. $13y - 5 = 73$

23. $14k - 21 = 21$ 24. $\frac{f}{24} + 3 = 15$ 25. $\frac{t}{11} - 11 = 11$

► Applications

26. Margaret brought her dogs to work one day. Altogether 38 feet walked into the building. How many dogs does Margaret have? Write an equation and then solve.

27. Roger rented a car for one day from Pro-Auto Rentals. His bill came to $72. Use the table below to find out how many miles Roger traveled. Write an equation and solve it.

Days	$ per Day	$ per Mile
1	24.00	0.08
2–4	20.00	0.075
5+	18.50	0.07

► Review (Lesson 1.5)

Compare. Use >, <, or =.

28. 659 ___ 699

29. 897 ___ 978

30. 1279 ___ 1257

31. 1956 ___ 1758

Pre-Algebra: Expressions and Sentences **367**

Chapter 16 Review

Complete each statement. (Lesson 16.1)

1. The letter d found in the ⬚ $24d$ is called a(n) ⬚ .

2. $5y - 14 = 36$ is called a(n) ⬚ .

Write as an algebraic expression. (Lesson 16.2)

3. five plus some number

4. some number minus twenty

5. the product of five and four

6. eighteen divided by some number

Write in words. (Lesson 16.2)

7. $5 + 18x$ **8.** $25 - b$ **9.** $4y + 8$ **10.** $18c$ **11.** $42 - 2q$

Find the value of each expression. (Lesson 16.3)

12. $15 + m$ when $m = 15$ **13.** $8c - 20$ when $c = 4$ **14.** $\frac{15b}{3}$ when $b = 12$

Write as an equation. (Lesson 16.4)

15. The quotient of some number and nine is ten.

16. The difference of some number and twenty is seventy-seven.

Solve. (Lesson 16.6)

17. $d - 8 = 32$ **18.** $25 + a = 72$ **19.** $a + 33 = 90$ **20.** $y - 56 = 19$

21. $e + 12 = 56$ **22.** $f - 11 = 39$ **23.** $h - 30 = 21$ **24.** $i + 29 = 42$

Solve. (Lesson 16.7)

25. $4c = 88$ **26.** $\frac{b}{3} = 17$ **27.** $12e = 72$ **28.** $\frac{m}{9} = 13$

29. $12a = 108$ **30.** $\frac{m}{30} = 120$ **31.** $\frac{n}{9} = 81$ **32.** $25b = 125$

Solve. (Lesson 16.8)

33. $5a - 11 = 114$ **34.** $\frac{r}{2} + 34 = 84$ **35.** $15c + 18 = 108$

36. $\frac{n}{10} - 36 = 27$ **37.** $12a + 21 = 81$ **38.** $\frac{d}{5} - 7 = 8$

39. $\frac{w}{3} + 92 = 305$ **40.** $44b - 68 = 416$ **41.** $10b + 12 = 82$

Chapter 16 Test

Find the value of each expression.

1. $14n$ when $n = 8$

2. $\frac{125}{b}$ when $b = 25$

3. $22n - 5$ when $n = 3$

4. $b - 42$ when $b = 70$

5. $\frac{b}{3}$ when $b = 18$

Solve.

6. $3b + 15 = 39$

7. $a + 62 = 109$

8. $40n - 123 = 77$

9. $\frac{108}{m} = 108$

10. $t - 58 = 83$

11. $18 + 2b = 106$

Write as an algebraic expression.

12. some number times fifteen

13. sixteen less than some number

14. thirty-three plus some number

15. the quotient of forty and four

Write as an equation.

16. Fifteen more than some number is seventy.

17. Five less than some number divided by eight is five.

18. The product of some number and fourteen is one hundred eighty-two.

Write in words.

19. $\frac{14}{y} + 13$

20. $12r - 3$

21. $\frac{370}{b} - 4$

22. $2z + 15$

"It was great when Molly came in with a chart that showed our band was finally making a profit. Since we know about integers, we can really understand how she keeps track of our band's money."

Pre-Algebra: Integers and Equations

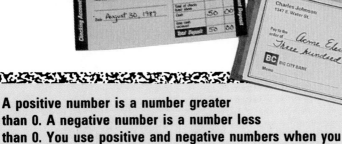

A positive number is a number greater
than 0. A negative number is a number less
than 0. You use positive and negative numbers when you
- record temperatures.
- watch a football game and keep track of gains and losses.
- deposit and withdraw money.

Deposit and Withdrawal

Charles opened a checking account with $250 that he
earned one summer. During the next week, he wrote
2 checks. Each check was for $50. Charles forgot to
keep a record of these checks. The following week, he
deposited $50. Two weeks later, Charles saw a compact
disk player on sale for $300. Charles thought he had
enough money, so he wrote a check for $300.

1. Make a line graph showing Charles's checking
   account balance after the first two weeks.
2. Did Charles have enough money to buy the
   CD player?
3. How would you graph Charles's balance after
   the $300 check?
4. How much money must Charles deposit so
   that his check for $300 will be cashed?

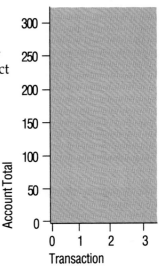

17.1 Introduction to Integers

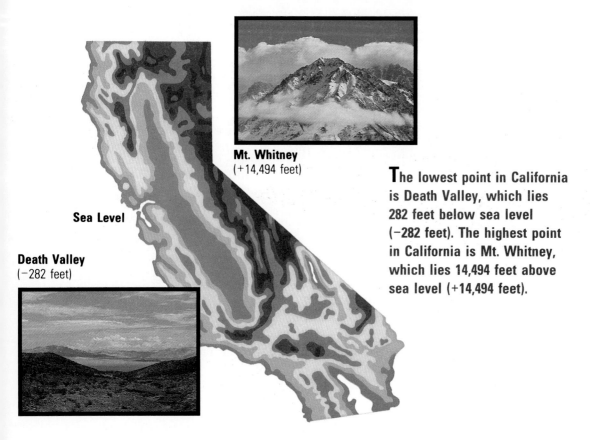

Mt. Whitney
(+14,494 feet)

Sea Level

Death Valley
(−282 feet)

The lowest point in California is Death Valley, which lies 282 feet below sea level (−282 feet). The highest point in California is Mt. Whitney, which lies 14,494 feet above sea level (+14,494 feet).

You have already met whole numbers like 6 and 28. On a number line, we can picture the whole numbers and their opposites. The opposites of 6 and 28 are −6 and −28. Together the whole numbers and their opposites are the **integers**.

$$\xleftarrow{\quad} \overset{\displaystyle -7 \quad -6 \quad -5 \quad -4 \quad -3 \quad -2 \quad -1 \quad 0 \quad +1 \quad +2 \quad +3 \quad +4 \quad +5 \quad +6 \quad +7}{|\quad|\quad|\quad|\quad|\quad|\quad|\quad|\quad|\quad|\quad|\quad|\quad|\quad|\quad|} \xrightarrow{\quad}$$

▶ Express Yourself

Here are some terms that will help you in this chapter.

   positive integer   a whole number greater than zero. Positive integers can be shown with a positive sign (+) and are to the right of zero on a number line.

negative integer an integer less than zero. Negative integers have a negative sign (−) and are to the left of zero on a number line.

opposite integers two integers with different signs that are equally distant from zero on a number line.

1. Describe two other ways you have heard the terms *positive, negative,* and *opposite* used.

2. What are the opposite integers for 14,494 and −282?

3. Refer to the introduction to this lesson. Find the difference in elevation between the top of Mt. Whitney and the lowest point in Death Valley.

▶ Practice What You Know

Add.

4. 49 + 28	**5.** 63 + 12	**6.** 15 + 1500	**7.** 325 + 4562
8. 238 + 671	**9.** 5553 + 228	**10.** 3857 + 2246	**11.** 9089 + 2378
12. 37 + 26	**13.** 444 + 93	**14.** 3286 + 992	**15.** 701 + 52

Subtract.

16. 12 − 6	**17.** 33 − 16	**18.** 45 − 9	**19.** 89 − 81
20. 875 − 87	**21.** 349 − 32	**22.** 4562 − 3729	**23.** 2600 − 34
24. 812 − 379	**25.** 64 − 32	**26.** 2709 − 84	**27.** 965 − 868

Write an equation and solve.

28. What is the distance between the highest point in Florida and the highest point in Wyoming?

29. The difference between the highest point in New York and the highest point in another state is 4538 feet. What is the height of the highest point of the other state? What is the other state?

30. The highest point in one state is 7514 feet higher than the highest point in Illinois. Identify the state.

State	Highest Point
Florida	345 ft.
Illinois	1235 ft.
Mississippi	806 ft.
New York	5344 ft.
South Dakota	7242 ft.
Texas	8749 ft.
Wyoming	13,804 ft.

17.2 Adding Integers

First Down! Jim throws a 5-yard forward pass to Juan. Juan runs 4 yards. Total gain is 9 yards.

Second Down! Bill is tackled 3 yards behind the line of scrimmage and he fumbles the ball. The ball rolls back 5 yards before his teammate falls on it. Total loss is 8 yards.

Third Down! Jim completes a 10-yard forward pass to Bob. Bob is forced back 2 yards. Total gain is 8 yards.

Adding integers can help you keep up with the total yards gained or lost in a football game.

Examples

To add integers, you can use a number line. Find zero on the number line. Then move to the right to add a positive (+) number or to the left to add a negative (−) number.

A Add. +5 + +4   Begin at 0. Move 5 units to the right. Then move 4 more units to the right. +5 + +4 = +9

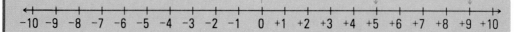

B Add. −3 + −5   Begin at 0. Move 3 units to the left. Then move five more units to the left. −3 + −5 = −8

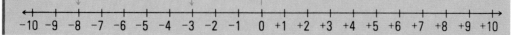

C Add. +10 + −2   Begin at 0. Move 10 units to the right. Then move 2 units to the left. +10 + −2 = +8

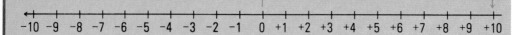

► **Think and Discuss**

1. If you are adding a negative number, do you move to the left or to the right?

2. Which number is farther from zero, $^-25$ or $^+19$?

3. Refer to the introduction to this lesson. What is the total loss or gain in the series of three downs?

4. Without adding, decide if $^-27 + {}^+14$ is positive. Explain.

Exercises

Add. (See Example A.)

5. $^+4 + {}^+4$ 6. $^+1 + {}^+5$ 7. $^+5 + {}^+8$ 8. $^+6 + 0$ 9. $^+8 + {}^+7$

10. $^+3 + {}^+4$ 11. $^+7 + {}^+2$ 12. $^+2 + {}^+3$ 13. $^+5 + {}^+3$ 14. $^+9 + {}^+8$

Add. (See Example B.)

15. $^-2 + {}^-2$ 16. $^-5 + {}^-5$ 17. $^-8 + 0$ 18. $^-4 + {}^-3$ 19. $^-7 + {}^-5$

20. $^-5 + {}^-2$ 21. $^-1 + {}^-4$ 22. $^-3 + {}^-3$ 23. $^-3 + {}^-6$ 24. $^-9 + {}^-7$

Add. (See Example C.)

25. $^-7 + {}^+8$ 26. $^+4 + {}^-9$ 27. $^-2 + {}^+1$ 28. $^+8 + {}^-4$ 29. $^-5 + {}^+9$

30. $^+6 + {}^-9$ 31. $^-7 + {}^+3$ 32. $^+6 + {}^-9$ 33. $^-5 + {}^+6$ 34. $^+8 + {}^-3$

► **Mixed Practice** (For more practice, see page 451.)

Add.

35. $^-5 + {}^-1$ 36. $^-9 + {}^+7$ 37. $^+8 + {}^-9$ 38. $^+4 + {}^+2$ 39. $^-7 + {}^-4$

40. $^+9 + {}^-9$ 41. $^-7 + {}^-2$ 42. $^-4 + {}^+7$ 43. $^+6 + {}^-8$ 44. $^+8 + {}^-5$

► **Applications**

45. The Bears lost thirteen yards on first down and twelve yards on second down. A pass was caught for a twenty-yard gain. What was the total gain or loss for the three downs?

46. Maria had $26 in her checking account. She wrote checks for $13 and $15. How much money does Maria need to deposit to make sure her checks can be cashed?

► **Review** (Lesson 11.3)

Solve each proportion.

47. $\frac{6}{9} = \frac{\text{▓}}{180}$ 48. $\frac{\text{▓}}{7} = \frac{25}{35}$ 49. $\frac{10}{12} = \frac{30}{\text{▓}}$ 50. $\frac{18}{\text{▓}} = \frac{12}{96}$

17.3 Subtracting Integers

Catherine usually listens to the radio in the morning to find out what the weather is like. This morning, she caught the tail end of the weather report. All she heard was . . . .

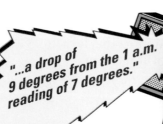

"...a drop of 9 degrees from the 1 a.m. reading of 7 degrees."

Catherine could subtract to find the temperature.

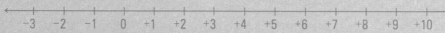

Examples

To subtract an integer, you can add its opposite.

A Subtract. +5 − −3

+5 − −3 = +5 + +3   Add the opposite of −3.
+5 − −3 = +8

B Subtract. +7 − +9

```
<----+----+----+----+----+----+----+----+----+----+----+----+----+----+--->
    -3   -2   -1    0   +1   +2   +3   +4   +5   +6   +7   +8   +9  +10
```

Use a number line. Begin at 0. Move 7 units to the right. Then move 9 units to the left.

+7 − +9 = −2

C Subtract. −6 − −4

```
<--+----+----+----+----+----+----+----+----+----+----+----+----+----+----+----+----+----+-->
  -9  -8  -7  -6  -5  -4  -3  -2  -1   0  +1  +2  +3  +4  +5  +6  +7  +8  +9
```

−6 − −4 = −6 + +4
Use a number line. Begin at 0. Move 6 units to the left. Then move 4 units to the right.
−6 − −4 = −2

▶ Think and Discuss

1. What is the opposite of +1? Of −1?
2. Refer to the introduction to this lesson. What was the temperature?

3. Subtract. $^-9 - {}^+4$

4. Complete the table below.

Working on a number line, you move:

to the right	to add a positive integer.
※	to add a negative integer.
※	to subtract a positive integer.
※	to subtract a negative integer.

Exercises

Subtract. (See Example A.)

5. $^+3 - {}^-2$ **6.** $^+5 - {}^-6$ **7.** $^+4 - {}^-3$ **8.** $^+7 - {}^-9$ **9.** $^+5 - {}^-4$

10. $0 - {}^-1$ **11.** $^+8 - {}^-8$ **12.** $^+3 - 0$ **13.** $^+6 - {}^-1$ **14.** $^+7 - {}^-8$

Subtract. (See Example B.)

15. $^+1 - {}^+4$ **16.** $^+5 - {}^+3$ **17.** $^+2 - {}^+7$ **18.** $^+8 - {}^+3$ **19.** $^+9 - {}^+2$

20. $^+9 - {}^+9$ **21.** $0 - {}^+2$ **22.** $^+7 - {}^+5$ **23.** $^+2 - {}^+8$ **24.** $^+6 - {}^+4$

Subtract. (See Example C.)

25. $^-6 - {}^-2$ **26.** $^-8 - {}^-9$ **27.** $^-5 - {}^-6$ **28.** $^-1 - {}^-4$ **29.** $^-3 - {}^-7$

30. $^-9 - {}^-4$ **31.** $^-1 - {}^-1$ **32.** $^-2 - {}^-6$ **33.** $^-3 - {}^-4$ **34.** $^-9 - {}^-6$

▶ **Mixed Practice** (For more practice, see page 452.)

Subtract.

35. $^-5 - {}^-11$ **36.** $^-6 - {}^-3$ **37.** $^+6 - {}^+9$ **38.** $^-11 - {}^+4$

39. $^+15 - {}^-9$ **40.** $^-7 - {}^+11$ **41.** $^-2 - {}^+1$ **42.** $^+9 - {}^+11$

43. $^+4 - {}^+12$ **44.** $^-3 - {}^-12$ **45.** $^+6 - {}^+10$ **46.** $0 - {}^+4$

47. $^-12 - {}^+3$ **48.** $^-12 - {}^-8$ **49.** $^+7 - {}^+3$ **50.** $^-9 - {}^+9$

▶ **Applications**

51. The temperature is $72°F$ inside and $^-12°F$ outside. What is the difference in temperatures?

52. How much warmer is it in Boston than it is in Portland?

Boston

Portland

▶ **Review** (Lesson 12.7)

Find each number.

53. 60% of 1570 **54.** $7\frac{1}{2}\%$ of 500 **55.** 475% of 80

17.4 Mapping and Coordinate Graphing

Jerome received a letter and a map from his cousin, Diane. "I'm looking forward to seeing you in Philadelphia during the holidays," Diane wrote. "Here's a map of central Philadelphia. Let me know where you want to go."

Jerome can use coordinate graphing to explain to Diane the locations of the places on the map he'd like to visit.

You have worked with number lines and have made line graphs. The grid on the right was made by drawing a horizontal number line and a vertical number line. The point where the two lines meet is called the **origin**. The origin is labeled with a zero. The grid is divided into four sections. You can name any point on the grid by using a pair of numbers called an **ordered pair**. Find the ordered pair (3, −2). The first number tells how far to move to the right or to the left on the horizontal number line. The second number tells how far to move up or down on the vertical number line.

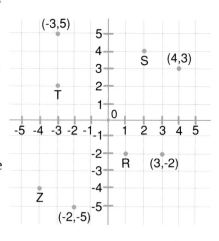

▶ Think and Discuss

1. Starting at the origin, what direction would you move on a horizontal number line for a negative number?

2. Starting at the origin, what direction would you move on a vertical number line for a negative number?

3. Write ordered pairs for the points S, T, R, and Z.

Look at the map of central Philadelphia below. You can use a grid to help you locate places on the map.

4. What street is on the horizontal number line?

5. What street is on the vertical number line?

6. Using the grid, how would you express the location of the corner of 11th Street and Cherry Street?

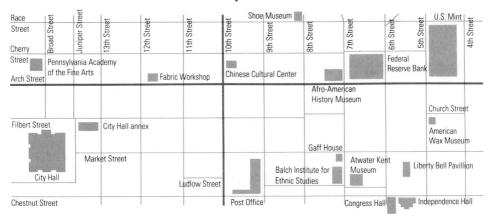

Exercises (For more practice, see page 452.)

Use the map to find the streets at these ordered pairs.

7. $(-2, 0)$ **8.** $(1, -2)$ **9.** $(-1, 2)$ **10.** $(-3, -3)$

Use the map to find the landmarks at these ordered pairs.

11. $(3, 0)$ **12.** $(4, -3)$ **13.** $(5, 2)$

▶ **Applications**

14. Diane and Jerome want to visit the Gaff House and the Shoe Museum. Write ordered pairs and name the streets that cross at each location.

15. Draw a grid on a sheet of paper. Label your school at the origin. Plot the locations and write ordered pairs for landmarks near your school.

▶ **Review** (Lesson 12.3)

Write as a percent.

16. $\frac{3}{5}$ **17.** $\frac{1}{2}$ **18.** $\frac{10}{10}$ **19.** $\frac{1}{10}$

20. $\frac{5}{6}$ **21.** $\frac{1}{5}$ **22.** $\frac{3}{8}$ **23.** $\frac{13}{16}$

17.5 Multiplying Integers

What does it mean to say, "My telephone number is not unlisted"?
What does it mean to say, "My telephone bill is not unpaid"? In ordinary
language, having the word *not* followed by a word beginning with the
prefix *un-* can be like multiplying two negative integers.

Examples

To multiply two integers, multiply as with whole numbers. If
the signs are the same, the product is positive. If the signs are
different, the product is negative.

A Multiply. +8 × +9	**B** Multiply. −4 × −7	**C** Multiply. −7 × +3
Same signs: positive product	Same signs: positive product	Different signs: negative product
+8 × +9 = +72	−4 × −7 = +28	−7 × +3 = −21

▶ Think and Discuss

1. Is the product of two negative numbers negative?

2. Multiply.   −9 × +5

3. What is the product of zero and any integer?

4. Does "not unfriendly" mean friendly? Explain.

5. If you multiply seven negative integers together, will the product be positive or negative?

6. Explain how the use of *not* followed by *un-* in ordinary language can be like multiplication of two negative integers.

Exercises

Multiply. (See Example A.)

7. +5 × +4   8. +7 × +7   9. +10 × +1   10. +4 × +20

11. +7 × +8   12. +7 × +11   13. +6 × +12   14. +14 × +7

15. +5 × +40   16. +9 × +16   17. +8 × +30   18. +15 × +4

Multiply. (See Example B.)

19. −6 × −3   20. −3 × −6   21. −9 × −11   22. −8 × −4   23. −7 × −4

24. −4 × −12   25. −2 × −24   26. −7 × −8   27. −9 × −9   28. −11 × −7

Multiply. (See Example C.)

29. −6 × +4   30. −1 × +10   31. +2 × −12   32. +5 × −20

33. −7 × +9   34. −5 × +6   35. +7 × −12   36. +8 × −11

▶ **Mixed Practice** (For more practice, see page 453.)

Multiply.

37. −3 × −5   38. +9 × −1   39. +9 × +1   40. −5 × −10   41. −7 × −15

42. −4 × +4   43. +8 × +10   44. −7 × −6   45. −9 × +10   46. −12 × +5

47. +5 × +5   48. −1 × −7   49. +8 × −7   50. +6 × +6   51. −20 × −4

▶ **Applications**

52. At 6 p.m. the temperature was 0°C. In the next three hours it dropped 3° each hour. What was the temperature at 9 p.m.?

53. A stock lost four points each day for four days. Then for two days it gained a point each day. Write the total change as an integer.

6 Day Results					
	M	T	W	T	F	M
PrgKn	-4	-4	-4	-4	+1	+1

▶ **Review** (Lesson 12.9)

Find each number.

54. 225% of what number is 81?

55. 16% of what number is 8?

56. 95% of what number is 285?

57. 4% of what number is 13.12?

58. 700% of what amount is $35,000?

59. 48% of what number is 5.76?

17.6 Dividing Integers

David, a railroad employee in Chicago, charts winter temperatures. When the temperature drops below freezing, David starts heaters so that the railroad switches do not freeze up. For four days in January, Chicago temperatures were −22°F, −15°F, −18°F, and −13°F. The sum of these temperatures is −68°F. What is the average temperature for those four days?

The rules for determining signs when dividing integers are exactly the same as the rules for determining signs when multiplying integers.

Examples

To divide integers, divide as with whole numbers. If the signs are the same, the quotient is positive. If the signs are different, the quotient is negative.

A Divide. +36 ÷ +4	**B** Divide. −56 ÷ −7	**C** Divide. −68 ÷ +4
Same signs: positive quotient	Same signs: positive quotient	Different signs: negative quotient
+36 ÷ +4 = +9	−56 ÷ −7 = +8	−68 ÷ +4 = −17

▶ Think and Discuss

1. Divide. −48 ÷ −6

2. Divide. −64 ÷ +8

3. Does a negative integer divided by a positive integer give a positive or a negative quotient?

4. Does a positive integer divided by a negative integer give a positive or a negative quotient?

5. What is the quotient of any integer divided by one?

6. Refer to the introduction to this lesson. What was the average temperature for the four days?

7. If you divide two negative integers, and then divide that quotient by a negative integer, will the final quotient be positive or negative?

Exercises

Divide. (See Example A.)

8. +81 ÷ +9   9. +25 ÷ +5   10. +8 ÷ +8   11. +45 ÷ +9   12. +21 ÷ +7

13. +56 ÷ +8   14. +10 ÷ +1   15. +48 ÷ +4   16. +63 ÷ +7   17. +72 ÷ 9

Divide. (See Example B.)

18. −90 ÷ −9   19. −54 ÷ −6   20. −36 ÷ −3   21. −12 ÷ −6   22. −80 ÷ −8

23. −7 ÷ −7   24. −72 ÷ −8   25. −22 ÷ −2   26. −45 ÷ −9   27. −63 ÷ −7

Divide. (See Example C.)

28. +16 ÷ −4   29. −20 ÷ +4   30. +25 ÷ −5   31. +36 ÷ −9   32. −49 ÷ +7

33. −50 ÷ +5   34. +56 ÷ −8   35. −81 ÷ +9   36. +72 ÷ −8   37. −63 ÷ +9

▶ **Mixed Practice** (For more practice, see page 453.)

Divide.

38. −60 ÷ −6   39. −10 ÷ +2   40. −24 ÷ +8   41. +72 ÷ +3   42. −64 ÷ −8

43. +15 ÷ −5   44. +9 ÷ −1   45. +13 ÷ +1   46. −18 ÷ +6   47. −42 ÷ −7

48. +27 ÷ −3   49. −56 ÷ −4   50. +48 ÷ +6   51. −12 ÷ +1   52. +64 ÷ −4

▶ **Applications**

53. The per-gallon cost of heating oil rose from $1.57 to $1.75 in two months. What was the average monthly change in the price per gallon?

54. The low in Bismarck, North Dakota, was −10 degrees Fahrenheit one day. The next day the low was −12. Find the average low temperature for these two days.

▶ **Review** (Lesson 16.6)

Solve and check each equation.

55. $a - 12 = 74$ 56. $21 + w = 98$ 57. $x + 3.6 = 8$ 58. $n - 47 = 9$

59. $b - 53 = 29$ 60. $c - 80 = 142$ 61. $78 + r = 123$ 62. $s + 16 = 43$

17.7 Solving Addition and Subtraction Equations Involving Integers

"I'm thinking of a number," Mr. Marvin, the mystical mathematician, told his class. "When you subtract −6 from it, you get −4. Can you guess what my number is?"

One way to find the number is to set up and solve an equation.

Examples

To solve an equation, isolate the variable on one side.

A Solve. $c + {}^-5 = {}^+4$

$$c + {}^-5 = {}^+4$$
$$c + {}^-5 - {}^-5 = {}^+4 - {}^-5 \qquad \text{Subtract } {}^-5 \text{ from both sides.}$$
$$c + 0 = {}^+4 - {}^-5 \qquad {}^-5 + {}^+5 = 0$$
$$c = 9$$

Check: $9 + {}^-5 \overset{?}{=} 4 \qquad$ Substitute 9 for c in the original equation.
$$4 = 4$$

B Solve. $b - {}^-6 = {}^-4$

$$b - {}^-6 = {}^-4$$
$$b - {}^-6 + {}^-6 = {}^-4 + {}^-6 \qquad \text{Add } {}^-6 \text{ to both sides.}$$
$$b + 0 = {}^-4 + {}^-6 \qquad {}^+6 + {}^-6 = 0$$
$$b = {}^-10$$

Check: ${}^-10 - {}^-6 \overset{?}{=} {}^-4 \qquad$ Substitute ${}^-10$ for b in the original equation.
$${}^-10 + {}^+6 \overset{?}{=} {}^-4$$
$${}^-4 = {}^-4$$

▶ Think and Discuss

1. Solve. $a + {}^-5 = {}^-8$

2. Solve. $b - {}^-12 = {}^+24$

3. Refer to the introduction to this lesson. What was Mr. Marvin's number?

4. Use an equation to solve "12 more than an integer is ${}^-8$."

5. Use an equation to solve "An integer decreased by 18 is ${}^-20$."

6. Try to solve these equations mentally.

   $x + {}^-3 = {}^+5$ $y + {}^+4 = {}^-1$ $z - {}^-2 = {}^+1$ $w - {}^+5 = {}^-1$

Exercises

Solve. (See Example A.) Check your answers.

7. $x + {}^+3 = {}^-1$ 8. $y + {}^+5 = {}^+2$ 9. $z + {}^+2 = {}^+1$ 10. $a + {}^+5 = {}^-5$

11. $e + {}^-6 = {}^+10$ 12. $j + {}^-1 = {}^-1$ 13. $b + {}^-4 = {}^+12$ 14. $m + {}^-8 = {}^-6$

Solve. (See Example B.) Check your answers.

15. $x - {}^-2 = {}^-3$ 16. $y - {}^-4 = {}^+2$ 17. $z - {}^-4 = {}^-8$ 18. $a - {}^-6 = {}^+2$

19. $a - {}^-1 = {}^+1$ 20. $b - {}^-1 = {}^-1$ 21. $c - {}^+5 = 0$ 22. $d - {}^+4 = {}^-3$

▶ Mixed Practice (For more practice, see page 454.)

Solve.

23. $e - {}^-2 = {}^+2$ 24. $f + {}^+5 = {}^-7$ 25. $g + {}^-9 = {}^-10$ 26. $h - {}^+6 = {}^+5$

27. $i + {}^-7 = {}^-4$ 28. $j - {}^-4 = {}^-2$ 29. $k + {}^+1 = {}^+1$ 30. $l + {}^+9 = {}^-4$

31. $m - {}^-10 = 0$ 32. $n + {}^+8 = {}^+6$ 33. $p + {}^-3 = {}^-1$ 34. $g - {}^+1 = {}^-10$

35. $r - {}^-6 = {}^+2$ 36. $s + {}^-5 = {}^+5$ 37. $v + {}^+9 = {}^+12$ 38. $w + {}^-8 = {}^-14$

▶ Applications

39. The temperature fell twelve degrees to three degrees below zero. How cold was it before? Set up an equation and solve.

40. When you add this number to ${}^-32$, the sum is ${}^-18$. Find the number using an equation.

▶ Review (Lesson 16.7)

Solve and check each equation.

41. $\frac{n}{8} = 14$ 42. $\frac{x}{3} = 55$ 43. $11s = 132$ 44. $7t = 266$

17.8 Solving Multiplication and Division Equations Involving Integers

Monica, a student in Mr. Marvin's class, raised her hand. "I played nine holes of golf and I scored below par. My score multiplied by negative 8 equals 16. Can you figure out my golf score?"

Mr. Marvin smiled. "For this problem I must solve an equation."

Examples

To solve an equation, isolate the variable on one side.

A Solve. $-8a = +16$

$$-8a = +16$$

$$\frac{-8a}{-8} = \frac{+16}{-8} \qquad \text{Divide both sides of the equation by } -8.$$

$$+1a = \frac{+16}{-8} \qquad \frac{-8}{-8} = +1$$

$$a = -2$$

Check: $-8 \times -2 \overset{?}{=} +16$ Substitute -2 for a in the original equation.

$$+16 = +16$$

B Solve. $\frac{b}{+6} = -7$

$$\frac{b}{+6} = -7$$

$$+6 \times \frac{b}{+6} = +6 \times -7 \qquad \text{Multiply both sides of the equation by } +6.$$

$$+1b = +6 \times -7 \qquad \frac{+6}{+6} = +1$$

$$b = -42$$

Check: $\frac{-42}{+6} \overset{?}{=} -7$ Substitute -42 for b in the original equation.

$$-7 = -7$$

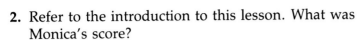

► Think and Discuss

1. Solve and then show the check. $-2k = -8$

2. Refer to the introduction to this lesson. What was Monica's score?

3. Try to solve these equations mentally.

   $-2k = -8$ $\frac{h}{+3} = +5$ $+4l = -12$ $\frac{m}{-4} = +3$

4. Why do you multiply both sides of the equation $\frac{t}{+8} = +16$ by 8 to solve it?

Exercises

Solve. (See Example A.)

5. $-3p = +9$ 6. $-2m = +10$ 7. $-1n = 0$ 8. $+5q = -5$

Solve. (See Example B.)

9. $\frac{p}{+2} = -4$ 10. $\frac{m}{+6} = +3$ 11. $\frac{n}{+6} = -3$ 12. $\frac{q}{-5} = -6$

► Mixed Practice (For more practice, see page 454.)

Solve.

13. $\frac{p}{-1} = -1$ 14. $-5m = +15$ 15. $+2n = 0$ 16. $\frac{q}{-4} = 20$

17. $\frac{a}{+3} = -3$ 18. $\frac{b}{-7} = +2$ 19. $+5c = -5$ 20. $-1d = -4$

21. $\frac{e}{+3} = -5$ 22. $-5f = +20$ 23. $+9q = -18$ 24. $+2h = +2$

► Applications

25. Monica played 3 rounds of golf. In all, she was 9 below par. What was her average score?

26. Alicia and Thomas sell homemade pastries. Every month they divide their profits equally. In June each got $457.50. What was their total profit?

► Review (Lesson 16.8)

Solve and check each equation.

27. $3a + 7 = 31$ 28. $\frac{y}{9} - 13 = 7$ 29. $16b - 48 = 192$

17.9 Solving Fermi Problems

Every year, in January, the President of the United States speaks to the Congress. The speech is called the State of the Union Address and is broadcast on television. The President usually speaks for about 45 minutes.

1. Estimate how many words the President says in this speech.

Did you have to guess in Question 1? Guessing is one way to estimate a quantity, but sometimes it is very difficult to make a reasonable guess. In this lesson, you will learn ways to improve your estimates as you learn about Fermi problems.

In a **Fermi problem**, you are asked a question and expected to give a reasonable estimate as your answer. To form your estimate, you are only allowed to use everyday facts you already know, as well as your common sense. The problem of the President's speech can be treated as a Fermi problem.

Enrico Fermi, physicist, (1901–1954)

▶ **Activity**

Work with 2 or 3 classmates. Try to solve the problem about the President's speech. The answers to the following questions use everyday facts and common sense. They will help you estimate the number of words the President spoke. In Fermi problems, you can estimate and round any of the numbers you use.

2. How many seconds are in a minute?

3. Discuss how many words the President might say each second. Decide on an estimate.

4. About how many words would the President say in 1 minute?

5. About how many words would the President say in 45 minutes?

Notice how the questions above allow you to organize the solution to the problem. You can sometimes make a reasonable estimate to solve a problem by breaking it down into smaller problems.

6. What was your group's solution to the problem? Compare it with the solution of another group. Compare the methods used by the two groups to solve the problem.

Now you are ready to solve some Fermi problems on your own.

Exercises

For each problem below, give a reasonable estimate. In addition, outline the steps by which you reached your estimate.

7. How many times do you inhale in the course of a day?

8. How many inflated balloons could you fit in your classroom?

9. How many teenagers live in the United States?

10. How many foul balls are hit into the stands during a major-league baseball season?

11. How many dentists are there in California?

▶ **Review** (Lessons 16.4, 16.6, 16.7, 16.8)

Write as an equation. Then solve.

12. Two times twelve equals some number.

13. Some number divided by 13 is 4.

14. Seven minus three is some number.

15. Twenty times some number is sixty.

16. The difference between nine and three is some number.

17. Eighteen plus two is some number.

18. Ten plus two times two is some number.

19. One plus one times some number is six.

20. Some number times three minus twelve is thirty.

17.10 Solving 2-Step Equations Involving Integers

Three girls started making birdhouses. Each contributed the same amount of money to buy supplies. The first week they sold 1 birdhouse for $9 but had a total loss of $15. How much did each contribute?

Examples

To solve two-step equations involving negative numbers, first add or subtract. Then multiply or divide.

A Solve. $3a + 9 = -15$

$$3a + 9 - 9 = -15 - 9 \qquad \text{Subtract 9 from both sides.}$$
$$3a = -24$$
$$\frac{3a}{3} = \frac{-24}{3} \qquad \text{Divide both sides by 3.}$$
$$a = -8$$

Check: $3 \times -8 + 9 \overset{?}{=} -15 \qquad$ Substitute -8 for a.
$$-24 + 9 \overset{?}{=} -15$$
$$-15 = -15$$

B Solve. $\frac{a}{+3} - {}^{+}9 = -36$

$$\frac{a}{+3} - {}^{+}9 + {}^{+}9 = -36 + {}^{+}9 \qquad \text{Add } {}^{+}9 \text{ to both sides.}$$
$$\frac{a}{+3} = -27$$
$${}^{+}3 \times \frac{a}{+3} = {}^{+}3 \times -27 \qquad \text{Multiply both sides by } {}^{+}3.$$
$$a = -81$$

Check: $\frac{-81}{+3} - {}^{+}9 \overset{?}{=} -36 \qquad$ Substitute -81 for a.
$$-27 - {}^{+}9 \overset{?}{=} -36$$
$$-36 = -36$$

▶ Think and Discuss

1. Describe how you would solve $-2m + -3 = +3$.

2. What steps would you use to solve $\frac{t}{+2} - +2 = -3$? Solve.

3. Refer to the introduction to this lesson. How much did the three girls each pay for supplies in their first week of business?

4. Solve $+6x + (-3) = +9$ by subtracting before dividing. Then try to solve the equation by dividing before subtracting. Which method is easier? Are both correct?

Exercises

Solve. (See Example A.)

5. $+2k + -3 = +5$ 6. $-3m + +1 = +10$ 7. $+6n - -4 = -8$

Solve. (See Example B.)

8. $\frac{k}{+2} + -6 = -1$ 9. $\frac{p}{-4} - -2 = +5$ 10. $\frac{q}{-3} + +4 = +2$

▶ Mixed Practice (For more practice, see page 455.)

Solve.

11. $\frac{k}{-1} + -5 = +2$ 12. $+2m + -3 = +13$ 13. $-5n - -5 = -10$

14. $+3p - +1 = -1$ 15. $\frac{r}{+4} + -4 = +4$ 16. $\frac{q}{-3} + +4 = -2$

17. $-1w + +6 = +2$ 18. $\frac{h}{+3} - -5 = -3$ 19. $+6j - +5 = +13$

▶ Applications

20. A business earned $1795 and had expenses of $262. How much did each of three partners receive if they divided the profits evenly?

21. Refer to the introduction to this lesson. One month, the three girls purchased $63 worth of supplies. They had a profit of $45. How many $9 birdhouses did they sell that month?

▶ Review (Lesson 3.9)

Simplify.

22. $8 \times 6 + 9 \times 6$ 23. $11 - 64 \div 8 + 2$ 24. $27 + 5 \times 8 \div 10$

25. $4 \times 4 \times 4 + 36$ 26. $9 + (2 \times 5 \times 3)$ 27. $11 \times (5 + 7)$

17.11 Using a Formula

Formulas are often used in sports. For example, to find a team's winning percentage you would use the formula:

$$\text{Winning Percentage (Pct.)} = \frac{\text{Number of wins}}{\text{Number of wins} + \text{number of losses}}$$

Look at the chart below.

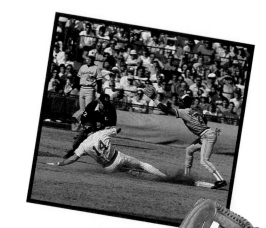

Southwest Division

Team	W	L	Pct	GB
Jackson	8	2	0.800	
Pelham	7	3	0.700	?
Kingsbridge	6	4	0.600	?
Polk	3	7	0.300	?
Flyport	1	9	0.100	?

1. How many games has Pelham High School won? How many games have they lost? How many games have they played?

2. Find the winning percentage for Pelham's team. Substitute information from the chart into the formula above.

Fans often want to know how far a team is from first place. In newspapers this is listed as GB, which stands for Games Behind. To find how many games behind the first-place team your team is, you can use the formula:

$$\text{GB} = \frac{(\text{wins} - \text{losses for first place team}) - (\text{wins} - \text{losses for your team})}{2}$$

For the Polk team,

$$\text{GB} = \frac{(8 - 2) - (3 - 7)}{2} = \frac{6 - (-4)}{2} = \frac{10}{2} = 5 \text{ games behind Jackson}$$

3. How many games behind Jackson is the Kingsbridge team?

4. How many games behind Jackson is the Flyport team?

5. How many games behind Jackson is the Pelham team?

6. How many games behind Jackson is Jackson?

7. Which of the four questions above were solved by using negative integers?

Shown below are the National League standings on August 12, 1987. Copy the table and complete the Pct and GB columns.

Eastern Division

	Team	W	L	Pct	GB
	St. Louis	69	43	0.616	0
8.	New York	64	49	0.566	▦
9.	Montreal	62	50	▦	7
10.	Philadelphia	57	55	0.509	▦
11.	Chicago	57	56	▦	12.5
12.	Pittsburgh	50	63	▦	19.5

Western Division

	Team	W	L	Pct	GB
	Cincinnati	59	55	0.518	0
13.	San Francisco	58	56	0.509	▦
14.	Houston	55	57	0.491	▦
15.	Los Angeles	50	62	▦	8
16.	Atlanta	49	63	▦	9
17.	San Diego	46	67	0.407	▦

To find a pitcher's earned-run average (ERA), use the formula:

$$\text{ERA} = \frac{\text{Number of earned runs allowed} \times 9}{\text{Number of innings pitched}}$$

Copy the chart below and complete the ERA column. Round to the nearest hundredth.

	Year	Pitcher	Team	Innings Pitched	Earned Runs	ERA
18.	1968	Bob Gibson	St. Louis	305	38	▦
19.	1978	Ron Guidry	New York	274	53	▦
20.	1984	Mike Boddicker	Baltimore	261	81	▦
21.	1984	Alejandro Pena	Los Angeles	199	55	▦

▶ **Review** (Lesson 8.9)

Copy and complete the table.

	Starting Time	Finishing Time	Elapsed Time
22.	6:50 a.m.	1:45 p.m.	▦
23.	8:15 p.m.	▦	10 hours, 52 minutes
24.	▦	7:40 a.m.	8 hours, 30 minutes

Chapter 17 Review

Write true or false. (Lesson 17.1)

1. Negative numbers are located to the right of zero on a number line.

2. The numbers 6 and 0 are opposite integers.

3. The sum of an integer and its opposite is zero.

Add. (Lesson 17.2)

4. $-3 + {}^+7$ 5. $-8 + {}^-12$ 6. ${}^+6 + {}^+16$ 7. ${}^+9 + {}^-15$

8. $-13 + {}^-5$ 9. $-4 + {}^+16$ 10. ${}^+17 + {}^+2$ 11. ${}^+1 + {}^-1$

Subtract. (Lesson 17.3)

12. $-7 - {}^+8$ 13. ${}^+14 - {}^+7$ 14. ${}^+11 - {}^-8$ 15. $-9 - {}^-1$

16. $-20 - {}^+10$ 17. ${}^+15 - {}^+5$ 18. $-16 - {}^+8$ 19. $-30 - {}^-40$

Multiply. (Lesson 17.5)

20. $-3 \times {}^-9$ 21. ${}^+9 \times {}^-8$ 22. ${}^+6 \times {}^+11$ 23. $-10 \times {}^+5$

24. $-4 \times {}^+15$ 25. $-2 \times {}^-13$ 26. ${}^+12 \times {}^-3$ 27. ${}^+7 \times {}^+7$

Divide. (Lesson 17.6)

28. $-64 \div {}^+8$ 29. ${}^+54 \div {}^+6$ 30. $-28 \div {}^-4$ 31. ${}^+40 \div {}^-8$

32. ${}^+35 \div {}^+5$ 33. ${}^+56 \div {}^-7$ 34. $-45 \div {}^-15$ 35. $-40 \div {}^+10$

Solve and check each equation. (Lessons 17.7, 17.8)

36. $b + {}^+5 = -4$ 37. $a - {}^-6 = -12$ 38. $x - {}^+3 = {}^+17$

39. $e + {}^-5 = {}^+25$ 40. $\frac{m}{-6} = {}^+6$ 41. $n \times {}^+8 = -32$

42. $-5 \times r = {}^+45$ 43. $\frac{z}{+9} = -9$ 44. $x - {}^+12 = -8$

Solve and check each equation. (Lesson 17.10)

45. $-6n + (-8) = -56$ 46. $\frac{q}{-2} - {}^+10 = -25$ 47. $\frac{r}{+4} - {}^-21 = {}^+12$

Chapter 17 Test

TEST

Multiply or divide.

1. $-5 \times +11$

2. $-93 \div -3$

3. $+85 \div -5$

4. $+16 \times -4$

5. $+48 \div +12$

6. $-77 \div +11$

7. -14×-3

8. $+22 \times +4$

9. $-32 \div -4$

10. $-8 \times +71$

11. -27×-18

12. $-63 \div +9$

13. $+19 \times -32$

14. $-11 \div -1$

15. $+90 \div +5$

Solve and check each equation.

16. $\frac{p}{-8} = -14$

17. $+3m - +9 = +30$

18. $c + -35 = -10$

19. $\frac{w}{20} + -4 = +4$

20. $y - -17 = -17$

21. $a + +29 = +70$

22. $\frac{r}{3} = -22$

23. $-5b - -50 = +200$

24. $12d = -144$

25. $e - +45 = +15$

26. $2s + -16 = -54$

27. $\frac{c}{-18} = +6$

28. $f - -18 = -27$

29. $\frac{12}{z} = -2$

30. $7x + +3 = -18$

Complete each statement.

31. On a number line, negative numbers are to the ___ of zero and positive numbers are to the ___ of zero.

32. When you multiply two negative numbers, the product is ___ .

33. When you add two negative numbers, the sum is ___ .

Add or subtract.

34. $+4 - -4$

35. $-16 + -16$

36. $-8 - +21$

37. $+17 + -7$

38. $+13 + +14$

39. $+25 - +25$

40. $-55 - -45$

41. $-24 + +48$

42. $-86 + +31$

43. $-86 - -47$

44. $+307 + -112$

45. $-99 + -27$

Cumulative Test   Chapters 1–17

TEST

▶ **Choose the letter that shows the correct answer.**

1. $3n = 36$
  a. $n = 18$
  b. $n = 12$
  c. $n = 9$
  d. not given

2. $50t + 8 = 13$
  a. $t = 5$
  b. $t = 25$
  c. $t = 11$
  d. not given

3. $-9 - {}^-6 =$ ___
  a. -15
  b. $+3$
  c. -3
  d. not given

4. $n + n \times 7 =$ ___
  when $n = 6$
  a. 48
  b. 84
  c. 49
  d. not given

▶ **Find the area of each figure.**

5. circle
  $r = 8$ in.
  a. 50.24 sq. in.
  b. 50.24 in.
  c. 200.96 sq. in.
  d. not given

6. parallelogram
  $b = 5$ in.
  $h = 9$ in.
  a. 14 sq. in.
  b. 28 in.
  c. 45 sq. in.
  d. not given

7. triangle
  $b = 10$ ft.
  $h = 15$ ft.
  a. 150 ft.
  b. 150 sq. ft.
  c. 75 ft.
  d. not given

8. square
  $s = 12$ ft.
  a. 48 sq. ft.
  b. 144 sq. ft.
  c. 16 sq. ft.
  d. not given

▶ **Choose the letter that shows the correct answer.**

9. 2 c. = ___ fl. oz.
  a. 32
  b. 12
  c. 8
  d. not given

10. $4\frac{1}{3}$ yd. = ___ in.
  a. 13
  b. 42
  c. 52
  d. not given

11. 12 oz. = ___ lb.
  a. $\frac{3}{4}$
  b. $1\frac{1}{2}$
  c. $\frac{2}{3}$
  d. not given

▶ **Write each of the following as a mathematical expression or equation.**

12. the product of thirty and five
  a. $\frac{5}{30}$
  b. $\frac{30}{5}$
  c. 30×5
  d. not given

13. the sum of 8 and 9 times some number
  a. $8 + 9 + n$
  b. $8 + 9n$
  c. $8 \times 9 \times n$
  d. not given

14. Eighty less than some number is fifty-four.
  a. $80 - n = 54$
  b. $n - 80 = 54$
  c. $80 - 54 = n$
  d. not given

► Compute.

15. -8×-7
 a. -56
 b. $+56$
 c. $+54$
 d. not given

16. $-54 \div +9$
 a. -6
 b. $+6$
 c. -7
 d. not given

17. $+25 \div -5$
 a. $+5$
 b. -5
 c. -1
 d. not given

18. $-4 + -5$
 a. $+20$
 b. $+9$
 c. -9
 d. not given

► Solve.

19. Jose bought a camera on sale for 25% off. He paid $86.25. What was the original price?
 a. $65.19
 b. $115
 c. $111.25
 d. not given

20. Ticket sales for a concert were 550 tickets at $6.25 each. How much money was raised by ticket sales?
 a. $3437.50
 b. $343,750
 c. $343.75
 d. not given

21. The temperatures for the week were: 66°, 37°, 47°, 45°, 51°, 39°, and 37°. What was the average temperature?
 a. 41°
 b. 54°
 c. 46°
 d. not given

22. Paul hiked $8\frac{1}{2}$ miles the first day and $7\frac{3}{4}$ miles the second day. How far did he hike in two days?
 a. $15\frac{1}{4}$ mi.
 b. $16\frac{1}{4}$ mi.
 c. $16\frac{1}{2}$ mi.
 d. not given

► Choose the best estimate.

23. 89×32
 a. ≈ 2400
 b. ≈ 2700
 c. ≈ 3600
 d. not given

24. $9.06 - 4.158$
 a. ≈ 5
 b. ≈ 3
 c. ≈ 6
 d. not given

25. $5,567 \div 70$
 a. ≈ 70
 b. ≈ 80
 c. ≈ 800
 d. not given

26. $3\frac{7}{8} + 9\frac{1}{12}$
 a. ≈ 12
 b. ≈ 13
 c. ≈ 14
 d. not given

► Compute.

27. $8 \times \$9.39$
 a. $72.12
 b. $75.04
 c. $74.42
 d. not given

28. $5 - 0.057$
 a. $75.04
 b. $3943
 c. 4.943
 d. not given

29. $91 \div 0.13$
 a. 700
 b. 70
 c. 7
 d. not given

30. $15 + 4.8 + 7.96$
 a. 859
 b. 26.66
 c. 27.76
 d. not given

▶ **Choose the letter that shows the correct answer.**

31. 5650 is 113% of what number?
a. 5000
b. 50,000
c. 500
d. not given

32. $\frac{1}{2}$% of 300 is
a. 15
b. 150
c. 1.5
d. not given

33. 80% of 570 is
a. 456
b. 45.6
c. 45,600
d. not given

▶ **Solve.**

34. Jane works 6 hours on Saturdays. She earns $4.45 an hour. How much will she have earned after four Saturdays?
a. $0.40
b. $81.92
c. $106.80
d. not given

35. Sam jogs 3 miles a day during the week and 4 miles a day on Saturdays and Sundays. How far does he jog in a week?
a. 100 miles
b. 19 miles
c. 20 miles
d. not given

▶ **Complete each statement. Choose the more reasonable measure.**

36. A college basketball player is 2 ___ tall.   cm   m

37. A rock has a mass of 8 ___.   mg   kg

38. Soda is sold in 2-___ bottles.   mL   L

39. The temperature on a hot summer day in Dallas might be 42 ___.   degrees Celsius   degrees Fahrenheit

40. An airmail letter weighs about $\frac{1}{2}$ ___.   oz.   lb.

41. A small carton of whipping cream contains 8 ___.   fl. oz.   c.

▶ **Solve.**

42. Jose earns $5.12 an hour. He works from 5 to 9 on Mondays and Wednesdays and from 11 to 4 on Saturdays. How much does he earn in a week?
a. $66.56
b. $61.44
c. $91.80
d. not given

43. Which is a better buy, $3.98 for 64 oz. of laundry soap or $2.19 for 32 oz.?
a. 32 oz. for $2.19
b. 64 oz. for $3.98
c. They are equal.
d. not given

1 ▸ Duck Down

Two fathers and two sons each shot a different duck at a carnival shooting gallery. Only three ducks were shot. How is this possible?

2 ▸ Letter Pattern

What is the next letter in this pattern?

Q.W.E.R.T.

3 ▸ Baseball Trades

A man bought an old baseball card for $1, sold it for $2, bought it back for $3, and sold it again for $4. How much profit, if any, did he make?

4 ▸ Worming Through

The volumes of an encyclopedia are arranged as shown. Each cover is $\frac{1}{8}$ inch thick. Volume 1 has 656 pages. Volume 2 has 832 pages. Each page is $\frac{1}{300}$ inch thick. A bookworm, starting at page 1 of Volume 1, eats its way to page 832 of Volume 2. How far did the worm travel?

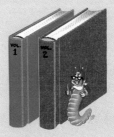

5 ▸ Before and After His Time

A man was born on the third day of 50 B.C. and died on the third day of A.D. 50. How many years did he live?

What's That?

Which is greatest, a *C-note*, a *grand*, or a *five-spot*?

Are you an *octogenarian*?

Do you suffer from *triskaidekaphobia*?

How many events are there in a *triathlon*?

Where is *seventh heaven*?

Extra Practice

▶ Lesson 1.1

Find the place value of the 4 in each number.

1. 4387   **2.** 46,322   **3.** 809,497   **4.** 461,003   **5.** 24,028,099

Write each number in words.

6. 9047   **7.** 81,502   **8.** 435,007   **9.** 600,536   **10.** 9,054,013

11. 755   **12.** 40,123   **13.** 109,066   **14.** 515,000   **15.** 3,330,001

Write the standard numeral.

16. 2 million, 75 thousand, 621

17. 73 million, 4 thousand, 25

18. 30 thousand, 700

19. 100 thousand, 42

20. 83 thousand, 427

21. 92 million, 761 thousand, 8

22. 6 billion, 122 thousand, 705

23. 2 thousand 19

24. 11 thousand, 900

25. 35 million, 54 thousand

26. 8 billion, 35 million, 67

27. 9 billion, 241

28. 627 million, 78 thousand, 5

29. 94 million, 375 thousand, 302

▶ Lesson 1.2

Find the place value of the 7 in each number.

1. 9.278   **2.** 2.6457   **3.** 4.752   **4.** 7.902   **5.** 0.0672

Write each decimal in words.

6. 4.91   **7.** 3.898   **8.** 1.0059   **9.** 0.066   **10.** 8.123

11. 9.7   **12.** 2.005   **13.** 12.17   **14.** 5.024   **15.** 3.6

Write the decimal.

16. eight and six hundredths

17. one and twelve hundredths

18. three and forty-two thousandths

19. ten and nine tenths

20. 54 hundredths

21. 78 and 91 thousandths

22. two thousandths

23. 14 and 503 thousandths

24. 338 thousandths

25. 100 and 7 hundredths

26. five and sixty-four thousandths

27. eighty-one and seven tenths

28. 3 thousand 607 and 14 ten thousandths

29. six hundred two and seventy-six thousandths

▶ Lesson 1.5

Compare. Use >, <, or = .

1. 4567 ___ 921
2. 56,423 ___ 65,423
3. 978,504 ___ 987,504
4. 7291 ___ 7219
5. 3,052,769 ___ 4,052,869
6. 71,434 ___ 71,432
7. 8919 ___ 9818
8. 190,655 ___ 190,566
9. 9762 ___ 10,439
10. 36,461 ___ 46,461
11. 8,003,758 ___ 803,785
12. 520,388 ___ 520,386
13. 89,052 ___ 98,025
14. 776,321 ___ 775,321
15. 144,648 ___ 144,486

Order from greatest to least.

16. 3467   3476   3477   3466
17. 75,965   7695   75,956   796
18. 50,531   50,513   50,135
19. 17,749   16,750   16,748
20. 429,672   429,267   429,726
21. 5,896,420   5,896,428   589,542
22. 2,078,399   2,068,993   4,079,393
23. 4,503,255   405,955   450,255
24. 8964   894   8846   9864
25. 69,304   690,304   69,403

26. Which is higher, Mt. Everest at 29,028 feet above sea level, or Mt. McKinley, at 20,322 feet?

27. Which river is shorter, the Yangtze in China (3437 miles) or the Yenisey in the USSR (3442 miles)?

▶ Lesson 1.6

Compare. Use >, <, or = .

1. 5.8 ___ 9.1
2. 99.05 ___ 99.50
3. 3.83 ___ 3.23
4. 2.08 ___ 2.0800
5. 4.897 ___ 4.987
6. 5.3094 ___ 5.3086
7. 7.452 ___ 4.725
8. 9.1 ___ 8.999
9. 8.700 ___ 8.7
10. 5.62 ___ 5.629
11. 30.7 ___ 3.077
12. 4.53 ___ 4.518
13. 0.02 ___ 0.0020
14. 2.49 ___ 2.398
15. 0.6 ___ 6.0

Order from least to greatest.

16. 3.9   3.891   3.7665   3.92
17. 0.051   0.0015   0.0515   0.01
18. 4.23   4.32   4.33   4.22
19. 7.58   7.8   7.5   7.588
20. 9.1   8.7   7.9   9.9
21. 0.11   1.19   0.19   0.91
22. 5.5   4.56   6.55   4.6
23. 13.0   13.72   13.726   13.7
24. 2.04   2.004   2.4   2.404
25. 0.83   0.8139   0.81   0.819

26. Which is shorter, a race track that is 29.095 miles, or one that is 28.9 miles?

27. Which route is longer, one that is 124.89 kilometers, or one that is 118.9 kilometers?

Lesson 1.7

Round to the nearest ten.

1. 39
2. 604
3. 8125
4. 19,663
5. 780,552

6. 7893
7. 35,647
8. 44,211
9. 500,478
10. 114,996

11. 84
12. 912
13. 2575
14. 58,409
15. 9,964,017

Round to the nearest hundred.

16. 94,283
17. 408,834
18. 55,751
19. 476,272
20. 99,499

21. 588
22. 953
23. 7425
24. 67,638
25. 932,171

26. 818
27. 51,339
28. 28,966
29. 44,742
30. 3,062,523

Round to the nearest thousand.

31. 2699
32. 44,735
33. 67,511
34. 98,463
35. 505,949

36. 8188
37. 717,351
38. 19,802
39. 376,590
40. 239,412

Round 89,415 to the nearest

41. ten.
42. ten thousand.
43. thousand.
44. hundred.

Round $44,572 to the nearest

45. thousand dollars.
46. hundred dollars.
47. ten dollars.

Lesson 1.8

Round to the nearest whole number or dollar.

1. 7.2
2. 3.088
3. 5.501
4. $19.63
5. 0.445

6. 9.83
7. $5.15
8. $39.49
9. 1.711
10. $2.75

11. $114.89
12. 34.5
13. 8.55
14. 805.19
15. $75.38

Round to the nearest tenth or ten cents.

16. $5.86
17. 6.945
18. 4.193
19. 11.025
20. $3.61

21. 1.39
22. $0.75
23. 0.052
24. $43.23
25. 2.66

26. 73.982
27. 115.536
28. 99.949
29. 8.291
30. 23.45

Round to the nearest hundredth.

31. 3.672
32. 0.325
33. 9.9992
34. 6.004
35. 5.0773

36. 2.816
37. 4.0537
38. 7.194
39. 18.0054
40. 8.982

Round 7.2853 to the nearest

41. tenth.
42. thousandth.
43. whole number.
44. hundredth.

Round $64.947 to the nearest

45. 10 dollars.
46. dollar.
47. 10 cents.
48. 100 dollars.

► Lesson 1.9

Add.

1.	23 $+$ 89	2.	0.654 $+$ 0.875	3.	926 $+$ 877	4.	4.57 $+$ 9.65	5.	8093 $+$ 5867
6.	97,635 $+$ 62,179	7.	9.352 $+$ 9.967	8.	5677 $+$ 5488	9.	35,608 $+$ 89,543	10.	46.727 $+$ 93.847
11.	5675 $+$ 8493	12.	77,563 $+$ 95,431	13.	3.9 $+$ 6.5	14.	9.86 $+$ 4.57	15.	7637 $+$ 899
16.	6.905 $+$ 4.865	17.	47.56 $+$ 38.75	18.	5.5087 $+$ 7.8952	19.	13,805 $+$ 978	20.	15.079 $+$ 7.951

21. $44 + 73 + 92$

22. $403 + 6795 + 9758$

23. $7.9 + 8.3 + 3.9$

24. $5.6 + 8.84 + 7.96$

25. $4.067 + 5 + 9.9$

26. $0.38 + 0.006 + 7$

27. $4.823 + 1.2295$

28. $0.765 + 0.84 + 3$

29. $6743 + 0.9685$

30. $75,403 + 9865$

31. $77,925 + 95,357$

32. $0.57 + 34 + 8.6$

33. $66 + 87 + 29 + 58$

34. $928 + 594 + 877$

35. $2.36 + 1.84 + 7$

36. $5624 + 19,408$

37. $0.07 + 7.77 + 77.07$

38. $8 + 5.6 + 4.13$

► Lesson 1.10

Subtract.

1.	276 $-$ 198	2.	5.24 $-$ 2.67	3.	8365 $-$ 4676	4.	9.631 $-$ 6.387	5.	45,728 $-$ 18,819
6.	4.16 $-$ 1.85	7.	9.5 $-$ 5.8	8.	8.43 $-$ 6.87	9.	6.561 $-$ 1.785	10.	7632 $-$ 3986
11.	19.2348 $-$ 13.8769	12.	9.24 $-$ 7.45	13.	81,423 $-$ 9,654	14.	9789 $-$ 6435	15.	7.6 $-$ 4.4
16.	3.91 $-$ 1.45	17.	33,824 $-$ 29,956	18.	5.6 $-$ 3.9	19.	99.72 $-$ 7.89	20.	4321 $-$ 892

21. $9.25 - 6.75$

22. $4359 - 2467$

23. $6.14 - 2.22$

24. $45,962 - 8973$

25. $63,248 - 41,459$

26. $3.4 - 2.5$

27. $54.762 - 38.883$

28. $15.61 - 7.85$

29. $9472 - 684$

30. $7.6 - 5.9$

31. $3255 - 699$

32. $427 - 99$

33. $8.1 - 4.7$

34. $452.8 - 339.9$

35. $8271 - 575$

36. $9.78 - 3.06$

37. $14.36 - 8.27$

38. $2467 - 895$

▶ Lesson 1.11

Subtract.

1.	502 − 389	**2.**	4005 − 1728	**3.**	8072 − 7388	**4.**	60,307 − 22,548	**5.**	70,043 − 49,165
6.	1.09 − 0.16	**7.**	8.0 − 2.6	**8.**	7.05 − 6.2	**9.**	2.002 − 0.74	**10.**	6.0306 − 3.5617
11.	10.8052 − 5.9678	**12.**	14 − 6.75	**13.**	407 − 205	**14.**	28.09 − 8.36	**15.**	2.2 − 1.689
16.	450.2 − 68.04	**17.**	80,060 − 33,457	**18.**	10 − 8.5	**19.**	4065 − 1873	**20.**	1.075 − 0.889

21. 9007 − 5328   **22.** 60,904 − 25,867   **23.** 90,530 − 44,682

24. 9 − 5.8   **25.** $34.07 − $13.75   **26.** 0.4001 − 0.1976

27. $900 − $145.89   **28.** 16 − 4.783   **29.** $5000 − $2618

30. 7064 − 895   **31.** 3.9 − 0.8675   **32.** 46 − 9.35

33. How much greater is the distance between the Earth and the sun (93,210,000 miles) than the distance between the Earth and the moon (238,906 miles)?

34. How much longer is the Nile River in Africa (4145 miles) than the Amazon River in South America (4006 miles)?

▶ Lesson 1.12

Estimate the sum or difference.

1.	566 + 714	**2.**	974 − 396	**3.**	814 + 852	**4.**	898 − 607	**5.**	165 + 672
6.	935 − 212	**7.**	9.45 − 5.78	**8.**	5.241 + 8.765	**9.**	73.85 − 15.92	**10.**	7.637 + 7.927
11.	93.37 − 42.57	**12.**	6.49 + 4.53	**13.**	3089 + 4455	**14.**	8376 + 7903	**15.**	8500 − 1298
16.	9833 − 3645	**17.**	6741 + 7499	**18.**	7457 − 3809	**19.**	$7.95 + 4.26	**20.**	$4.07 − 1.59
21.	8.67 + 9.18	**22.**	36.992 − 27.873	**23.**	73.506 + 68.747	**24.**	9.639 − 8.75	**25.**	17,726 − 9,254
26.	33,465 + 21,599	**27.**	987 − 76	**28.**	$45.66 − 16.25	**29.**	439 565 + 64	**30.**	$3.74 8.89 + 2.31

▶ Lesson 2.2

Multiply.

1. 9862
   × 5

2. 63
   × 8

3. 777
   × 4

4. 369
   × 9

5. 3878
   × 6

6. 534
   × 7

7. 2976
   × 5

8. 95
   × 3

9. 4895
   × 9

10. 6048
    × 8

11. 37
    × 4

12. 307
    × 7

13. 5814
    × 2

14. 88
    × 9

15. 7009
    × 0

16. 919
    × 6

17. 1954
    × 3

18. 69
    × 9

19. 4075
    × 8

20. 566
    × 7

21. 942×8
22. 68×7
23. 535×4
24. 999×9
25. 4386×6
26. 97×3
27. 442×8
28. 7000×5
29. 63×2
30. 7096×7

31. One box contains 356 pencils. How many pencils are in 9 boxes?

32. One package contains 575 pins. How many pins are in 7 boxes?

▶ Lesson 2.3

Multiply mentally.

1. 50×300
2. 7000×700
3. 15×400
4. 9×4000
5. 70×60
6. 100×50
7. 32×10
8. 800×800
9. 1000×78
10. 20×500
11. 80×60
12. 10×996
13. 900×3000
14. 500×900
15. 600×500
16. 43×1000
17. 10×6589
18. 800×50
19. 100×9700
20. 400×400
21. 1000×27
22. $9 \times 90,000$
23. 7000×90
24. 70×80
25. 100×84
26. 30×400
27. 8000×200
28. 763×1000
29. $56,432 \times 10$
30. 2000×50
31. 400×800
32. 4000×400

33. A truck carries 576 cartons. Another truck carries 100 times that many. How many does the second truck carry?

34. There are 50 envelopes in a box and there are 9 boxes. How many envelopes are there in all?

35. A page of stamps contains 100 stamps. How many stamps are there on 75 pages?

36. Terri drove 98 miles. Kim drove 10 times as far. How far did Kim drive?

► Lesson 2.4

Multiply.

1.	81 × 62	2.	75 × 93	3.	54 × 84	4.	237 × 29	5.	856 × 91
6.	525 × 407	7.	306 × 505	8.	186 × 890	9.	771 × 573	10.	8029 × 609
11.	90 × 84	12.	603 × 72	13.	5007 × 15	14.	791 × 40	15.	33 × 88
16.	520 × 42	17.	178 × 81	18.	6000 × 46	19.	53 × 13	20.	805 × 409

21. 23 × 418　22. 639 × 107　23. 75 × 905　24. 92 × 981　25. 41 × 552

26. 60 × 346　27. 500 × 649　28. 88 × 94　29. 207 × 117　30. 81 × 235

31. A box contains 5250 cards. How many cards are in 18 boxes?

32. An airplane seats 338 passengers. If every seat is taken, how many people fly in 410 trips?

► Lesson 2.6

Estimate each product.

1.	45 × 56	2.	83 × 91	3.	76 × 3	4.	97 × 3	5.	62 × 74
6.	185 × 43	7.	450 × 8	8.	917 × 66	9.	872 × 49	10.	833 × 6
11.	49 × 6	12.	72 × 6	13.	93 × 9	14.	329 × 43	15.	816 × 32
16.	390 × 74	17.	17 × 29	18.	86 × 92	19.	72 × 9	20.	38 × 7

21. Carl saves $89 a month. About how much will he save in 18 months?

22. JoJo earns $167 a week during the summer. About how much can she earn in 11 weeks?

23. There are 65 classes at King High School. Each class has about 27 students. About how many students are in the school?

24. There are 894 packages of paper in each carton. About how many packages are there in 511 cartons?

► **Lesson 2.7**

Multiply.

1. 98
× 4.1

2. 462
× 0.5

3. 707
× 9.2

4. $5.39
× 7

5. $26.98
× 3

6. 8
× 0.4

7. 12.6
× 2

8. $19.54
× 8

9. 2
× 0.259

10. 2.873
× 5

11. 2.75
× 7

12. 903
× 0.06

13. $8.49
× 17

14. 68
× 4.9

15. 5.5
× 82

16. 3 × $55.75

17. 1.14 × 9

18. 6 × 4.3

19. 0.209 × 7

20. 21 × $4.06

21. 8.5 × 35

22. 99 × $1.19

23. 7 × 3.691

24. 0.026 × 6

25. 54 × 9.03

26. 5.5 × 317

27. $8.23 × 4

28. 7 × 0.008

29. 452 × 3.3

30. 2 × $89.06

31. 0.095 × 4

32. 16 × $0.78

33. 5 × 52.4

34. 4.367 × 3

35. $19.95 × 6

► **Lesson 2.8**

Multiply.

1. 76.5 × 100

2. 1000 × 4.5

3. 40 × 0.007

4. 0.08 × 1000

5. 10 × 0.532

6. 1.46 × 50

7. 8.9 × 700

8. 100 × 6.789

9. 72.09 × 100

10. 0.06 × 10

11. 4.9 × 1000

12. 60 × 5.5

13. 800 × $0.50

14. 1000 × $42.75

15. $99.99 × 100

16. 300 × 8.6

17. 0.326 × 100

18. 77.9 × 40

19. $0.62 × 900

20. 1000 × 3.91

21. $0.07 × 1000

22. 100 × 8.906

23. 400 × $0.40

24. 800 × 3.9

25. Yahna wants to buy 250 balloons at $0.03 each. What is the total cost?

26. Colored paper costs $0.08. How much will 1000 sheets cost?

27. A box of paper clips contains 100 clips. Each box cost $0.79. How much will 10 boxes cost?

28. Fabric is on sale for $3.99 a yard. Kim buys 20 yards. How much does she have to pay?

29. Henry's old car costs him $0.24 a mile to operate. How much does it cost Henry to drive 1000 miles?

30. Aluminum for recycling is bought for $0.38 a pound. How much money can you make if you sell 100 pounds?

► Lesson 2.10

Multiply.

1. 9.3×2.1	2. 0.006×0.05	3. 1.8×0.9	4. 7.02×1.95	5. 0.88×7.69
6. 5.55×4.4	7. 0.091×6.7	8. 25.3×1.9	9. 32.79×0.06	10. 4.4×2.9
11. 8.005×0.4	12. 0.381×1.07	13. 0.7×0.9	14. 7.0×0.9	15. 7.007×9.09
16. 7.0×9.0	17. 9.07×4.009	18. 8.5×0.072	19. 4.17×0.07	20. 2.4×8.2
21. 0.905×2.06	22. 3.8×0.8	23. 67.13×0.08	24. 1.46×0.56	25. 5.75×1.25

26. 9.3×2.1 27. 0.25×5.7 28. 8.8×0.8 29. 2.305×9.7

30. 0.6×73.9 31. 1.14×2.36 32. 8.2×0.5 33. 3.92×0.178

34. Joel needs a piece of wood that is 2.5 times as long as a pole 1.9 meters long. How long a piece does he need?

35. Jenny makes $4.78 an hour. How much does she make if she works 6.5 hours?

► Lesson 3.2

Divide.

1. $6)\overline{594}$ 2. $9)\overline{3082}$ 3. $8)\overline{978}$ 4. $3)\overline{5343}$ 5. $5)\overline{8985}$

6. $2)\overline{145}$ 7. $4)\overline{652}$ 8. $7)\overline{2349}$ 9. $6)\overline{4186}$ 10. $3)\overline{936}$

11. $9)\overline{4572}$ 12. $5)\overline{623}$ 13. $8)\overline{632}$ 14. $2)\overline{12,810}$ 15. $6)\overline{872}$

16. $4)\overline{32,200}$ 17. $7)\overline{1295}$ 18. $3)\overline{2763}$ 19. $5)\overline{10,700}$ 20. $8)\overline{904}$

21. $98 \div 4$ 22. $86 \div 2$ 23. $400 \div 8$ 24. $3192 \div 6$

25. $3065 \div 9$ 26. $609 \div 5$ 27. $976 \div 7$ 28. $8864 \div 2$

29. $99 \div 3$ 30. $2832 \div 4$ 31. $565 \div 6$ 32. $7123 \div 8$

33. Derrick uses 2 washers for each nut and bolt he uses. How many nuts and bolts can he tighten if he has 503 washers?

34. Jonita is making packages of cards. Each package contains 8 cards. How many packages can she make with 456 cards?

▶ Lesson 3.3

Divide.

1. $6.89 \div 100$
2. $0.9 \div 1000$
3. $4752 \div 10$
4. $0.56 \div 100$

5. $58{,}932 \div 1000$
6. $751 \div 10$
7. $123{,}495 \div 100$
8. $9000 \div 30$

9. $400 \div 200$
10. $0.16 \div 100$
11. $8000 \div 500$
12. $7.698 \div 100$

13. $63{,}000 \div 90$
14. $456{,}322 \div 10$
15. $4900 \div 70$
16. $25{,}500 \div 10$

17. $320{,}000 \div 80$
18. $7835 \div 1000$
19. $3.6 \div 1000$
20. $56{,}000 \div 800$

21. $450 \div 90$
22. $3600 \div 600$
23. $9.83 \div 100$
24. $0.075 \div 100$

25. $20{,}000 \div 500$
26. $1800 \div 90$
27. $5.4 \div 10$
28. $54{,}000 \div 60$

29. For the school dance, Lamar bought 3500 tickets in rolls of 50 each. How many rolls did he buy?

30. It is Cary's job to unload the boxes of books. There are 850 books. How many stacks of ten books can he make?

▶ Lesson 3.4

Estimate.

1. $199 \div 19$
2. $732 \div 8$
3. $3456 \div 97$
4. $689 \div 32$

5. $60{,}438 \div 525$
6. $470 \div 77$
7. $399 \div 5$
8. $9875 \div 249$

9. $624 \div 7$
10. $7936 \div 38$
11. $9119 \div 29$
12. $32{,}228 \div 436$

13. $15{,}987 \div 387$
14. $808 \div 9$
15. $356 \div 41$
16. $839 \div 778$

17. $425 \div 58$
18. $2811 \div 92$
19. $742 \div 13$
20. $6399 \div 78$

21. $277 \div 18$
22. $6219 \div 329$
23. $555 \div 7$
24. $3978 \div 83$

25. $62{,}649 \div 888$
26. $448 \div 52$
27. $7234 \div 86$
28. $604 \div 35$

29. On a trip the Jimenez family drove 2389 miles at an average speed of 55 miles per hour. Estimate the number of driving hours it took.

30. On a map each centimeter represents 46 kilometers. The distance between two cities is 987 kilometers. About how many centimeters would that be on the map?

▶ Lesson 3.5

Divide.

1. $80 \overline{)567}$ 2. $11 \overline{)9744}$ 3. $60 \overline{)960}$ 4. $25 \overline{)925}$ 5. $31 \overline{)378}$

6. $58 \overline{)600}$ 7. $20 \overline{)6423}$ 8. $99 \overline{)431}$ 9. $40 \overline{)620}$ 10. $12 \overline{)840}$

11. $66 \overline{)6666}$ 12. $75 \overline{)1500}$ 13. $30 \overline{)8492}$ 14. $70 \overline{)635}$ 15. $50 \overline{)9533}$

16. $24 \overline{)1680}$ 17. $14 \overline{)1190}$ 18. $90 \overline{)873}$ 19. $20 \overline{)660}$ 20. $81 \overline{)794}$

21. $412 \div 39$ 22. $3105 \div 69$ 23. $1080 \div 72$ 24. $250 \div 40$

25. $1050 \div 25$ 26. $1960 \div 70$ 27. $794 \div 60$ 28. $3060 \div 30$

29. $2250 \div 18$ 30. $5000 \div 42$ 31. $998 \div 20$ 32. $4850 \div 37$

33. Marbles are packaged with 35 marbles per package. How many packages can be made from 3360 marbles?

34. Tracy collects quarters. How many quarters are in 78 dollars?

▶ Lesson 3.6

Divide.

1. $125 \overline{)3875}$ 2. $903 \overline{)17,157}$ 3. $600 \overline{)31,598}$ 4. $805 \overline{)4830}$

5. $550 \overline{)12,660}$ 6. $799 \overline{)15,980}$ 7. $186 \overline{)959}$ 8. $400 \overline{)21,742}$

9. $491 \overline{)17,185}$ 10. $310 \overline{)9300}$ 11. $212 \overline{)1698}$ 12. $890 \overline{)44,500}$

13. $700 \overline{)28,855}$ 14. $298 \overline{)3278}$ 15. $103 \overline{)721}$ 16. $500 \overline{)8590}$

17. $86,945 \div 900$ 18. $6494 \div 191$ 19. $7555 \div 150$ 20. $72,912 \div 300$

21. Rubber bands are boxed in groups of 500. How many full boxes can be made from 25,125 rubber bands?

22. Tickets are sold in lots of 800. How many full lots are needed for a crowd of 53,629?

23. Prize money of $1,625,000 is divided evenly among 125 winners. How much money does each winner receive?

24. Colored paper comes in packages of 250 sheets. How many packages can be made from 60,000 sheets?

► Lesson 3.9

Simplify.

1. $4 \times (2 + 3)$
2. $18 - 2 \times 3 + 4$
3. $(4 \times 6) - (5 \times 2)$
4. $(72 \div 8) \times 3$
5. $2 \times 2 + (64 \div 8)$
6. $20 \div 4 + 6 \times 6$
7. $9 \times 2 + 6 - 20$
8. $50 - (5 \times 5) - 1$
9. $3 \times 8 + (24 \div 8)$
10. $5 + (5 \times 5) - 5$
11. $(5 + 5) \times 5 - 5$
12. $5 - 5 + 5 \times 5$
13. $19 - (16 - 8) \times 2$
14. $54 \div 6 \times 8 + 3$
15. $90 \div 9 + 80 \div 8$
16. $14 - 4 \times (0 \times 9)$
17. $2 \times 2 \times 5 \div (5 + 5)$
18. $(24 + 36) \div 12$
19. $32 \div 4 + 6 \times 5$
20. $3 + 2 \times 5 - 7$
21. $29 - 7 \times 3 + 4$
22. $8 \times 7 + 15 \div 3$
23. $3 \times 3 \times 3 - 14$
24. $0 \times (12 \div 6 + 3)$
25. $80 - 2 \times 2 \times 5$
26. $9 \div 9 \times 9 - 9$
27. $42 \div 6 + (4 - 2)$
28. $(60 - 30) \div 5 + 4$
29. $44 - 28 \div 4 \times 2$
30. $12 + 35 \div 5 - 9$

► Lesson 3.10

Divide.

1. $16\overline{)0.48}$
2. $7\overline{)2.10}$
3. $25\overline{)52.5}$
4. $35\overline{)0.315}$
5. $8\overline{)167.2}$
6. $42\overline{)37.8}$
7. $5\overline{)\$45.95}$
8. $11\overline{)0.814}$
9. $3\overline{)6.9}$
10. $66\overline{)12.54}$
11. $9\overline{)0.081}$
12. $93\overline{)6.51}$
13. $4\overline{)0.96}$
14. $2\overline{)21.8}$
15. $15\overline{)141.0}$
16. $8\overline{)\$56.32}$
17. $51\overline{)0.663}$
18. $7\overline{)0.049}$
19. $36\overline{)14.4}$
20. $21\overline{)134.4}$
21. $109.2 \div 6$
22. $0.748 \div 44$
23. $98.1 \div 3$
24. $0.207 \div 23$
25. $0.0056 \div 7$
26. $10.50 \div 5$
27. $0.666 \div 18$
28. $294.5 \div 95$
29. $0.0054 \div 2$
30. $\$74.22 \div 3$
31. $0.008 \div 2$
32. $1.891 \div 31$

33. Five friends had dinner out together. The bill was $48.25. What was each person's equal share of the bill?

34. Twelve relay runners each ran an equal lap of a 57.6 mile race. How many miles did each person run?

▶ Lesson 3.11

Divide. Round the quotient to the nearest tenth or nearest cent.

1. $46\overline{)89}$
2. $13\overline{)5.8}$
3. $5\overline{)62.7}$
4. $3\overline{)74}$
5. $19\overline{)557}$

6. $4\overline{)\$34.91}$
7. $7\overline{)88}$
8. $4\overline{)3.21}$
9. $11\overline{)\$56.24}$
10. $3\overline{)\$97.63}$

11. $25\overline{)86}$
12. $18\overline{)488}$
13. $34\overline{)\$560}$
14. $21\overline{)8.7}$
15. $3\overline{)368}$

16. $74\overline{)\$90}$
17. $49\overline{)39}$
18. $8\overline{)700}$
19. $5\overline{)0.813}$
20. $5\overline{)\$15.72}$

21. $\$9.89 \div 2$
22. $3 \div 16$
23. $4.59 \div 4$
24. $892 \div 22$

25. $\$89.17 \div 5$
26. $777 \div 4$
27. $9.41 \div 6$
28. $\$7.79 \div 8$

29. $38.6 \div 12$
30. $\$46.45 \div 3$
31. $73 \div 4$
32. $4.96 \div 5$

33. Seven girls chipped in together to buy a large bag of cosmetics on sale for $33.57. How much did each girl have to contribute?

34. On the class trip to Washington, 45 people ride in each bus. If 837 people have signed up for the trip, how many buses are needed?

▶ Lesson 3.12

Divide. Round to the nearest tenth, if needed.

1. $1.7\overline{)16.83}$
2. $1.11\overline{)6.549}$
3. $0.02\overline{)8.36}$
4. $0.5\overline{)1.265}$

5. $4.5\overline{)9.0}$
6. $1.3\overline{)16.9}$
7. $0.4\overline{)0.324}$
8. $0.08\overline{)0.3568}$

9. $0.19\overline{)1.52}$
10. $7.8\overline{)47.58}$
11. $0.03\overline{)90.6}$
12. $0.25\overline{)1.975}$

13. $6.4\overline{)569.6}$
14. $0.6\overline{)0.738}$
15. $1.2\overline{)66.48}$
16. $0.81\overline{)38.07}$

17. $0.01\overline{)453.8}$
18. $4.3\overline{)4.128}$
19. $22.2\overline{)1.554}$
20. $0.7\overline{)26.67}$

21. $61.1 \div 6.5$
22. $2.925 \div 0.75$
23. $17.69 \div 2.9$
24. $98.65 \div 0.005$

25. $0.872 \div 0.2$
26. $157.5 \div 3.5$
27. $79.2 \div 1.8$
28. $388.6 \div 0.58$

29. $12.5 \div 0.5$
30. $8.136 \div 4.52$
31. $8.1 \div 6.75$
32. $36.96 \div 0.06$

33. Hilary runs 10.75 miles every five days. How far does she run each day?

34. Henry has $33.75 in quarters. How many quarters does he have?

► **Lesson 3.13**

Divide. Round to the nearest tenth, if needed.

1. $0.08\overline{)64}$ 2. $1.5\overline{)6}$ 3. $8.2\overline{)41}$ 4. $0.45\overline{)3}$ 5. $2.4\overline{)12}$

6. $1.05\overline{)84}$ 7. $0.03\overline{)9}$ 8. $2.5\overline{)5}$ 9. $1.5\overline{)555}$ 10. $0.12\overline{)48}$

11. $0.7\overline{)147}$ 12. $6.3\overline{)315}$ 13. $0.01\overline{)8}$ 14. $4.57\overline{)914}$ 15. $0.75\overline{)2}$

16. $0.6\overline{)18}$ 17. $0.85\overline{)255}$ 18. $9.1\overline{)8190}$ 19. $0.5\overline{)9}$ 20. $3.3\overline{)528}$

21. $45 \div 0.2$ 22. $16 \div 8.35$ 23. $9 \div 0.06$ 24. $90 \div 0.4$

25. $112 \div 2.8$ 26. $12 \div 2.36$ 27. $843 \div 2.81$ 28. $1356 \div 4.52$

29. $15{,}536 \div 15.5$ 30. $74{,}856 \div 9.02$ 31. $400 \div 78.9$

32. Pete drove 182 miles in his car. He used 6.5 gallons of gas. How many miles did he average per gallon of gas?

33. The distance between two cities on a map is 540 kilometers. On a map this measures 4.5 centimeters. How many kilometers does each centimeter represent on the map?

► **Lesson 4.2**

Convert each measure.

1. 73 m to mm 2. 9.9 km to cm 3. 256 mm to cm

4. 4.6 cm to m 5. 8732 m to km 6. 19.8 m to cm

7. 65.4 cm to mm 8. 0.328 km to m 9. 7243 mm to m

10. 16 m to km 11. 508 cm to mm 12. 1.89 m to mm

13. 9 cm to mm 14. 9087 mm to m 15. 5 km to cm

16. 5280 m to km 17. 0.85 m to cm 18. 5136 mm to m

19. 181 cm to km 20. 1.9 mm to cm 21. 2.4 km to cm

22. 680 m to km 23. 3737 mm to m 24. 85.4 m to cm

25. 29 cm to mm 26. 6 km to cm 27. 739 m to mm

28. 4526 cm to km 29. 97.4 cm to mm 30. 0.997 m to km

31. 164 mm to cm 32. 1000 mm to m 33. 39.5 km to cm

34. 8762 cm to mm 35. 8.05 mm to m 36. 5.51 km to m

37. 709 cm to km 38. 113 cm to m 39. 5.67 mm to cm

▶ Lesson 4.3

Measure the length of each of the following to the nearest centimeter and the nearest millimeter.

1. your math book
2. your thumb
3. an earring
4. your pencil
5. a piece of hair
6. a carrot
7. an eraser
8. a belt
9. a shoestring
10. an envelope
11. a stapler
12. a calculator

Complete each statement. Choose the more reasonable measure.

13. A light switch is about 11 ___ long.   m   cm
14. A chair is about 1 ___ tall.   mm   m
15. A paper clip is about 3 ___ long.   mm   cm
16. A one-foot ruler has about 300 ___ on it.   cm   mm
17. A refrigerator is about 2 ___ tall.   cm   m
18. A box of cereal is about 26 ___ tall.   m   cm
19. A hockey stick is about 1.5 ___ long.   mm   m

▶ Lesson 4.5

Convert each measure.

1. 0.97 mg to g
2. 0.17 mg to g
3. 899 g to kg
4. 3 mg to g
5. 250 g to mg
6. 75 g to kg
7. 190 mg to g
8. 60.8 kg to mg
9. 8.98 g to kg
10. 56 g to kg
11. 117.4 kg to g
12. 0.25 mg to g
13. 400 mg to g
14. 50.4 g to mg
15. 4.75 kg to g

Complete each statement. Choose the more reasonable measure.

16. The mass of an apple is about 250 ___.   g   kg
17. The mass of a personal computer is about 18 ___.   g   kg
18. The mass of a car is about 2 ___.   kg   T
19. The mass of a dust particle is measured in ___.   g   mg
20. The mass of a dime is about 2 ___.   g   mg
21. The mass of a dollar bill is about 1 ___.   g   mg
22. The mass of a large stone statue is about 3 ___.   T   mg

► Lesson 4.6

Convert each measure.

1. 6.5 L to mL
2. 887 mL to L
3. 40 L to mL
4. 99 mL to L
5. 2.5 L to mL
6. 7.85 L to mL
7. 4.7 L to mL
8. 15 L to mL
9. 4.25 mL to L
10. 400 mL to L
11. 37 L to mL
12. 0.98 L to mL
13. 4.18 L to mL
14. 20.88 mL to L
15. 100.6 mL to L

Complete each statement. Choose the more reasonable measure.

16. Harvey bought 30 ▨ of gas for his car. mL L

17. A graduated cylinder contains 100 ▨ of liquid. mL L

18. The chef added 4 ▨ of anise flavoring to the cookie dough. mL L

19. The decorator bought 75 ▨ of paint for the new building. mL L

20. Each hiker carried a ▨ of water in her canteen. mL L

21. Joan bought a 1 ▨ container of orange juice. mL L

22. He gave her 50 ▨ of perfume for her birthday. mL L

23. A large bottle of soy sauce contains 592 ▨. mL L

► Lesson 5.2

Find 3 fractions equivalent to each fraction.

1. $\frac{3}{4}$
2. $\frac{11}{12}$
3. $\frac{7}{9}$
4. $\frac{12}{8}$
5. $\frac{2}{5}$
6. $\frac{3}{16}$
7. $\frac{4}{15}$
8. $\frac{10}{9}$
9. $\frac{3}{8}$
10. $\frac{5}{2}$
11. $\frac{5}{16}$
12. $\frac{9}{10}$

Find a fraction equivalent to each fraction.

13. $\frac{5}{10}$
14. $\frac{15}{25}$
15. $\frac{6}{9}$
16. $\frac{8}{12}$
17. $\frac{3}{12}$
18. $\frac{21}{30}$
19. $\frac{15}{45}$
20. $\frac{20}{32}$
21. $\frac{20}{60}$
22. $\frac{15}{20}$
23. $\frac{18}{15}$
24. $\frac{56}{64}$
25. $\frac{3}{9}$
26. $\frac{18}{36}$
27. $\frac{25}{40}$
28. $\frac{9}{27}$
29. $\frac{40}{30}$
30. $\frac{9}{15}$
31. $\frac{20}{50}$
32. $\frac{24}{18}$
33. $\frac{30}{100}$
34. $\frac{6}{24}$
35. $\frac{5}{30}$
36. $\frac{15}{50}$
37. $\frac{12}{9}$
38. $\frac{15}{30}$
39. $\frac{18}{10}$
40. $\frac{25}{30}$
41. $\frac{21}{24}$
42. $\frac{2}{16}$
43. $\frac{16}{14}$
44. $\frac{10}{16}$
45. $\frac{6}{10}$
46. $\frac{9}{18}$
47. $\frac{45}{25}$
48. $\frac{9}{6}$

Write each improper fraction as a whole or mixed number.

1. $\frac{9}{8}$ 2. $\frac{13}{2}$ 3. $\frac{5}{5}$ 4. $\frac{61}{8}$ 5. $\frac{4}{4}$ 6. $\frac{23}{4}$

7. $\frac{57}{8}$ 8. $\frac{21}{12}$ 9. $\frac{19}{8}$ 10. $\frac{3}{3}$ 11. $\frac{17}{10}$ 12. $\frac{15}{6}$

13. $\frac{10}{10}$ 14. $\frac{15}{4}$ 15. $\frac{18}{6}$ 16. $\frac{12}{5}$ 17. $\frac{82}{9}$ 18. $\frac{30}{5}$

19. $\frac{9}{9}$ 20. $\frac{16}{2}$ 21. $\frac{11}{3}$ 22. $\frac{20}{4}$ 23. $\frac{19}{19}$ 24. $\frac{22}{3}$

Write each mixed or whole number as an improper fraction.

25. $4\frac{1}{6}$ 26. 8   27. 6   28. $8\frac{1}{9}$ 29. $6\frac{7}{10}$ 30. $25\frac{3}{4}$

31. $2\frac{7}{9}$ 32. $4\frac{2}{3}$ 33. $3\frac{5}{6}$ 34. 22   35. $3\frac{3}{8}$ 36. $10\frac{7}{9}$

37. 12   38. $5\frac{2}{9}$ 39. $3\frac{5}{12}$ 40. $5\frac{9}{10}$ 41. $2\frac{5}{6}$ 42. $6\frac{1}{6}$

43. $2\frac{1}{4}$ 44. 20   45. $4\frac{2}{5}$ 46. $1\frac{3}{16}$ 47. 15   48. $2\frac{8}{9}$

Compare. Use $<$, $>$, or $=$.

1. $\frac{5}{8}$ ▨ $\frac{2}{3}$ 2. $\frac{1}{2}$ ▨ $\frac{10}{20}$ 3. $\frac{9}{10}$ ▨ $\frac{7}{10}$ 4. $\frac{1}{8}$ ▨ $\frac{1}{5}$

5. $\frac{2}{3}$ ▨ $\frac{10}{15}$ 6. $\frac{3}{10}$ ▨ $\frac{2}{5}$ 7. $2\frac{1}{2}$ ▨ $\frac{3}{2}$ 8. $3\frac{3}{4}$ ▨ $2\frac{3}{4}$

9. $\frac{1}{3}$ ▨ $\frac{5}{6}$ 10. $\frac{7}{8}$ ▨ $\frac{14}{16}$ 11. $9\frac{2}{9}$ ▨ $5\frac{7}{9}$ 12. $\frac{3}{16}$ ▨ $\frac{11}{16}$

13. $\frac{4}{9}$ ▨ $\frac{1}{9}$ 14. $3\frac{1}{3}$ ▨ $3\frac{7}{10}$ 15. $\frac{15}{4}$ ▨ $\frac{15}{16}$ 16. $\frac{8}{8}$ ▨ $\frac{3}{3}$

17. $2\frac{1}{6}$ ▨ $2\frac{1}{5}$ 18. $\frac{16}{8}$ ▨ $\frac{10}{20}$ 19. $\frac{4}{5}$ ▨ $\frac{7}{10}$ 20. $\frac{5}{20}$ ▨ $\frac{1}{4}$

21. $\frac{15}{25}$ ▨ $\frac{2}{5}$ 22. $\frac{7}{12}$ ▨ $\frac{5}{12}$ 23. $\frac{5}{8}$ ▨ $1\frac{3}{8}$ 24. $\frac{14}{20}$ ▨ $\frac{9}{10}$

25. $3\frac{6}{9}$ ▨ $3\frac{1}{3}$ 26. $\frac{7}{8}$ ▨ $\frac{3}{8}$ 27. $6\frac{1}{6}$ ▨ $6\frac{5}{6}$ 28. $4\frac{1}{2}$ ▨ $3\frac{7}{9}$

29. $\frac{7}{16}$ ▨ $\frac{15}{16}$ 30. $\frac{5}{5}$ ▨ $\frac{9}{9}$ 31. $\frac{25}{3}$ ▨ $8\frac{1}{3}$ 32. $\frac{5}{6}$ ▨ $\frac{4}{5}$

33. 4 ▨ $\frac{16}{4}$ 34. $2\frac{1}{8}$ ▨ $2\frac{1}{6}$ 35. $1\frac{4}{5}$ ▨ $\frac{9}{6}$ 36. $\frac{11}{12}$ ▨ $\frac{9}{8}$

37. $2\frac{3}{4}$ ▨ $\frac{11}{3}$ 38. $7\frac{2}{3}$ ▨ $9\frac{2}{3}$ 39. $\frac{15}{5}$ ▨ 3 40. $\frac{3}{4}$ ▨ $\frac{9}{12}$

▶ Lesson 5.5

List the factors.

1. 9   2. 21   3. 30   4. 25   5. 7   6. 24

7. 19   8. 5   9. 45   10. 50   11. 28   12. 11

Find the GCF of each pair of numbers.

13. 3 6   14. 10 25   15. 8 24   16. 45 10   17. 9 30

18. 15 5   19. 40 60   20. 7 21   21. 5 4   22. 13 39

23. 8 16   24. 8 18   25. 18 6   26. 21 28   27. 50 30

28. 14 7   29. 17 2   30. 55 11   31. 10 26   32. 42 14

33. 5 30   34. 8 9   35. 18 15   36. 20 19   37. 12 9

38. 32 64   39. 24 60   40. 49 63   41. 25 100   42. 19 29

43. 30 45   44. 34 17   45. 12 15   46. 35 85   47. 29 43

▶ Lesson 5.6

Write in lowest terms.

1. $\frac{10}{15}$ 2. $\frac{30}{40}$ 3. $\frac{25}{125}$ 4. $\frac{50}{75}$ 5. $\frac{18}{22}$ 6. $\frac{35}{45}$

7. $\frac{25}{30}$ 8. $\frac{6}{36}$ 9. $\frac{10}{50}$ 10. $\frac{55}{10}$ 11. $\frac{9}{30}$ 12. $\frac{9}{24}$

13. $\frac{15}{12}$ 14. $\frac{9}{6}$ 15. $\frac{6}{8}$ 16. $\frac{10}{4}$ 17. $\frac{12}{9}$ 18. $\frac{5}{15}$

19. $\frac{12}{48}$ 20. $\frac{35}{25}$ 21. $\frac{25}{75}$ 22. $\frac{14}{16}$ 23. $\frac{20}{45}$ 24. $\frac{14}{12}$

25. $\frac{14}{4}$ 26. $\frac{20}{30}$ 27. $\frac{32}{36}$ 28. $\frac{13}{26}$ 29. $\frac{50}{6}$ 30. $\frac{16}{24}$

31. $\frac{36}{30}$ 32. $\frac{14}{21}$ 33. $\frac{8}{48}$ 34. $\frac{6}{48}$ 35. $\frac{20}{25}$ 36. $\frac{40}{36}$

List the factors and write in lowest terms.

37. $\frac{4}{9}$ 38. $\frac{13}{15}$ 39. $\frac{7}{8}$ 40. $\frac{11}{25}$ 41. $\frac{5}{28}$ 42. $\frac{3}{16}$

43. $\frac{15}{19}$ 44. $\frac{5}{12}$ 45. $\frac{3}{40}$ 46. $\frac{16}{25}$ 47. $\frac{9}{8}$ 48. $\frac{14}{17}$

49. $\frac{30}{11}$ 50. $\frac{7}{24}$ 51. $\frac{9}{4}$ 52. $\frac{6}{5}$ 53. $\frac{27}{35}$ 54. $\frac{15}{32}$

► Lesson 5.7

Convert each fraction to a decimal and each decimal to a fraction.

1. $\frac{13}{15}$
2. 0.6
3. 0.035
4. $\frac{13}{20}$
5. $\frac{4}{9}$
6. 0.04

7. $\frac{2}{15}$
8. $\frac{21}{100}$
9. $\frac{1}{6}$
10. $\frac{1}{30}$
11. 0.8
12. 0.08

13. $\frac{3}{8}$
14. $\frac{19}{20}$
15. 0.85
16. 0.1
17. $\frac{6}{11}$
18. $\frac{1}{5}$

19. $\frac{5}{8}$
20. $\frac{1}{8}$
21. 0.875
22. 0.98
23. 0.255
24. 0.148

25. 0.011
26. 0.002
27. $\frac{3}{11}$
28. $\frac{5}{16}$
29. 0.2
30. $\frac{5}{9}$

31. $\frac{10}{11}$
32. $\frac{5}{12}$
33. 0.76
34. 0.041
35. 0.199
36. $\frac{2}{9}$

37. 0.4
38. 0.007
39. $\frac{7}{15}$
40. $\frac{1}{16}$
41. 0.097
42. $\frac{1}{12}$

43. 0.03
44. $\frac{1}{9}$
45. 0.3
46. $\frac{11}{12}$
47. 0.33
48. 0.44

49. $\frac{11}{15}$
50. $\frac{13}{50}$
51. 0.009
52. 0.28
53. $\frac{14}{15}$
54. 0.999

► Lesson 6.1

Multiply. Write the answers in lowest terms.

1. $5 \times \frac{2}{3}$
2. $\frac{2}{3} \times \frac{2}{3}$
3. $\frac{5}{6} \times \frac{1}{8}$
4. $2 \times \frac{7}{8}$

5. $11 \times \frac{2}{9}$
6. $\frac{1}{6} \times \frac{3}{4}$
7. $\frac{2}{5} \times \frac{3}{4}$
8. $8 \times \frac{5}{6}$

9. $\frac{1}{8} \times \frac{1}{9}$
10. $4 \times \frac{5}{12}$
11. $12 \times \frac{1}{5}$
12. $\frac{4}{5} \times \frac{4}{9}$

13. $\frac{1}{16} \times \frac{1}{2}$
14. $7 \times \frac{4}{7}$
15. $\frac{3}{8} \times \frac{2}{3}$
16. $\frac{5}{6} \times \frac{2}{3}$

17. $3 \times \frac{1}{9}$
18. $\frac{3}{5} \times \frac{3}{8}$
19. $\frac{3}{4} \times \frac{1}{5}$
20. $6 \times \frac{7}{8}$

21. $\frac{1}{8} \times \frac{5}{9}$
22. $\frac{3}{16} \times \frac{1}{4}$
23. $\frac{7}{9} \times \frac{2}{3}$
24. $2 \times \frac{4}{5}$

25. $3 \times \frac{5}{8}$
26. $\frac{2}{3} \times \frac{5}{12}$
27. $\frac{3}{4} \times \frac{1}{8}$
28. $13 \times \frac{1}{4}$

29. $\frac{7}{12} \times \frac{1}{10}$
30. $\frac{3}{10} \times \frac{1}{5}$
31. $10 \times \frac{7}{9}$
32. $\frac{1}{9} \times \frac{2}{5}$

33. A rice salad recipe uses 3 cups of rice and serves 12 people. How much rice would you use if you made $\frac{1}{2}$ the recipe?

34. Sam is a terrific basketball player. He makes $\frac{2}{3}$ of all the baskets he attempts. How many baskets does he make if he attempts 36 shots?

► Lesson 6.2

Multiply. Write the answers in lowest terms.

1. $\frac{7}{8} \times \frac{6}{9}$ 2. $\frac{5}{8} \times \frac{3}{10}$ 3. $\frac{4}{5} \times \frac{1}{6}$ 4. $\frac{9}{10} \times \frac{5}{12}$

5. $\frac{6}{10} \times \frac{7}{14}$ 6. $\frac{1}{8} \times \frac{4}{5}$ 7. $\frac{2}{9} \times \frac{3}{16}$ 8. $\frac{1}{6} \times \frac{3}{8}$

9. $\frac{7}{10} \times \frac{5}{12}$ 10. $\frac{5}{9} \times \frac{3}{5}$ 11. $\frac{1}{12} \times \frac{9}{16}$ 12. $\frac{11}{16} \times \frac{8}{9}$

13. $\frac{5}{6} \times \frac{1}{5}$ 14. $\frac{4}{9} \times \frac{1}{2}$ 15. $\frac{5}{16} \times \frac{4}{5}$ 16. $\frac{2}{3} \times \frac{9}{16}$

17. $\frac{1}{12} \times \frac{4}{9}$ 18. $\frac{3}{5} \times \frac{15}{16}$ 19. $\frac{1}{4} \times \frac{4}{9}$ 20. $\frac{5}{6} \times \frac{3}{10}$

21. $\frac{11}{12} \times \frac{4}{5}$ 22. $\frac{5}{16} \times \frac{2}{5}$ 23. $\frac{5}{8} \times \frac{9}{10}$ 24. $\frac{4}{5} \times \frac{7}{16}$

25. $\frac{6}{8} \times \frac{3}{12}$ 26. $\frac{14}{16} \times \frac{8}{28}$ 27. $\frac{5}{7} \times \frac{35}{45}$ 28. $\frac{10}{15} \times \frac{20}{50}$

29. Pablo sold $\frac{4}{5}$ of the tickets to the school carnival. Students bought $\frac{5}{8}$ of the tickets he sold. What fraction of the tickets were bought by students?

30. Half of the students at North High School work on Saturdays. Two-fifths of those who work also play on a school team. What fraction of the students work and play on a team?

► Lesson 6.4

Multiply. Write the answers in lowest terms.

1. $2\frac{2}{9} \times 1\frac{4}{5}$ 2. $9\frac{1}{2} \times \frac{1}{8}$ 3. $6 \times 2\frac{3}{10}$ 4. $4\frac{1}{8} \times 2\frac{2}{3}$

5. $1\frac{15}{16} \times 8$ 6. $2\frac{5}{8} \times 1\frac{1}{5}$ 7. $\frac{7}{12} \times 3\frac{3}{4}$ 8. $5\frac{1}{3} \times \frac{7}{8}$

9. $1\frac{1}{10} \times \frac{14}{15}$ 10. $3\frac{1}{9} \times 1\frac{3}{4}$ 11. $8\frac{5}{8} \times 1\frac{1}{3}$ 12. $\frac{9}{16} \times 2\frac{3}{4}$

13. $5\frac{5}{6} \times 3$ 14. $5 \times 4\frac{9}{10}$ 15. $\frac{4}{5} \times 7\frac{2}{9}$ 16. $\frac{11}{12} \times 3\frac{3}{10}$

17. $1\frac{1}{16} \times 3\frac{1}{5}$ 18. $3\frac{5}{9} \times \frac{3}{8}$ 19. $6\frac{2}{9} \times 1\frac{4}{5}$ 20. $1\frac{1}{3} \times \frac{3}{16}$

21. $4\frac{1}{2} \times \frac{2}{9}$ 22. $3\frac{3}{5} \times 1\frac{2}{3}$ 23. $2\frac{1}{6} \times 2$ 24. $\frac{2}{5} \times 2\frac{1}{12}$

25. $5\frac{1}{3} \times 2\frac{7}{16}$ 26. $4\frac{3}{4} \times 3\frac{1}{5}$ 27. $10\frac{2}{3} \times 1\frac{1}{2}$ 28. $14 \times 1\frac{3}{8}$

29. A pizza recipe uses $1\frac{2}{3}$ cups of sliced mushrooms. How many cups do you need for 8 pizzas?

30. Kiri uses $2\frac{1}{2}$ yards of ribbon for each yard of fabric. How much ribbon does she need for $4\frac{2}{3}$ yards of fabric?

► Lesson 6.6

Find the reciprocal.

1. $\frac{3}{4}$ 2. 7   3. 30   4. $\frac{7}{8}$ 5. $\frac{4}{9}$ 6. $\frac{13}{15}$

7. $\frac{5}{12}$ 8. 19   9. $\frac{7}{10}$ 10. 3   11. $\frac{4}{5}$ 12. 8

Divide.

13. $\frac{1}{3} \div \frac{4}{9}$ 14. $\frac{15}{16} \div 5$ 15. $\frac{11}{12} \div \frac{3}{4}$ 16. $\frac{5}{6} \div 9$

17. $\frac{3}{8} \div \frac{3}{4}$ 18. $\frac{9}{10} \div \frac{7}{8}$ 19. $\frac{3}{4} \div 6$ 20. $\frac{13}{16} \div 4$

21. $\frac{2}{9} \div \frac{5}{6}$ 22. $\frac{1}{3} \div 15$ 23. $\frac{14}{15} \div \frac{3}{5}$ 24. $\frac{9}{16} \div 9$

25. $\frac{5}{16} \div \frac{5}{9}$ 26. $\frac{11}{12} \div 3$ 27. $\frac{7}{8} \div \frac{3}{4}$ 28. $\frac{7}{10} \div \frac{7}{9}$

29. $\frac{4}{5} \div 10$ 30. $\frac{3}{10} \div 6$ 31. $\frac{3}{16} \div \frac{3}{5}$ 32. $\frac{4}{5} \div \frac{1}{16}$

33. Ben wants to divide a 9-foot board into pieces $\frac{3}{4}$ of a foot long. How many pieces will there be?

34. A piece of wire is $\frac{11}{12}$ of a yard in length. Paul needs wire pieces $\frac{1}{3}$ of a yard in length. How many pieces of wire can be cut from the wire?

► Lesson 6.7

Divide. Write the answers in lowest terms.

1. $1\frac{2}{3} \div 4\frac{1}{2}$ 2. $\frac{3}{8} \div 1\frac{1}{6}$ 3. $6\frac{5}{6} \div \frac{5}{6}$ 4. $\frac{3}{16} \div 5\frac{1}{8}$

5. $2\frac{3}{4} \div \frac{4}{5}$ 6. $\frac{9}{10} \div 1\frac{1}{2}$ 7. $3\frac{3}{5} \div 3\frac{3}{5}$ 8. $4\frac{2}{9} \div 1\frac{7}{12}$

9. $2\frac{1}{16} \div 1\frac{5}{6}$ 10. $3\frac{5}{9} \div 2$ 11. $11\frac{1}{4} \div \frac{3}{16}$ 12. $\frac{5}{6} \div 3\frac{1}{9}$

13. $7\frac{1}{2} \div \frac{3}{5}$ 14. $1\frac{3}{8} \div 3\frac{2}{3}$ 15. $\frac{4}{5} \div 2\frac{1}{2}$ 16. $\frac{5}{12} \div 2\frac{1}{4}$

17. $5\frac{5}{6} \div 2\frac{11}{12}$ 18. $\frac{2}{3} \div 1\frac{1}{3}$ 19. $1\frac{1}{4} \div \frac{5}{16}$ 20. $2\frac{1}{8} \div \frac{3}{4}$

21. $3\frac{7}{8} \div \frac{7}{8}$ 22. $1\frac{9}{16} \div 2\frac{5}{8}$ 23. $\frac{7}{15} \div 1\frac{1}{9}$ 24. $\frac{15}{16} \div 3\frac{3}{4}$

25. $\frac{11}{20} \div 5\frac{1}{2}$ 26. $6\frac{1}{9} \div 1\frac{3}{8}$ 27. $\frac{7}{8} \div 5\frac{3}{5}$ 28. $9\frac{7}{9} \div 1\frac{5}{6}$

29. A desk phone is $4\frac{3}{4}$ inches wide. How many phones can fit on a desk that is $30\frac{1}{2}$ inches wide?

30. A rope is $16\frac{1}{2}$ feet long. How many pieces $\frac{3}{4}$ of a foot long can be cut from the rope?

▶ Lesson 7.1

Add or subtract. Write the answers in lowest terms.

1. $\dfrac{5}{8}$ $-\dfrac{1}{8}$　　2. $\dfrac{11}{16}$ $+\dfrac{9}{16}$　　3. $\dfrac{5}{6}$ $-\dfrac{1}{6}$　　4. $\dfrac{7}{9}$ $+\dfrac{5}{9}$　　5. $\dfrac{7}{8}$ $-\dfrac{1}{8}$　　6. $\dfrac{9}{10}$ $-\dfrac{4}{10}$

7. $\dfrac{1}{2}$ $\dfrac{1}{2}$ $+\dfrac{1}{2}$　　8. $\dfrac{8}{9}$ $\dfrac{5}{9}$ $+\dfrac{2}{9}$　　9. $\dfrac{11}{12}$ $\dfrac{5}{12}$ $+\dfrac{7}{12}$　　10. $\dfrac{5}{16}$ $\dfrac{9}{16}$ $+\dfrac{15}{16}$　　11. $\dfrac{4}{5}$ $\dfrac{2}{5}$ $+\dfrac{3}{5}$　　12. $\dfrac{2}{3}$ $\dfrac{1}{3}$ $+\dfrac{2}{3}$

13. $\dfrac{15}{16}$ $-\dfrac{9}{16}$　　14. $\dfrac{5}{12}$ $-\dfrac{1}{12}$　　15. $\dfrac{14}{15}$ $-\dfrac{8}{15}$　　16. $\dfrac{4}{9}$ $+\dfrac{5}{9}$　　17. $\dfrac{7}{12}$ $-\dfrac{5}{12}$　　18. $\dfrac{7}{9}$ $+\dfrac{8}{9}$

19. $\dfrac{5}{9} - \dfrac{4}{9}$　　20. $\dfrac{11}{12} + \dfrac{5}{12}$　　21. $\dfrac{9}{16} - \dfrac{1}{16}$　　22. $\dfrac{3}{10} - \dfrac{1}{10}$

23. $\dfrac{3}{4} + \dfrac{1}{4}$　　24. $\dfrac{7}{8} + \dfrac{3}{8}$　　25. $\dfrac{11}{15} - \dfrac{8}{15}$　　26. $\dfrac{13}{16} - \dfrac{7}{16}$

27. Juan swam $\dfrac{11}{16}$ of a mile. Maria swam $\dfrac{9}{16}$ of a mile. Who swam the shorter distance? How much shorter?

28. Kim lives $\dfrac{7}{10}$ of a mile east of the school. Carla lives $\dfrac{7}{10}$ of a mile west of the school. How far is it from Kim's house to Carla's house?

▶ Lesson 7.2

List the first five multiples.

1. 9　　2. 14　　3. 24　　4. 40　　5. 16　　6. 36

Find the LCD.

7. $\dfrac{2}{15}$ $\dfrac{11}{12}$　　8. $\dfrac{1}{2}$ $\dfrac{3}{4}$ $\dfrac{5}{6}$　　9. $\dfrac{3}{8}$ $\dfrac{4}{5}$　　10. $\dfrac{3}{10}$ $\dfrac{4}{5}$

11. $\dfrac{2}{3}$ $\dfrac{7}{8}$　　12. $\dfrac{3}{4}$ $\dfrac{5}{8}$ $\dfrac{1}{2}$　　13. $\dfrac{2}{5}$ $\dfrac{2}{3}$　　14. $\dfrac{1}{4}$ $\dfrac{2}{5}$ $\dfrac{7}{10}$

15. $\dfrac{9}{16}$ $\dfrac{1}{2}$ $\dfrac{7}{8}$　　16. $\dfrac{3}{4}$ $\dfrac{5}{12}$　　17. $\dfrac{4}{5}$ $\dfrac{5}{9}$　　18. $\dfrac{1}{3}$ $\dfrac{5}{6}$

19. $\dfrac{1}{8}$ $\dfrac{5}{12}$　　20. $\dfrac{5}{6}$ $\dfrac{4}{5}$　　21. $\dfrac{1}{3}$ $\dfrac{1}{4}$　　22. $\dfrac{1}{6}$ $\dfrac{7}{8}$

23. $\dfrac{7}{15}$ $\dfrac{2}{3}$　　24. $\dfrac{7}{8}$ $\dfrac{2}{9}$　　25. $\dfrac{5}{8}$ $\dfrac{4}{9}$ $\dfrac{1}{2}$　　26. $\dfrac{3}{10}$ $\dfrac{1}{15}$

27. $\dfrac{5}{6}$ $\dfrac{2}{3}$ $\dfrac{1}{2}$　　28. $\dfrac{4}{5}$ $\dfrac{9}{10}$ $\dfrac{8}{15}$　　29. $\dfrac{1}{12}$ $\dfrac{5}{9}$ $\dfrac{1}{6}$　　30. $\dfrac{1}{3}$ $\dfrac{7}{9}$

31. $\dfrac{7}{10}$ $\dfrac{1}{2}$　　32. $\dfrac{3}{5}$ $\dfrac{1}{10}$ $\dfrac{1}{8}$　　33. $\dfrac{1}{6}$ $\dfrac{5}{9}$　　34. $\dfrac{1}{2}$ $\dfrac{5}{12}$ $\dfrac{2}{9}$

Lesson 7.3

Add. Write the answers in lowest terms.

1. $\dfrac{2}{3}$ $+\dfrac{1}{9}$
2. $\dfrac{7}{8}$ $+\dfrac{3}{4}$
3. $\dfrac{4}{5}$ $+\dfrac{3}{4}$
4. $\dfrac{5}{16}$ $+\dfrac{1}{8}$
5. $\dfrac{1}{3}$ $+\dfrac{1}{8}$
6. $\dfrac{7}{10}$ $+\dfrac{1}{5}$

7. $\dfrac{5}{6}$ $+\dfrac{1}{4}$
8. $\dfrac{2}{9}$ $+\dfrac{1}{3}$
9. $\dfrac{1}{4}$ $+\dfrac{9}{16}$
10. $\dfrac{5}{8}$ $+\dfrac{1}{5}$
11. $\dfrac{1}{2}$ $+\dfrac{9}{10}$
12. $\dfrac{5}{12}$ $+\dfrac{5}{6}$

13. $\dfrac{3}{4}$ $+\dfrac{1}{6}$
14. $\dfrac{7}{8}$ $+\dfrac{3}{16}$
15. $\dfrac{2}{3}$ $+\dfrac{3}{10}$
16. $\dfrac{3}{8}$ $+\dfrac{11}{12}$
17. $\dfrac{11}{16}$ $+\dfrac{1}{2}$
18. $\dfrac{4}{5}$ $+\dfrac{7}{10}$

19. $\dfrac{2}{3}+\dfrac{4}{9}$
20. $\dfrac{7}{16}+\dfrac{1}{4}$
21. $\dfrac{1}{10}+\dfrac{1}{6}$
22. $\dfrac{5}{12}+\dfrac{1}{4}$

23. $\dfrac{1}{6}+\dfrac{1}{3}+\dfrac{3}{4}$
24. $\dfrac{1}{2}+\dfrac{5}{6}+\dfrac{4}{9}$
25. $\dfrac{7}{8}+\dfrac{2}{3}+\dfrac{1}{6}$
26. $\dfrac{3}{4}+\dfrac{1}{2}+\dfrac{4}{5}$

27. Sandra picked $\dfrac{1}{3}$ of the apples on one day and $\dfrac{3}{8}$ of the apples the next day. What fraction of the apples did she pick in all?

28. Greg sprinted $\dfrac{1}{4}$ of a mile on Tuesday, $\dfrac{1}{5}$ of a mile on Wednesday, and $\dfrac{1}{8}$ of a mile on Thursday. How far did he sprint altogether?

Lesson 7.4

Add. Write the answers in lowest terms.

1. $1\dfrac{3}{8}$ $+2\dfrac{1}{2}$
2. $10\dfrac{1}{8}$ $+6\dfrac{7}{8}$
3. $5\dfrac{3}{8}$ $+2\dfrac{2}{5}$
4. $3\dfrac{9}{16}$ $+2\dfrac{11}{16}$
5. $8\dfrac{2}{3}$ $+4\dfrac{2}{3}$

6. $1\dfrac{9}{10}$ $+3\dfrac{4}{5}$
7. $4\dfrac{1}{8}$ $+3\dfrac{3}{16}$
8. $1\dfrac{5}{8}$ $+2\dfrac{7}{8}$
9. $5\dfrac{7}{9}$ $+4\dfrac{7}{9}$
10. $6\dfrac{1}{2}$ $+4\dfrac{3}{4}$

11. $3\dfrac{1}{6}+2\dfrac{4}{9}$
12. $1\dfrac{3}{4}+4\dfrac{2}{5}$
13. $8\dfrac{1}{9}+7\dfrac{5}{12}$
14. $9\dfrac{7}{15}+3\dfrac{2}{15}$

15. $6\dfrac{2}{3}+1\dfrac{8}{9}$
16. $1\dfrac{1}{12}+1\dfrac{1}{6}$
17. $4\dfrac{7}{8}+9\dfrac{3}{10}$
18. $5\dfrac{2}{9}+3\dfrac{7}{9}$

19. $1\dfrac{5}{6}+4\dfrac{5}{6}$
20. $3\dfrac{9}{10}+3\dfrac{5}{6}$
21. $4\dfrac{4}{15}+8\dfrac{1}{15}$
22. $7\dfrac{1}{2}+6\dfrac{1}{2}$

23. Jo has $3\dfrac{3}{16}$ yards of silk. Ina has $4\dfrac{3}{4}$ yards of silk. How much silk do they have in all?

24. Terry rode his bike $9\dfrac{7}{10}$ miles. Jim rode his bike $8\dfrac{2}{3}$ miles. How far did the boys ride in all?

► Lesson 7.5

Subtract. Write the answers in lowest terms.

1. $\dfrac{5}{8}$ $-\dfrac{1}{3}$
2. $\dfrac{3}{4}$ $-\dfrac{1}{16}$
3. $\dfrac{7}{10}$ $-\dfrac{2}{5}$
4. $\dfrac{11}{12}$ $-\dfrac{1}{6}$
5. $\dfrac{7}{8}$ $-\dfrac{1}{4}$
6. $\dfrac{3}{4}$ $-\dfrac{1}{5}$

7. $\dfrac{8}{9}$ $-\dfrac{1}{3}$
8. $\dfrac{1}{2}$ $-\dfrac{3}{16}$
9. $\dfrac{9}{10}$ $-\dfrac{3}{4}$
10. $\dfrac{5}{8}$ $-\dfrac{1}{12}$
11. $\dfrac{3}{4}$ $-\dfrac{1}{6}$
12. $\dfrac{7}{12}$ $-\dfrac{1}{3}$

13. $\dfrac{5}{6} - \dfrac{2}{3}$
14. $\dfrac{7}{10} - \dfrac{2}{3}$
15. $\dfrac{5}{9} - \dfrac{1}{6}$
16. $\dfrac{1}{2} - \dfrac{1}{10}$
17. $\dfrac{3}{5} - \dfrac{1}{3}$

18. $\dfrac{3}{4} - \dfrac{2}{5}$
19. $\dfrac{5}{6} - \dfrac{3}{8}$
20. $\dfrac{11}{12} - \dfrac{7}{8}$
21. $\dfrac{9}{10} - \dfrac{1}{3}$
22. $\dfrac{15}{16} - \dfrac{3}{4}$

23. $\dfrac{3}{8} - \dfrac{1}{5}$
24. $\dfrac{8}{9} - \dfrac{1}{6}$
25. $\dfrac{4}{5} - \dfrac{1}{8}$
26. $\dfrac{7}{12} - \dfrac{1}{5}$
27. $\dfrac{5}{6} - \dfrac{4}{15}$

28. Gabe finished $\dfrac{7}{10}$ of his job. Jim finished $\dfrac{3}{4}$ of his job. Who completed more of his job? By what fraction?

29. Zoe collected $\dfrac{7}{12}$ of the junior class fund. Nicole collected $\dfrac{3}{8}$ of the fund. Who collected more? By what fraction?

► Lesson 7.7

Rename each whole number.

1. $8 = 7\dfrac{\blacksquare}{20}$
2. $4 = 3\dfrac{\blacksquare}{2}$
3. $20 = 19\dfrac{\blacksquare}{16}$
4. $9 = 8\dfrac{\blacksquare}{5}$

Subtract. Write the answers in lowest terms.

5. 9 $- 2\dfrac{1}{6}$
6. 4 $- 3\dfrac{1}{2}$
7. 15 $- 6\dfrac{1}{3}$
8. 7 $- 1\dfrac{1}{10}$
9. 3 $- 1\dfrac{3}{5}$

10. $6 - 3\dfrac{5}{12}$
11. $20 - 5\dfrac{1}{4}$
12. $8 - 2\dfrac{2}{5}$
13. $5 - 2\dfrac{7}{8}$

14. $25 - 5\dfrac{9}{10}$
15. $12 - 4\dfrac{5}{16}$
16. $12 - 9\dfrac{2}{9}$
17. $40 - 25\dfrac{7}{20}$

18. $66 - 23\dfrac{1}{2}$
19. $39 - 13\dfrac{4}{9}$
20. $50 - 25\dfrac{1}{8}$
21. $38 - 11\dfrac{8}{9}$

22. Meagan hiked 9 miles on the first day and $7\dfrac{1}{10}$ miles on the second day. How much farther did she hike the first day?

23. Dave worked $3\dfrac{2}{3}$ hours Friday and 8 hours Saturday. How much longer did he work on Saturday?

Rename each mixed number.

1. $6\frac{1}{2} = 5\frac{\blacksquare}{2}$
2. $4\frac{3}{4} = 3\frac{\blacksquare}{4}$
3. $2\frac{5}{8} = 1\frac{\blacksquare}{8}$
4. $8\frac{9}{10} = 7\frac{\blacksquare}{10}$

Subtract. Write the answers in lowest terms.

5. $\begin{array}{r} 3\frac{1}{4} \\ -\ 1\frac{3}{4} \\ \hline \end{array}$
6. $\begin{array}{r} 8\frac{7}{8} \\ -\ 2\frac{1}{6} \\ \hline \end{array}$
7. $\begin{array}{r} 5\frac{11}{12} \\ -\ 3\frac{1}{6} \\ \hline \end{array}$
8. $\begin{array}{r} 6\frac{1}{9} \\ -\ 2\frac{1}{8} \\ \hline \end{array}$
9. $\begin{array}{r} 9\frac{5}{9} \\ -\ 1\frac{1}{3} \\ \hline \end{array}$

10. $\begin{array}{r} 7\frac{2}{15} \\ -\ 3\frac{4}{15} \\ \hline \end{array}$
11. $\begin{array}{r} 6\frac{11}{16} \\ -\ 4\frac{3}{16} \\ \hline \end{array}$
12. $\begin{array}{r} 10\frac{1}{8} \\ -\ 6\frac{5}{8} \\ \hline \end{array}$
13. $\begin{array}{r} 12\frac{1}{4} \\ -\ 6\frac{1}{3} \\ \hline \end{array}$
14. $\begin{array}{r} 15\frac{4}{5} \\ -\ 5\frac{3}{5} \\ \hline \end{array}$

15. $11\frac{1}{10} - 3\frac{4}{5}$
16. $9\frac{7}{12} - 4\frac{1}{12}$
17. $8\frac{1}{6} - 4\frac{1}{3}$
18. $4\frac{1}{4} - 1\frac{1}{2}$

19. $17\frac{8}{9} - 9\frac{4}{9}$
20. $20\frac{3}{5} - 4\frac{2}{5}$
21. $13\frac{5}{12} - 6\frac{11}{12}$
22. $16\frac{7}{8} - 2\frac{1}{16}$

23. $18\frac{5}{6} - 9\frac{1}{4}$
24. $24\frac{3}{10} - 14\frac{3}{5}$
25. $15\frac{5}{12} - 8\frac{1}{12}$
26. $7\frac{2}{3} - 5\frac{1}{2}$

27. $12\frac{5}{9} - 6\frac{7}{9}$
28. $14\frac{9}{10} - 6\frac{1}{10}$
29. $19\frac{3}{4} - 17\frac{7}{8}$
30. $21\frac{11}{12} - 8\frac{3}{8}$

31. Reg skated $6\frac{3}{4}$ miles. Rich skated $7\frac{5}{6}$ miles. How much farther did Rich skate?

32. Claire watched the Superbowl for $1\frac{5}{12}$ hours. Jed watched it for $2\frac{1}{12}$ hours. What was the difference in their viewing time?

Estimate to the nearest whole number and to the nearest $\frac{1}{2}$.

1. $4\frac{5}{8} + 3\frac{3}{8}$
2. $12\frac{9}{10} - 5\frac{3}{4}$
3. $5\frac{5}{6} + 4\frac{4}{9}$
4. $7\frac{1}{9} - 4\frac{1}{6}$

5. $3\frac{7}{16} - 2\frac{1}{3}$
6. $9\frac{5}{15} - 3\frac{9}{16}$
7. $4\frac{1}{4} + 7\frac{5}{8}$
8. $3\frac{5}{12} + 2\frac{8}{9}$

9. $1\frac{5}{9} + 2\frac{2}{3}$
10. $7\frac{1}{3} - 5\frac{2}{5}$
11. $6\frac{2}{9} - 1\frac{5}{9}$
12. $4\frac{3}{4} + 4\frac{7}{8}$

13. $3\frac{1}{5} - 1\frac{4}{5}$
14. $5\frac{7}{16} + 1\frac{1}{8}$
15. $7\frac{5}{6} - 2\frac{2}{3}$
16. $5\frac{3}{5} + 6\frac{11}{12}$

17. $2\frac{1}{9} + 2\frac{8}{9}$
18. $8\frac{1}{4} - 2\frac{5}{8}$
19. $3\frac{2}{3} - 1\frac{7}{12}$
20. $7\frac{7}{9} + 8\frac{9}{16}$

21. Karl bought $3\frac{7}{8}$ pounds of peanuts and $2\frac{1}{3}$ pounds of walnuts. About how many pounds of nuts did he buy?

22. Sandi bought $6\frac{1}{2}$ pounds of apples and $4\frac{3}{4}$ pounds of bananas. About how many pounds of fruit did she buy?

Lesson 8.2

Convert each measure.

1. 489 ft. to yd.
2. 132 in. to ft.
3. 2 mi. to ft.
4. 5280 ft. to mi.
5. 7 mi. to ft.
6. 99 ft. to yd.
7. 216 in. to yd.
8. 3520 yd. to mi.
9. 48 ft. to yd.
10. 168 in. to ft.
11. 440 ft. to yd.
12. 8 yd. to in.
13. 732 in. to ft.
14. 3 mi. to ft.
15. 72 in. to ft.
16. 4 mi. to ft.
17. 663 yd. to ft.
18. 36 in. to yd.
19. 11 mi. to yd.
20. 133 yd. to ft.
21. 576 ft. to yd.
22. 77 yd. to ft.
23. 204 in. to ft.
24. 15 yd. to in.
25. 168 yd. to in.
26. 12 mi. to ft.
27. 648 in. to yd.
28. 456 ft. to yd.
29. 918 yd. to ft.
30. 31,680 ft. to mi.
31. 156 in. to ft.
32. 3 mi. to yd.
33. 14,080 yd. to mi.

Lesson 8.3

Measure each item to the nearest inch, $\frac{1}{2}$ inch, and $\frac{1}{4}$ inch.

1. _____

2. _____

3. _____

4. _____

5. the length of a calculator
6. the length of your shoe
7. the length of your pencil
8. the length of an envelope
9. the length of a cassette tape
10. the width of a book

Complete each statement. Choose the more reasonable measure.

11. A horse has legs about ___ long.   3 ft.   15 in.
12. A rabbit has ears about ___ long.   4 in.   2 ft.
13. The length of a whistle is about ___ .   6 ft.   2.5 in.
14. Socks are about ___ long.   12 in.   4 ft.
15. The height of a quart bottle of milk is about ___ .   3 ft.   10 in.
16. The distance between Milwaukee and Chicago
    is about ___ .   90,000 ft.   90 mi.

▶ Lesson 8.4

Convert each measure.

1. 490 oz. to lb.
2. 17 lb. to oz.
3. 5 T. to lb.
4. 8 lb. 5 oz. to oz.
5. 19,000 lb. to T.
6. 336 oz. to lb.
7. 4 lb. 7 oz. to oz.
8. 55 lb. to oz.
9. 23 T. to lb.
10. 91 oz. to lb.
11. 25 lb. to oz.
12. 44,000 lb. to T.
13. 788 oz. to lb.
14. 18 oz. to lb.
15. 297 oz. to lb.
16. 100 lb. to oz.
17. 200 oz. to lb.
18. 923 oz. to lb.
19. 16 lb. to oz.
20. 44 oz. to lb.
21. 8 T. to lb.
22. 31,000 lb. to T.
23. 81 lb. to oz.
24. 800 oz. to lb.

Complete each statement. Choose the more reasonable measure.

25. A postcard weighs about $\frac{1}{2}$ ___.   oz.   lb.

26. A full-grown dog weighs about 75 ___.   lb.   T.
27. A truck weighs about 3 ___.   lb.   T.
28. A baby elephant weighs about 300 ___.   lb.   T.

▶ Lesson 8.6

Convert each measure.

1. 12 qt. to gal.
2. 43 c. to qt.
3. 56 pt. to qt.
4. 60 fl. oz. to c.
5. 18 gal. to pt.
6. 21 qt. to c.
7. 9 c. to fl. oz.
8. 66 pt. to gal.
9. 16 gal. to qt.
10. 72 c. to pt.
11. 114 qt. to gal.
12. 100 pt. to c.
13. 15 qt. to c.
14. 88 fl. oz. to c.
15. 72 c. to qt.
16. 46 gal. to qt.
17. 35 c. to fl. oz.
18. 26 pt. to qt.
19. 27 gal. to pt.
20. 155 c. to qt.
21. 18 qt. to pt.
22. 68 fl. oz. to c.
23. 97 qt. to gal.
24. 25 c. to qt.

Complete each statement. Choose the more reasonable measure.

25. A small bottle of cough medicine contains 4 ___.   fl. oz.   qt.
26. The cake recipe uses 2 ___ of milk.   c.   gal.
27. The bottle of shampoo contains 15 ___.   fl. oz.   c.
28. Sam bought 3 ___ of paint.   fl. oz.   gal.

▶ Lesson 8.7

Add or subtract.

1. 19 lb. 11 oz.
    + 7 lb. 13 oz.

2. 4 yd. 1 ft. 8 in.
    + 5 yd. 2 ft. 9 in.

3. 7 gal. 1 qt.
    − 3 gal. 3 qt.

4. 15 yd. 2 ft. 2 in.
    − 12 yd. 2 ft. 7 in.

5. 6 gal. 3 qt.
    + 2 gal. 2 qt.

6. 15 T.   100 lb.
    − 9 T. 1,200 lb.

7. 18 min. 45 sec.
    + 32 min. 55 sec.

8. 6 yd. 2 ft. 10 in.
    + 8 yd. 2 ft. 11 in.

9. 16 lb. 14 oz.
    + 13 lb. 15 oz.

10. 100 lb. 10 oz.
    − 75 lb. 14 oz.

11. 9 min.
    − 2 min. 32 sec.

12. 4 T. 1,772 lb.
    + 1 T.   968 lb.

13. 6 hr. 17 min.
    − 3 hr. 25 min.

14. 5 hr. 33 min.
    + 3 hr. 48 min.

15. 34 lb.
    − 29 lb. 7 oz.

16. 10 yd. 2 ft. 10 in.
    + 15 yd. 2 ft.

17. 10 gal. 1 qt.
    + 7 gal. 3 qt.

18. 88 lb. 8 oz.
    + 55 lb. 8 oz.

▶ Lesson 8.8

Use the thermometer to determine each temperature.

1. a spring day in Cleveland in degrees Fahrenheit

2. a very cold winter day in Colorado in degrees Celsius

3. a very cold winter day in New Hampshire in degrees Fahrenheit

4. a very hot summer day in Dallas in degrees Celsius

5. a person with a slight fever in degrees Fahrenheit

6. a person with a slight fever in degrees Celsius

7. a point just above when water turns to ice in degrees Fahrenheit

8. a point just below when water turns to ice in degrees Celsius

9. a point just below when water boils in degrees Celsius

10. a point just above when water boils in degrees Fahrenheit

11. a cool oven, set to let bread rise in degrees Celsius

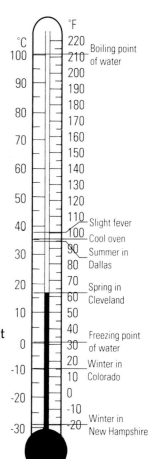

427

▶ **Lesson 8.10**

Determine the total hours worked each week. Determine the gross earnings.

1. First week
   Sunday: 1–5
   Tuesday: 5–9
   Thursday: 5–9
   Saturday: 12–5
   Hourly rate: $4.75

2. Second week
   Monday: 4–6
   Tuesday: 4–6
   Thursday: 4–6
   Friday: 4–9
   Hourly rate: $5.25

3. Third week
   Tuesday: 6–9
   Friday: 5–9
   Saturday: 10–2
   Sunday: 2–5
   Hourly rate: $4.13

4. Fourth week
   Sunday: 12–5
   Wednesday: 4–8
   Friday: 4–9
   Saturday: 9–12
   Hourly rate: $4.81

5. Fifth week
   Monday: 4–7
   Tuesday: 5–8
   Wednesday: 4–7
   Thursday: 5–8
   Hourly rate: $5.85

6. Sixth week
   Monday: 6–10
   Wednesday: 7–9
   Thursday: 7–9
   Friday: 6–10
   Hourly rate: $4.95

▶ **Lesson 9.2**

Use the graph below to answer each question.

1. Which class is the most popular?

2. Which class is the least popular?

3. Which class has about twice as many students as Chinese cooking?

4. Which two classes have about the same number of students?

5. Which class has about $\frac{1}{3}$ as many students as Chinese cooking?

6. Estimate the total number of students.

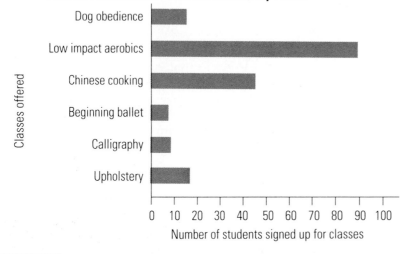

Classes Offered at the Riverside Community Center

▶ Lesson 9.3

Use the graph on the right to answer the following questions.

1. Twice as many boys as girls prefer to watch which sport?

2. More girls than boys prefer to watch which sports?

3. Which sport do both boys and girls like least to watch?

4. About $\frac{2}{3}$ as many boys as girls prefer to watch which sport?

5. About $\frac{1}{3}$ as many boys as girls prefer to watch which sport?

6. What three sports do most students prefer to watch?

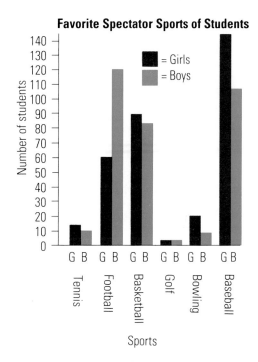

Favorite Spectator Sports of Students

▶ Lesson 9.4

1. Construct a bar graph from the table below. Describe the results.

1988 Sales for Mona's Pet Palace

Type of Animal	Number Sold
Dogs	50
Cats	80
Birds	42
Rodents	170
Reptiles	17

2. Construct a bar graph from the table below. Describe the results.

Caloric Content of Foods

Type of Food	Number of Calories per Serving
Egg	80
8-oz. yogurt with fruit	240
$\frac{2}{3}$ c. of rice	120
$\frac{1}{8}$ of a cherry pie	310
Orange	75
3-oz. candy bar with nuts	450

▶ **Lesson 9.5**

Use the graph on the right to answer the following questions.

1. What was the price of a first-class stamp in 1978?

2. In what year did the first-class stamp double the 1952 cost?

3. What was the shortest time between stamp price increases?

4. What was the largest jump in first-class stamp prices?

5. Which price of stamp was used for the longest period of time?

6. What is the price of a first-class stamp today?

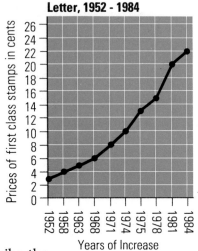

Cost of Sending a First Class Letter, 1952 - 1984

Prices of first class stamps in cents

Years of Increase

▶ **Lesson 9.6**

1. Construct a line graph from the table below. Describe the results.

Profits for Success-Bright Company from 1983–1988

Year	Profits (in millions of dollars)
1983	0.5
1984	0.9
1985	1.8
1986	2.5
1987	3.6
1988	5.5

2. Construct a line graph from the table below. Describe the results.

Gabe's Plant Experiment

After Week:	Height of Plant in Inches
#1	$\frac{1}{4}$ in.
#2	$\frac{1}{2}$ in.
#3	$\frac{3}{4}$ in.
#4	$1\frac{1}{2}$ in.
#5	$2\frac{1}{2}$ in.
#6	$3\frac{1}{2}$ in.
#7	$4\frac{3}{4}$ in.

► Lesson 9.7

Use the graph on the right to answer the following questions.

1. About how many television sets are there in Brazil?

2. Which country has about half the number of sets as the U.S.?

3. About how many sets are there in Italy, Great Britain, and China combined?

4. Which two countries have about the same number of sets?

5. Use the table below to construct a pictograph. Use one coin to represent $50.

Number of Television Sets in Some Countries in 1986

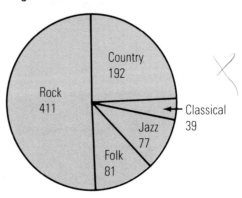

United States
Soviet Union
China
Japan
Italy
Great Britain
Brazil

= 100 television sets for each 1000 people.

Amounts Earned in the Freshman Class Candy Sale by Homerooms

Homeroom	Dollars Earned
206	$175
213	$65
301	$250
309	$105
310	$385

► Lesson 9.8

Use the graph on the right to answer the following questions.

1. What type of music do most students prefer?

2. What type of music is the second choice among students?

3. What 2 types of music are preferred by about the same number of students?

4. What type of music is chosen by the fewest number of students?

5. What 3 types together make up about $\frac{1}{4}$ of the graph?

Music Survey of Students at Kimberly High School

Rock 411
Country 192
Classical 39
Jazz 77
Folk 81

► Lesson 10.2

Make a frequency table for each set of data below. Use your
table to answer the questions at the right.

Cost of blank VCR tapes

$3	$2	$5	$4	$6
$3	$3	$4	$4	$6
$2	$3	$6		

1. How much does the most expensive tape cost?

2. What is the most common cost for a tape?

3. What is the difference in cost between the least and the most expensive tape?

Number of gymnasts from individual schools taking part in the state meet

2	1	1	9	6	1
3	5	3	9	5	10
6	2	3	1	1	3
4	5	3	3	3	2

4. How many schools participated in the state gymnastics meet?

5. How many schools entered only one gymnast in the meet?

6. How many schools entered 7 or 8 gymnasts in the meet?

7. How many schools entered 5 or more gymnasts in the meet?

► Lesson 10.3

Find the range, the mode, and the median.

1. 5   6   4.5   5.5   5.5

2. 9   6.5   5.5   8.5   7.5

3. 18,933   24,635   33,941   32,627

4. 26,225   28,317   27,934   27,878

5. 65   78   85   78   85   91

6. 79   83   83   79   89   83

7. 13.2   8.9   7.9   12.4

8. 6.8   8.6   8.7   6.8   7.4

9. $1285 $4360 $945 $2850

10. $3650 $725 $1495 $3650

11. 380   425   650   535   425   619

12. 978   1237   1205   999   1327

13. 400   385   400   376   400   385

14. 265   310   275   275   266   305

15. 22.3   18.9   17.5   18.1   11.8

16. $9.89 $5.65 $4.25 $4.25 $13.41

17. $77 $61 $98 $101 $98 $98

18. 99   94   96   96   94   99   94   94

19. 12   13   14   14   12   12   13

20. $6 $5 $9 $6 $5 $9 $5 $5

▶ Lesson 10.4

Find the mean.

1. $2.50 $4 $1.25 $5.75 $4.75
2. $6.50 $4.75 $3.75 $5.50 $4.25
3. 65   61   80   94   78   79   86   82
4. 71   76   82   88   91   93   77
5. 11   17   8   19   21   3   6   5   4
6. 18   18   26   12   18   20   22   13
7. $0.98 $1.82 $0.58 $0.44 $1.26
8. $0.51 $0.69 $1.55 $2.55 $1.10
9. 128   191   204   255   341   363
10. 385   250   95   118   192   99   98
11. 9   3   4   0   12   11   2   15   13   8
12. 1   0   1   7   22   6   0   5   18   3
13. $2.85 $3.90 $1.75 $1.20 $2.35
14. $5.95 $1.85 $0.95 $4.30 $4.85
15. 2550   3175   2100   1900   3725   2890   2760
16. $1525 $1675 $1950 $2950 $1400 $2400 $1700
17. $20,500 $45,900 $35,400 $68,700 $44,500 $35,400
18. 12,385   21,344   22,895   26,890   26,789   16,542

▶ Lesson 10.6

Use the information in each table to construct a histogram.

1. Freshman students at Main High School with summer jobs

Hours worked per week	0–4	5–8	9–12	13–16	17–20	21–24	25–28	29–36	36+
Number of students	6	12	45	62	35	37	16	15	9

Divide the number of hours into categories, 0–12, 13–24, 25–36+.

2. Snowblower sales at Harvey's Hardware Store

Month	Jan.	Feb.	Mar.	Apr.	May	June	July	Aug.	Sep.	Oct.	Nov.	Dec.
Number of Snowblowers	35	26	15	8	0	0	5	15	9	18	42	53

Divide the year into 2-month periods.

► **Lesson 11.1**

Write each ratio in 3 different ways.

1. 6 puppies to 4 kittens

2. 4 computers for 10 people

3. 8 gloves for 4 hands

4. $12.95 for 3 posters

5. $10 for 4 tickets

6. 6 goals in 2 games

7. 2 runs for 13 times at bat

8. $21.50 for 2 pairs of shoes

9. 1 cake with 14 candles

10. 3 phones for 9 lines

11. 30 tests for 30 students

12. 1 calendar for 365 days

Find 3 ratios equivalent to each ratio.

13. 2 cages for 5 snakes

14. 4 tokens for 6 boys

15. 8 radios for 1 plug

16. 14 maps for 2 globes

17. 8 flowers for 32 leaves

18. 3 mountains for 120 hikers

19. 8 laces for 9 skates

20. 1 train for 256 people

21. 10 buses for 320 riders

22. 6 dogs for 12 leashes

► **Lesson 11.2**

Tell whether each statement is a proportion. Use = or ≠.

1. $\frac{24}{1}$ ※ ___ $\frac{96}{4}$

2. $\frac{16}{2}$ ※ ___ $\frac{80}{5}$

3. $\frac{3}{1}$ ※ ___ $\frac{9}{3}$

4. $\frac{12}{1}$ ※ ___ $\frac{2}{24}$

5. $\frac{30}{1}$ ※ ___ $\frac{300}{10}$

6. $\frac{364}{7}$ ※ ___ $\frac{52}{2}$

7. $\frac{2}{1}$ ※ ___ $\frac{4}{8}$

8. $\frac{32}{8}$ ※ ___ $\frac{8}{2}$

9. $\frac{20}{6}$ ※ ___ $\frac{50}{10}$

10. $\frac{5}{2}$ ※ ___ $\frac{15}{4}$

11. $\frac{192}{3}$ ※ ___ $\frac{64}{1}$

12. $\frac{1}{7}$ ※ ___ $\frac{3}{24}$

13. $\frac{6}{16}$ ※ ___ $\frac{12}{32}$

14. $\frac{30}{5}$ ※ ___ $\frac{6}{1}$

15. $\frac{13}{169}$ ※ ___ $\frac{12}{144}$

16. $\frac{17}{20}$ ※ ___ $\frac{40}{34}$

17. $\frac{34}{40}$ ※ ___ $\frac{17}{20}$

18. $\frac{88}{22}$ ※ ___ $\frac{5}{4}$

19. $\frac{5}{10}$ ※ ___ $\frac{10}{50}$

20. $\frac{24}{8}$ ※ ___ $\frac{48}{16}$

21. $\frac{8}{9}$ ※ ___ $\frac{64}{81}$

22. $\frac{9}{4}$ ※ ___ $\frac{18}{8}$

23. $\frac{50}{45}$ ※ ___ $\frac{10}{9}$

24. $\frac{6}{36}$ ※ ___ $\frac{12}{48}$

25. $\frac{9}{1}$ ※ ___ $\frac{2}{18}$

26. $\frac{3}{4}$ ※ ___ $\frac{75}{100}$

27. $\frac{10}{6}$ ※ ___ $\frac{30}{18}$

28. $\frac{12}{48}$ ※ ___ $\frac{5}{10}$

29. $\frac{17}{3}$ ※ ___ $\frac{51}{9}$

30. $\frac{24}{9}$ ※ ___ $\frac{36}{12}$

31. $\frac{8}{7}$ ※ ___ $\frac{56}{64}$

32. $\frac{33}{11}$ ※ ___ $\frac{66}{22}$

33. $\frac{12}{4}$ ※ ___ $\frac{6}{2}$

34. $\frac{81}{9}$ ※ ___ $\frac{49}{7}$

35. $\frac{5}{3}$ ※ ___ $\frac{15}{9}$

36. $\frac{8}{9}$ ※ ___ $\frac{16}{19}$

37. $\frac{11}{2}$ ※ ___ $\frac{24}{3}$

38. $\frac{20}{30}$ ※ ___ $\frac{3}{2}$

39. $\frac{55}{5}$ ※ ___ $\frac{88}{8}$

40. $\frac{12}{7}$ ※ ___ $\frac{24}{14}$

► Lesson 11.3

Solve each proportion.

1. $\frac{8}{6} = \frac{\blacksquare}{60}$

2. $\frac{4}{\blacksquare} = \frac{12}{36}$

3. $\frac{3}{\blacksquare} = \frac{39}{169}$

4. $\frac{22}{11} = \frac{\blacksquare}{49}$

5. $\frac{3}{5} = \frac{\blacksquare}{25}$

6. $\frac{125}{5} = \frac{625}{\blacksquare}$

7. $\frac{\blacksquare}{9} = \frac{6}{18}$

8. $\frac{14}{\blacksquare} = \frac{21}{15}$

9. $\frac{\blacksquare}{3} = \frac{150}{225}$

10. $\frac{24}{32} = \frac{\blacksquare}{4}$

11. $\frac{91}{7} = \frac{65}{\blacksquare}$

12. $\frac{200}{350} = \frac{\blacksquare}{70}$

13. $\frac{4}{2} = \frac{16}{\blacksquare}$

14. $\frac{9}{10} = \frac{9}{\blacksquare}$

15. $\frac{\blacksquare}{39} = \frac{4}{52}$

16. $\frac{6}{1} = \frac{\blacksquare}{300}$

17. $\frac{\blacksquare}{1} = \frac{378}{14}$

18. $\frac{8}{20} = \frac{30}{\blacksquare}$

19. $\frac{\blacksquare}{7} = \frac{18}{42}$

20. $\frac{16}{\blacksquare} = \frac{17}{17}$

21. $\frac{8}{14} = \frac{\blacksquare}{35}$

22. $\frac{\blacksquare}{20} = \frac{15}{12}$

23. $\frac{40}{\blacksquare} = \frac{90}{63}$

24. $\frac{18}{5} = \frac{36}{\blacksquare}$

25. $\frac{\blacksquare}{5} = \frac{104}{20}$

26. $\frac{19}{\blacksquare} = \frac{57}{6}$

27. $\frac{28}{42} = \frac{\blacksquare}{39}$

28. $\frac{30}{18} = \frac{\blacksquare}{30}$

29. $\frac{24}{8} = \frac{\blacksquare}{4}$

30. $\frac{\blacksquare}{9} = \frac{6}{54}$

31. $\frac{21}{\blacksquare} = \frac{49}{14}$

32. $\frac{58}{29} = \frac{\blacksquare}{54}$

33. $\frac{90}{\blacksquare} = \frac{10}{3}$

34. $\frac{\blacksquare}{3} = \frac{64}{12}$

35. $\frac{10}{\blacksquare} = \frac{50}{5}$

36. $\frac{75}{2} = \frac{\blacksquare}{6}$

37. $\frac{11}{8} = \frac{22}{\blacksquare}$

38. $\frac{9}{\blacksquare} = \frac{27}{3}$

39. $\frac{65}{\blacksquare} = \frac{13}{1}$

40. $\frac{34}{17} = \frac{\blacksquare}{1}$

► Lesson 11.4

Find the unit price.

1. $0.98 for 11 oz. of hand lotion

2. $1.69 for 4 qt. of milk

3. $1.98 for 2.5 lb. of asparagus

4. $0.45 for 1.5 lb. of bananas

5. $15.65 for 3 posters

6. $9.78 for 4 pairs of socks

7. $5.15 for 4 rolls of tape

8. $17.66 for 5 rolls of film

9. $8.75 for 2 records

10. $11.89 for 10 gallons of gas

Find the better buy.

11. $4.46 for 5 lb. of oranges or $2.71 for 3 lb. of oranges

12. $6.59 for $\frac{1}{2}$ lb. of nuts or $13.08 for a lb. of nuts

13. $1.10 for 9 oz. of soy sauce or $1.97 for 15 oz. of soy sauce

14. $1.76 for 3 lb. of potatoes or $2.88 for 5 lb. of potatoes

15. $3.87 for a gallon of juice or $0.99 for a quart of juice

▶ **Lesson 11.5**

Find the actual length for each scale length given.

1. 5 in. scale: 1 in. = 25 mi.

2. 20.4 in. scale: 1 in. = 15 yd.

3. 9.5 in. scale: 1 in. = 15 mi.

4. $2\frac{7}{8}$ in. scale: $\frac{1}{8}$ in. = 30 mi.

5. 4.6 in. scale: 1 in. = 1.5 yd.

6. $3\frac{3}{4}$ in. scale: $\frac{1}{4}$ in. = 10 mi.

7. $\frac{1}{4}$ in. scale: $2\frac{1}{2}$ in. = 10 mi.

8. 3 in. scale: 5 in. = 7.5 yd.

9. $1\frac{5}{8}$ in. scale: $\frac{1}{4}$ in. = 40 mi.

10. 5.7 in. scale: 5 in. = 50 ft.

Find the scale length for each actual length given.

11. 15.5 yd. scale: 1 in. = 2 yd.

12. 30 yd. scale: 1 in. = 5 yd.

13. 76 ft. scale: $\frac{1}{2}$ in. = 2 ft.

14. 550 yd. scale: 0.5 in. = 50 yd.

15. 225 mi. scale: 1 in. = 50 mi.

16. 880 mi. scale: 1 in. = 20 mi.

17. 475 mi. scale: 2 in. = 100 mi.

18. 10 yd. scale: $\frac{1}{4}$ in. = 1 yd.

19. 32 mi. scale: 5 in. = 160 mi.

20. 1350 mi. scale: 3 in. = 450 mi.

▶ **Lesson 11.7**

Each pair of polygons is similar. Find the length of the unknown sides.

1.
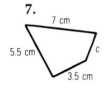

2.
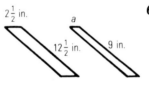

3.
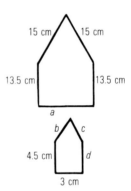

4.
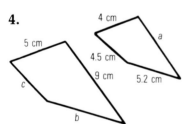

5.
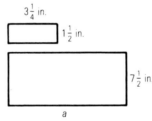

6.
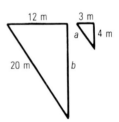

7.
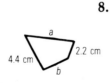

8.
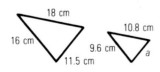

▶ Lesson 12.2

Write each decimal as a percent.

1. 0.1 **2.** 0.86 **3.** 0.33 **4.** 0.175 **5.** 0.858 **6.** $0.66\overline{6}$

7. 0.97 **8.** 0.206 **9.** 0.41 **10.** 0.06 **11.** 0.29 **12.** 0.8925

13. 0.7 **14.** 0.0225 **15.** 0.674 **16.** 0.625 **17.** 0.8 **18.** $0.77\overline{3}$

19. 0.09 **20.** 0.11 **21.** 0.028 **22.** 0.59 **23.** 0.035 **24.** 0.555

25. 0.084 **26.** 0.48 **27.** 0.915 **28.** 0.30 **29.** 0.73 **30.** 0.059

Write each percent as a decimal.

31. 69% **32.** 3% **33.** 19% **34.** 14.5% **35.** 40% **36.** $87\frac{1}{2}\%$

37. $8\frac{1}{2}\%$ **38.** 95% **39.** 11.8% **40.** 29% **41.** $13\frac{1}{2}\%$ **42.** $6\frac{1}{4}\%$

43. 85% **44.** 6% **45.** $23\frac{3}{4}\%$ **46.** 81% **47.** 65% **48.** $16\frac{2}{3}\%$

49. $42\frac{1}{2}\%$ **50.** 21.4% **51.** 38.4% **52.** 76% **53.** $5\frac{1}{2}\%$ **54.** 22.6%

55. $1\frac{1}{2}\%$ **56.** 2% **57.** 12% **58.** 43% **59.** 7.2% **60.** $83\frac{1}{3}\%$

▶ Lesson 12.3

Write as a percent.

1. $\frac{1}{10}$ **2.** $\frac{16}{20}$ **3.** $\frac{18}{27}$ **4.** $\frac{9}{24}$ **5.** $\frac{9}{20}$ **6.** $\frac{1}{12}$

7. $\frac{25}{50}$ **8.** $\frac{50}{60}$ **9.** $\frac{11}{12}$ **10.** $\frac{8}{8}$ **11.** $\frac{9}{10}$ **12.** $\frac{11}{25}$

13. $\frac{2}{25}$ **14.** $\frac{35}{45}$ **15.** $\frac{12}{20}$ **16.** $\frac{3}{12}$ **17.** $\frac{15}{60}$ **18.** $\frac{1}{6}$

19. $\frac{7}{12}$ **20.** $\frac{13}{20}$ **21.** $\frac{13}{50}$ **22.** $\frac{16}{25}$ **23.** $\frac{13}{16}$ **24.** $\frac{6}{18}$

25. $\frac{1}{20}$ **26.** $\frac{7}{25}$ **27.** $\frac{2}{16}$ **28.** $\frac{18}{30}$ **29.** $\frac{25}{75}$ **30.** $\frac{5}{12}$

31. $\frac{39}{50}$ **32.** $\frac{4}{5}$ **33.** $\frac{17}{20}$ **34.** $\frac{6}{8}$ **35.** $\frac{1}{50}$ **36.** $\frac{10}{10}$

37. Susan made 33 out of 50 free throw attempts. What was her free throw percentage?

38. Yolanda made 12 out of 15 free throw attempts. What was her free throw percentage?

39. Paul made 8 out of 40 baskets during the last game. What was his shooting percentage?

40. Jan made 28 out of 32 baskets during the same game. What was his shooting percentage?

Write as a fraction in lowest terms.

1. 60% 2. $8\frac{1}{3}\%$ 3. 2% 4. 93.75% 5. 19% 6. $91\frac{2}{3}\%$

7. $13\frac{1}{2}\%$ 8. 6% 9. 74% 10. 6.8% 11. 3% 12. 8%

13. 80% 14. 29% 15. 35% 16. 50% 17. 44% 18. $6\frac{2}{3}\%$

19. 7% 20. $88\frac{1}{2}\%$ 21. $56\frac{1}{4}\%$ 22. 10.5% 23. 68% 24. 96%

25. $35\frac{1}{4}\%$ 26. $87\frac{1}{2}\%$ 27. 9% 28. 70% 29. $12\frac{1}{2}\%$ 30. 15%

31. In the mountains it snowed 78% of the days last year. What fraction is that?

32. Last winter the sun was out 64% of the time. What fraction is that?

33. It rained 25% of the time last summer. What fraction is that?

34. In New York City, 10 percent of the temperatures were above 95 degrees last summer. What fraction is that?

► **Lesson 12.6**

Write each fraction or decimal as a percent.

1. 8 2. $\frac{3}{800}$ 3. $1\frac{3}{4}$ 4. $\frac{9}{5}$ 5. $\frac{2}{250}$ 6. 2.2

7. $6\frac{1}{2}$ 8. 11 9. $\frac{6}{1000}$ 10. $5\frac{2}{3}$ 11. $\frac{45}{15}$ 12. $5\frac{9}{10}$

13. 2.25 14. $\frac{35}{7}$ 15. 10.25 16. 4.4 17. $\frac{7}{800}$ 18. $\frac{5}{1000}$

19. 1.8 20. 4.6 21. 7.5 22. $\frac{5}{800}$ 23. $\frac{14}{7}$ 24. $\frac{5}{600}$

25. 19 26. $\frac{10}{3}$ 27. 12.5 28. $\frac{2}{1000}$ 29. 9.75 30. $\frac{2}{500}$

Write each percent as a fraction and as a decimal.

31. $\frac{7}{10}\%$ 32. 525% 33. $\frac{1}{20}\%$ 34. $\frac{1}{3}\%$ 35. $\frac{7}{8}\%$ 36. $\frac{4}{5}\%$

37. $\frac{1}{12}\%$ 38. $\frac{7}{20}\%$ 39. $\frac{1}{8}\%$ 40. $\frac{2}{5}\%$ 41. $\frac{3}{10}\%$ 42. $\frac{2}{3}\%$

43. $887\frac{1}{2}\%$ 44. $\frac{19}{20}\%$ 45. $166\frac{2}{3}\%$ 46. $\frac{1}{10}\%$ 47. $\frac{1}{16}\%$ 48. $\frac{1}{5}\%$

▶ Lesson 12.7

Find each number.

1. 40% of $32
2. 15% of 2600
3. 300% of $250
4. $66\frac{2}{3}$% of 1800

5. 50% of 9
6. $33\frac{1}{3}$% of 66
7. 10% of 4321
8. 75% of 984

9. 64% of 8764
10. 8% of 1790
11. 12% of $26,800
12. 9.5% of 784

13. 35% of $90
14. 24% of 650
15. 300% of 430
16. 63% of 78,500

17. 4% of 7890
18. 80% of $6400
19. $\frac{1}{4}$% of 240
20. 150% of 96

21. 0.5% of 40
22. 90% of $9840
23. 8% of $49
24. 100% of 659

25. 50% of 964
26. 99% of 1700
27. 12% of 992
28. 82% of 3750

29. 136% of 225
30. 95% of 500
31. 37.5% of 96
32. 0.8% of 55

33. In a recent poll, 19% of the 400 people questioned said they preferred winter to summer. How many people was that?

34. In another poll, $\frac{1}{2}$% of 1000 people said they liked rainy days better than sunny days. How many people was that?

▶ Lesson 12.8

Find each percent.

1. 38 is what percent of 950?
2. What percent of 88 is 528?

3. 285 is what percent of 475?
4. 322 is what percent of 2300?

5. 616 is what percent of 1400?
6. 0.28 is what percent of 35?

7. 33 is what percent of 264?
8. 16 is what percent of 800?

9. 1110 is what percent of 1200?
10. What percent of 900 is 666?

11. What percent of 5 is $\frac{1}{2}$?
12. 504 is what percent of 56?

13. 84 is what percent of 400?
14. What percent of 80 is 0.4?

15. 57 is what percent of 152?
16. 2.5 is what percent of 125?

17. What percent of $10,000 is $750?
18. 4.38 is what percent of 4.38?

19. 150 is what percent of 60?
20. 3466 is what percent of 13,864?

21. Last year Company A made $30,000 in profits. This year the profits were $129,000. What percent of $30,000 is $129,000?

22. Of the 900 people who work for the city, 702 of them received pay raises this year. What percent received pay raises?

▶ Lesson 12.9

Find the number.

1. 25% of what number is 69?
2. 2895 is 50% of what number?
3. 100% of what number is 9.8?
4. 99% of what number is $19,800?
5. 49% of what number is 3332?
6. 24 is 80% of what number?
7. $695 is 8% of what number?
8. 300% of what number is 12?
9. 50 is 40% of what number?
10. 6 is 20% of what number?
11. 0.6% of what number is 3?
12. 87.5% of what number is 301?
13. $\frac{1}{2}$% of what number is 0.9?
14. $49.80 is 83% of what number?
15. 45 is 5% of what number?
16. 396 is 90% of what number?
17. 18% of what number is $720?
18. 750% of what number is 7350?
19. $\frac{1}{4}$% of what number is 2.2?
20. $108 is 400% of what number?

21. Seventy percent, or 266 students in the freshman class, eat fresh fruit for breakfast. How many students are there in the freshman class?

22. Thirty students, or 2.5% of the student body, are on the gymnastics team. How many students are there in all?

▶ Lesson 12.10

Find the percent of change.

1. $25 to $30
2. $150 to $75
3. 16 to 18
4. 10 to 45
5. 27 to 36
6. 144 to 168
7. $33 to $22
8. $800 to $560
9. 8 to 10
10. 75 to 45
11. $88 to $176
12. $475 to $427.50
13. 21 to 105
14. $9.90 to $6.93
15. 45,000 to 50,000
16. $36,000 to $20,160
17. $15.50 to $3.10
18. 300 to 500
19. 12 to 84
20. 635 to 508
21. 10 to 11
22. $700 to $1400
23. 169 to 152.1
24. 80 to 24
25. $150 to $60
26. 4 to 36
27. $45 to $42.75

28. The original cost of a computer was $1200. The price was lowered to $960. What was the percent of change?

29. A modem for a computer cost $575. The price was lowered to $529. What was the percent of change?

▶ Lesson 13.2

One marble is drawn from a sack without looking. There are
5 blue marbles, 2 yellow marbles, and 3 white marbles. Find the
probability of drawing the following.

1. a blue marble

2. a yellow marble

3. a blue or white marble

4. a blue or yellow marble

A currency exchange sold the following license plates one day:
MTH 458, MTH 459, MTH 460, MTH 461, MTH 462, 123 456,
123 457, and 123 458. You bought a license plate that day from
the place and had no choice of plates. Find the probability that
you received a license plate with the following.

5. an M

6. numbers only

7. a 6

8. a T and a 5

9. a 4

10. a 4 and a 9

11. 2, 4, and 6

12. an A

▶ Lesson 13.3

Before a skating contest 12 skaters drew numbers to determine
the performance order. Find the following odds.

1. of skating first

2. against skating last

3. of skating in the first 6

4. of skating first or second

5. against skating in the last 5

6. of skating in the first 10

The 26 company mailboxes were lettered A to Z and were
assigned in no particular order. Find the following odds.

7. of getting the letter Q

8. against getting the letter A

9. against getting a vowel
(A, E, I, O, or U)

10. of getting a vowel
(A, E, I, O, or U)

11. of getting a letter in the first half of the alphabet

12. against getting one of the first 10 letters

▶ Lesson 13.5

Draw a tree diagram listing all possible outcomes for each of the following.

1. making a cheese pizza, sausage pizza, mushroom pizza, or combination pizza with a thin crust or a deep-dish crust

2. choosing a 4-ounce, 8-ounce, or 12-ounce container of juice, soda, milk, or lemonade

3. choosing one sport from soccer, football, wrestling, baseball, and track and one club from Spanish club, pep band, and student council

Draw a tree diagram. Then find the probability of the underlined event. The order is not important.

4. a coin is tossed twice and a number cube is rolled; *2 heads and a 5*

5. a coin is tossed 4 times; *2 heads and 2 tails*

6. a number cube is rolled twice; *a 1 and a 2*

7. a number cube is rolled 3 times; *three 6s*

▶ Lesson 13.6

Find the number of possible outcomes. What is the probability that a computer would randomly select each of the underlined prizes?

1. vacations in Atlanta, Omaha, Detroit, Sacramento, or Buffalo staying at a hotel, motel, condominium, or resort; *resort in Atlanta*

2. $100, $500, $1000, or $5000 and a visit to the White House, the Grand Canyon, the Empire State Building, Cypress Gardens, the Grand Old Opry, or Disneyland; *$500 and a trip to the White House*

3. tickets to 4 different concerts, 3 different plays, or 5 different sporting events and dinner at one of 10 different restaurants including La Diner; *a concert followed by dinner at La Diner*

▶ Lesson 14.2

Find the perimeter of each polygon.

1.
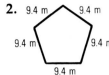
8.5 cm 8.5 cm
6.5 cm

2.
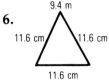
9.4 m 9.4 m
9.4 m 9.4 m
9.4 m

3.

$3\frac{1}{2}$ in.
$2\frac{3}{4}$ in.

4.
12 m
12 m

5.
2.8 cm 8.9 cm
2.5 cm
5.8 cm

6.
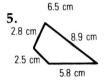
11.6 cm 11.6 cm
11.6 cm

7.
6.7 m
7.4 m

8.
$1\frac{1}{2}$ in.
11 in.

9. Rectangle
$l = 12.7$ cm
$w = 2.9$ cm

10. Rectangle
$l = 6.2$ cm
$w = 3.9$ cm

11. Rectangle
$l = 5\frac{1}{2}$ ft.
$w = 8\frac{1}{2}$ ft.

12. Rectangle
$l = 9\frac{1}{2}$ in.
$w = 6\frac{3}{4}$ in.

13. 14.8 m   14.8 m   14.8 m   14.8 m

14. 9.2 cm   8.7 cm   6.3 cm

15. 7 in.   7 in.   7 in.   5 in.   5 in.

16. 12 ft.   13 ft.   8 ft.   9 ft.

▶ Lesson 14.3

Estimate the perimeter of each figure.

1.
2.2 cm
1.3 cm
2.6 cm
1.3 cm

2.
4 cm
2 cm
3.2 cm

3.
2 cm 2.2 cm
3.5 cm

4.
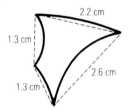
2.8 cm
1.3 cm
1.6 cm
2.5 cm

5.
2 cm
1.6 cm 2 cm
2.2 cm

6.
2.8 cm
2.4 cm
1.6 cm

7.
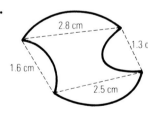
0.5 cm
2.2 cm
0.8 cm 1.5 cm
1.8 cm
2 cm

8.
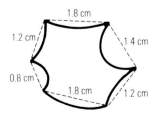
1.8 cm
1.2 cm 1.4 cm
0.8 cm
1.8 cm 1.2 cm

9.
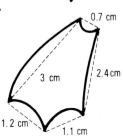
0.7 cm
3 cm 2.4 cm
1.2 cm 1.1 cm

▶ Lesson 14.4

Find each circumference.

1. $r = 0.9$ m
2. $d = 7$ cm
3. $d = 100$ ft.
4. $r = 3.2$ cm

5. $d = 15$ mm
6. $r = 4.4$ m
7. $r = 1.7$ cm
8. $d = 0.5$ m

9. $r = 16$ m
10. $r = 13$ cm
11. $d = 20.2$ m
12. $d = 40$ ft.

13. $d = 3.6$ cm
14. $r = 19$ m
15. $r = 16.5$ cm
16. $d = 17$ m

17. $r = 0.3$ m
18. $d = 20$ cm
19. $d = 5.1$ m
20. $r = 150$ ft.

21. $d = 1.1$ m
22. $r = 32$ mm
23. $d = 11$ cm
24. $r = 25$ in.

25. $r = 55$ ft.
26. $d = 80$ mm
27. $d = 260$ ft.
28. $r = 22$ cm

29. $d = 0.2$ m
30. $d = 1.4$ cm
31. $r = 5.7$ cm
32. $r = 43$ m

▶ Lesson 14.5

Find the area of each parallelogram or rectangle.

1.

2.6 cm
2.6 cm

2.
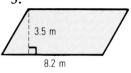
6.5 m
7.5 m

3.
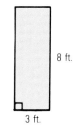
8 ft.
3 ft.

4.

1.7 cm
3.9 cm

5.
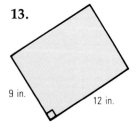
3.5 m
8.2 m

6.
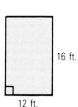
16 ft.
12 ft.

7.
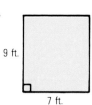
5.6 cm
5.6 cm

8.
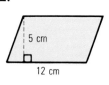
4.8 m
1.9 m

9.

$2\frac{3}{4}$ in.
3 in.

10.

14.5 m
14.5 m

11.
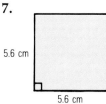
9 ft.
7 ft.

12.
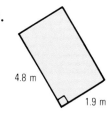
5 cm
12 cm

13.
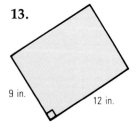
9 in.
12 in.

14.

15 ft.
5 ft.

15.

13.6 cm
13.6 cm

16.
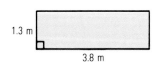
1.3 m
3.8 m

► **Lesson 14.6**

Find the area of each triangle.

1.
3 in.
4 in.

2.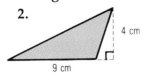
4 cm
9 cm

3.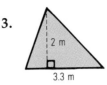
2 m
3.3 m

4.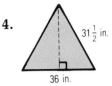
$31\frac{1}{2}$ in.
36 in.

5.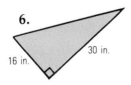
12 cm
10 cm

6.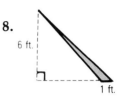
30 in.
16 in.

7.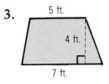
12 cm
6 cm

8.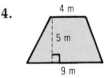
6 ft.
1 ft.

9. $b = 9$ ft.
$h = 8$ ft.

10. $b = 25$ ft.
$h = 5\frac{1}{2}$ ft.

11. $b = 12$ cm
$h = 20$ cm

12. $b = 3.5$ m
$h = 15$ m

13. $b = 1.8$ m
$h = 7$ m

14. $b = 55$ in.
$h = 12$ in.

15. $b = 11.2$ cm
$h = 4.3$ cm

16. $b = 66$ mm
$h = 24$ mm

17. $b = 18$ m
$h = 20$ m

18. $b = 8\frac{1}{4}$ ft.
$h = 12\frac{2}{3}$ ft.

19. $b = 2.8$ m
$h = 3.2$ m

20. $b = 20$ cm
$h = 15$ cm

► **Lesson 14.7**

Find the area of each trapezoid.

1.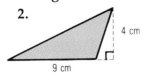
13 in.
8 in.
21 in.

2.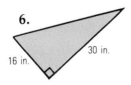
4.5 cm
3.0 cm
2.5 cm

3.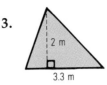
5 ft.
4 ft.
7 ft.

4.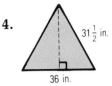
4 m
5 m
9 m

5.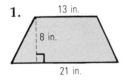
26 mm
9 mm
34 mm

6.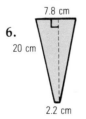
7.8 cm
20 cm
2.2 cm

7.
3.1 m
4 m
5.2 m

8.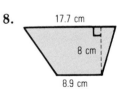
17.7 cm
8 cm
8.9 cm

9. $a = 1.6$ m
$b = 8.6$ m
$h = 2$ m

10. $a = 4$ in.
$b = 6$ in.
$h = 11$ in.

11. $a = 2.4$ cm
$b = 1.6$ cm
$h = 0.7$ cm

12. $a = 10\frac{1}{2}$ in.
$b = 15\frac{1}{2}$ in.
$h = 25$ in.

13. $a = 13$ in.
$b = 7$ in.
$h = 30$ in.

14. $a = 31$ in.
$b = 9\frac{1}{8}$ in.
$h = 8$ in.

15. $a = 9.7$ cm
$b = 15.1$ cm
$h = 5$ cm

16. $a = 18\frac{3}{4}$ in.
$b = 19\frac{1}{4}$ in.
$h = 16$ in.

▶ Lesson 14.8

Find the area. Round answers to the nearest tenth as needed.

1.   2 m

2.   200 ft.

3.   1.3 cm

4.   7 m

5.   60 ft.

6.   14 ft.

7.   9 m

8.   42 cm

9. $r = 60$ ft. 10. $d = 4.8$ cm 11. $d = 80$ ft. 12. $r = 22$ in.

13. $d = 18.4$ cm 14. $r = 1.8$ m 15. $d = 42$ mm 16. $r = 80$ in.

17. $r = 19$ cm 18. $d = 150$ ft. 19. $d = 65$ m 20. $r = 17$ cm

21. $r = 6.2$ m 22. $d = 9.4$ m 23. $r = 0.3$ m 24. $d = 3.3$ cm

25. $d = 600$ ft. 26. $d = 75$ mm 27. $r = 99$ cm 28. $r = 430$ ft.

▶ Lesson 15.3

Find the surface area of each figure.

1. 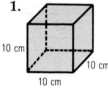 10 cm, 10 cm, 10 cm

2.   0.5 m, 0.8 m, 0.4 m

3. 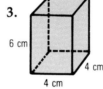 6 cm, 4 cm, 4 cm

4.   2 m, 3 m, 5 m

5. Prism
   $l = 12$ in.
   $w = 8$ in.
   $h = 4$ in.

6. Cube
   $s = 3.3$ m

7. Cube
   $s = 8.1$ cm

8. Cube
   $s = 3\frac{3}{4}$ ft.

9. Prism
   $l = 4$ in.
   $w = 3$ in.
   $h = 6$ in.

10. Cube
    $s = 9.8$ m

11. Cube
    $s = 13$ cm

12. Prism
    $l = 6.7$ cm
    $w = 4.1$ cm
    $h = 2.3$ cm

13. Prism
    $l = 3$ ft.
    $w = 1\frac{1}{2}$ ft.
    $h = 5\frac{1}{2}$ ft.

14. Cube
    $s = 25$ in.

15. Prism
    $l = 9$ in.
    $w = 7$ in.
    $h = 6$ in.

16. Cube
    $s = 12.2$ cm

▶ **Lesson 15.4**

Find the surface area of each figure.

1.
10 in.
4 in.

2.
3 m
6 m

3.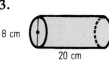
8 cm
20 cm

4.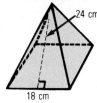
24 cm
18 cm

5. Cylinder
   $h = 7$ cm
   $r = 3$ cm

6. Pyramid
   $s = 3$ m
   $h = 6.5$ m

7. Cylinder
   $h = 12$ m
   $d = 10$ m

8. Cylinder
   $h = 21$ m
   $r = 8$ m

9. Cylinder
   $h = 10$ ft.
   $d = 20$ ft.

10. Pyramid
    $s = 8$ ft.
    $h = 6$ ft.

11. Cylinder
    $h = 9$ m
    $r = 10$ m

12. Cylinder
    $h = 4$ cm
    $r = 4$ cm

13. Pyramid
    $s = 6$ in.
    $h = 5$ in.

14. Cylinder
    $h = 11$ m
    $r = 30$ m

15. Pyramid
    $s = 11.5$ cm
    $h = 9$ cm

16. Cylinder
    $h = 6$ m
    $d = 4$ m

▶ **Lesson 15.5**

Find the volume of each figure.

1.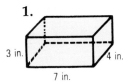
3 in.
4 in.
7 in.

2.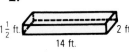
$1\frac{1}{2}$ ft.
2 ft.
14 ft.

3.
10 cm
8 cm

4.
5 m
5 m
5 m

5. Cube
   $s = 8.5$ m

6. Cube
   $s = 14$ cm

7. Prism
   $l = 3$ ft.
   $w = 3$ ft.
   $h = 4$ ft.

8. Pyramid
   $s = 5$ m
   $a = 7$ m

9. Cube
   $s = 30$ in.

10. Prism
    $l = 9$ ft.
    $w = 7$ ft.
    $h = 20$ ft.

11. Pyramid
    $s = 11$ cm
    $a = 8$ cm

12. Prism
    $l = 0.3$ m
    $w = 0.6$ m
    $h = 0.8$ m

13. Pyramid
    $s = 25$ cm
    $a = 22$ cm

14. Cube
    $s = 6.4$ cm

15. Prism
    $l = 8$ in.
    $w = 2$ in.
    $h = 14$ in.

16. Prism
    $l = 48$ in.
    $w = 12$ in.
    $h = 2$ in.

▶ Lesson 15.7

Find the volume of each figure.

1.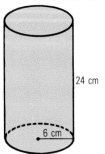
24 cm
6 cm

2.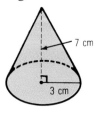
7 cm
3 cm

3.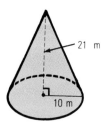
21 m
10 m

4.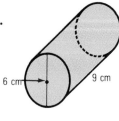
6 cm
9 cm

5. Cone
   $r = 2$ cm
   $a = 6$ cm

6. Cylinder
   $r = 5$ cm
   $h = 8$ cm

7. Cylinder
   $d = 18$ m
   $h = 7$ m

8. Cone
   $r = 1.5$ m
   $a = 5$ m

9. Cone
   $d = 10$ cm
   $a = 10$ cm

10. Cone
    $r = 1.5$ m
    $a = 1.5$ m

11. Cylinder
    $r = 8$ cm
    $h = 21$ cm

12. Cylinder
    $r = 3$ cm
    $h = 7$ cm

13. Cone
    $d = 10.4$ cm
    $a = 4$ cm

14. Cylinder
    $r = 3$ m
    $h = 12$ m

15. Cone
    $r = 10$ cm
    $a = 5$ cm

16. Cylinder
    $d = 5$ m
    $h = 3.7$ m

▶ Lesson 16.2

Write as an algebraic expression.

1. two times some number plus eight
2. six times some number
3. some number more than thirteen
4. ten less than forty-three
5. fifty minus three
6. the sum of some number and forty
7. sixty-five minus some number
8. twenty-eight times some number
9. five times four
10. the sum of three and some number
11. ninety divided by some number
12. fourteen less than some number
13. fifteen more than twenty-five
14. nine minus six
15. ten divided by five
16. thirteen minus seven
17. seven times nine
18. the sum of seven and twenty

Write in words.

19. $5d - 6$
20. $12 + 3e$
21. $\frac{40}{c} - 7$
22. $11 + b$
23. $100 - z$
24. $6 + \frac{a}{9}$
25. $\frac{y}{30}$
26. $6t + 10$

▶ Lesson 16.3

Find the value of each expression.

1. $8 + e$ when $e = 37$ **2.** $\frac{x}{4} - 3$ when $x = 40$ **3.** $9a + 5$ when $a = 2$

4. $5y$ when $y = 15$ **5.** $f - 12$ when $f = 68$ **6.** $\frac{77}{7} - b$ when $b = 5$

7. $\frac{169}{c}$ when $c = 13$ **8.** $a + \frac{45}{a}$ when $a = 9$ **9.** $187 - r$ when $r = 49$

10. $8g + g$ when $g = 7$ **11.** $s + 35$ when $s = 18$ **12.** $\frac{65}{n}$ when $n = 5$

13. $\frac{86}{q} + q$ when $q = 2$ **14.** $\frac{16}{x} + 3$ when $x = 1$ **15.** $50r + r$ when $r = 10$

16. $a + 12a$ when $a = 12$ **17.** $129 - b$ when $b = 17$ **18.** $2c + 42$ when $c = 7$

19. $\frac{64}{t} - t$ when $t = 8$ **20.** $\frac{x}{3}$ when $x = 96$ **21.** $\frac{84}{r} + 3$ when $r = 2$

22. $4f + 9$ when $f = 3$ **23.** $91 - b$ when $b = 54$ **24.** $c + \frac{c}{8}$ when $c = 32$

▶ Lesson 16.4

Write as an equation.

1. Three times seven is some number.

2. The quotient of twelve and some number is two.

3. Some number less than thirteen is five.

4. The difference between fifty-eight and some number is four.

5. The product of some number and nine minus twenty is fifty-two.

6. The sum of thirty-eight and some number is ninety-four.

7. Sixteen minus some number is eleven.

8. Forty-eight divided by some number is three.

9. Some number divided by five plus twenty is thirty-eight.

10. Thirty-one less than some number is fifty-two.

11. Five plus five times some number is seventy.

12. The sum of six and some number is twenty-six.

► Lesson 16.6

Solve.

1. $b + 6 = 18$
2. $n - 9 = 15$
3. $z + 12 = 61$
4. $a - 2 = 50$

5. $b - 5 = 42$
6. $m - 3 = 19$
7. $c + 16 = 30$
8. $a - 17 = 3$

9. $y - 34 = 9$
10. $m + 18 = 54$
11. $n - 23 = 2$
12. $s + 26 = 53$

13. $y - 30 = 12$
14. $h + 55 = 65$
15. $b + 4 = 19$
16. $n + 12 = 20$

17. $s + 7 = 16$
18. $c + 21 = 29$
19. $a - 32 = 14$
20. $q + 14 = 28$

21. $y - 13 = 23$
22. $d + 19 = 38$
23. $t - 25 = 55$
24. $w - 59 = 31$

25. $r - 12 = 48$
26. $e + 75 = 100$
27. $c - 15 = 40$
28. $s - 9 = 35$

29. $f - 8 = 42$
30. $p + 44 = 66$
31. $q + 5 = 35$
32. $w - 16 = 32$

33. $a - 33 = 11$
34. $r + 21 = 46$
35. $d + 9 = 54$
36. $c - 6 = 24$

37. $m - 25 = 8$
38. $n + 13 = 39$
39. $f - 11 = 88$
40. $g + 42 = 80$

► Lesson 16.7

Solve.

1. $25t = 125$
2. $\frac{r}{12} = 12$
3. $13x = 169$
4. $\frac{b}{4} = 24$

5. $\frac{x}{9} = 7$
6. $\frac{m}{50} = 9$
7. $6y = 72$
8. $9s = 162$

9. $\frac{r}{3} = 37$
10. $7n = 105$
11. $\frac{p}{8} = 32$
12. $12w = 240$

13. $\frac{y}{5} = 19$
14. $10x = 440$
15. $\frac{s}{6} = 90$
16. $2c = 38$

17. $3n = 45$
18. $\frac{t}{2} = 75$
19. $4x = 64$
20. $\frac{y}{15} = 5$

21. $\frac{z}{18} = 2$
22. $9n = 216$
23. $7r = 119$
24. $\frac{q}{11} = 11$

25. $8y = 120$
26. $\frac{b}{14} = 5$
27. $\frac{c}{22} = 30$
28. $35r = 105$

29. $16s = 96$
30. $\frac{y}{8} = 18$
31. $25d = 625$
32. $\frac{c}{2} = 57$

33. $40m = 280$
34. $16b = 80$
35. $\frac{r}{90} = 8$
36. $\frac{q}{19} = 5$

37. $\frac{y}{25} = 9$
38. $12n = 600$
39. $\frac{x}{18} = 4$
40. $45p = 90$

► Lesson 16.8

Solve.

1. $4x - 80 = 40$
2. $8a + 55 = 127$
3. $\frac{c}{10} - 3 = 9$

4. $\frac{d}{5} - 8 = 1$
5. $3e - 4 = 23$
6. $\frac{p}{6} - 2 = 6$

7. $4s - 2 = 26$
8. $8x - 10 = 54$
9. $\frac{a}{7} + 8 = 15$

10. $2d + 6 = 16$
11. $\frac{w}{2} + 13 = 88$
12. $6y + 27 = 99$

13. $\frac{b}{9} - 5 = 7$
14. $\frac{m}{3} + 15 = 25$
15. $\frac{q}{4} + 27 = 47$

16. $3c - 5 = 7$
17. $\frac{r}{2} + 5 = 19$
18. $5a - 8 = 12$

19. $8c - 13 = 43$
20. $\frac{x}{11} + 10 = 15$
21. $7n + 4 = 46$

22. $10t + 9 = 129$
23. $\frac{y}{5} - 5 = 5$
24. $13s + 7 = 72$

25. $\frac{a}{8} - 12 = 38$
26. $9x + 23 = 104$
27. $\frac{e}{2} + 47 = 71$

28. $14r - 19 = 65$
29. $25a + 15 = 240$
30. $\frac{s}{3} - 41 = 17$

► Lesson 17.2

Add.

1. $+9 + -4$
2. $-8 + -13$
3. $-6 + +15$
4. $+7 + +23$

5. $-9 + -12$
6. $+15 + +35$
7. $+17 + -11$
8. $-4 + +4$

9. $-31 + +37$
10. $+55 + +20$
11. $+14 + -28$
12. $-18 + -19$

13. $+63 + -63$
14. $-20 + -38$
15. $+22 + -74$
16. $-42 + -16$

17. $+36 + +54$
18. $+88 + -11$
19. $-51 + +69$
20. $-75 + -150$

21. $-29 + +29$
22. $-80 + -46$
23. $+92 + +33$
24. $-48 + +17$

25. $-95 + -67$
26. $+37 + -66$
27. $+11 + +72$
28. $-31 + -52$

29. $-79 + -49$
30. $-65 + +13$
31. $+83 + -14$
32. $+25 + +56$

33. $+94 + -87$
34. $-62 + +17$
35. $+51 + -51$
36. $-24 + -74$

37. $+57 + +39$
38. $-76 + +42$
39. $-35 + -85$
40. $+71 + -66$

► **Lesson 17.3**

Subtract.

1. +6 − −12
2. −5 − −20
3. +8 − +17
4. −4 − +11

5. −13 − +9
6. +10 − −25
7. −3 − −2
8. +6 − −6

9. +30 − +24
10. −23 − −18
11. +5 − −17
12. −8 − +19

13. +16 − −4
14. +21 − −9
15. −14 − −14
16. +31 − +35

17. −45 − +35
18. +21 − +21
19. +19 − +12
20. −26 − +13

21. +50 − −25
22. −33 − −11
23. +10 − −40
24. −12 − +24

25. −95 − −10
26. −95 − +10
27. +95 − +10
28. +95 − −10

29. +4 − −18
30. −72 − +28
31. +5 − −25
32. −16 − −64

33. −7 − +30
34. +50 − +25
35. −33 − −99
36. +12 − +36

37. +35 − +6
38. −51 − −17
39. −75 − +25
40. +21 − −7

► **Lesson 17.4**

Use the graph at the right. Write the ordered pair for each point.

1. E
2. A
3. H

4. C
5. G
6. B

7. D
8. F
9. K

10. M
11. P
12. T

Name the point for each ordered pair.

13. (−4, 0)
14. (5, 4)
15. (−2, 1)

16. (−6, −3)
17. (5, −1)
18. (−5, 6)

19. (0, −3)
20. (4, 5)
21. (3, −3)

22. (−2, 4)
23. (2, −5)
24. (−6, 2)

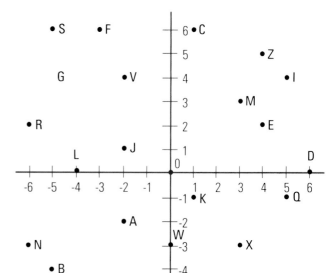

▶ Lesson 17.5

Multiply.

1. -9×-10
2. $+15 \times +4$
3. $+13 \times -13$
4. $-25 \times +5$

5. $+14 \times -2$
6. -8×-20
7. $+5 \times +60$
8. -12×-3

9. $+10 \times -40$
10. -33×-2
11. -16×-6
12. $-3 \times +15$

13. $+18 \times +2$
14. -6×-40
15. $+19 \times -4$
16. -8×-21

17. -29×-5
18. $-66 \times +2$
19. $+37 \times 0$
20. -1×-99

21. $+44 \times -7$
22. $+17 \times -6$
23. $+88 \times +3$
24. $-36 \times +3$

25. $-21 \times +8$
26. $+50 \times -4$
27. -42×-10
28. $+93 \times -6$

29. -38×-5
30. $+18 \times -7$
31. $-9 \times +24$
32. $+34 \times +9$

33. $+28 \times -4$
34. -67×-1
35. 0×-89
36. $-95 \times +5$

37. -8×-26
38. $+4 \times -49$
39. $+12 \times -20$
40. $+15 \times +15$

▶ Lesson 17.6

Divide.

1. $-88 \div -11$
2. $+40 \div -4$
3. $+48 \div +12$
4. $-65 \div +5$

5. $-50 \div +25$
6. $-84 \div -2$
7. $+39 \div -13$
8. $+93 \div +3$

9. $+100 \div +10$
10. $-66 \div +33$
11. $-42 \div -3$
12. $+81 \div -3$

13. $-125 \div +5$
14. $-60 \div -4$
15. $+96 \div +16$
16. $+108 \div -9$

17. $-400 \div -50$
18. $+320 \div +80$
19. $+280 \div -70$
20. $-630 \div +90$

21. $+128 \div -2$
22. $-195 \div -5$
23. $-90 \div +9$
24. $+121 \div +11$

25. $-84 \div -4$
26. $+120 \div -6$
27. $-168 \div +8$
28. $+225 \div +25$

29. $+444 \div -4$
30. $-98 \div +2$
31. $-75 \div -1$
32. $+169 \div -13$

33. $-130 \div -10$
34. $+225 \div -15$
35. $-300 \div +30$
36. $+96 \div +3$

37. $+70 \div -5$
38. $-55 \div +5$
39. $-200 \div -4$
40. $+99 \div +9$

Lesson 17.7

Solve. Check your answers.

1. $m + -8 = -4$ 2. $n - -7 = +28$ 3. $c + +15 = -45$ 4. $d - +6 = -26$

5. $x - -10 = +25$ 6. $b + +13 = -3$ 7. $e - -7 = +38$ 8. $r + -21 = +15$

9. $y + +5 = -62$ 10. $w - -12 = +6$ 11. $f + -30 = -9$ 12. $b - +4 = +48$

13. $s + -3 = -13$ 14. $d - -25 = +7$ 15. $m - +54 = -1$ 16. $c + +12 = +36$

17. $c - +17 = -8$ 18. $n + -38 = -5$ 19. $e - -27 = +2$ 20. $r + -33 = -44$

21. $a + -23 = -9$ 22. $r - -51 = -8$ 23. $m - +27 = -3$ 24. $z + +48 = +75$

25. $b - +16 = +4$ 26. $x + -57 = -1$ 27. $n - -12 = +3$ 28. $c + -66 = -17$

29. $e + +25 = +2$ 30. $t - +11 = -8$ 31. $s + -8 = +72$ 32. $m - -8 = -40$

33. $y - -13 = +9$ 34. $b + +7 = -30$ 35. $c - +16 = +9$ 36. $n + -7 = +24$

37. $t + -45 = -6$ 38. $r - -90 = +8$ 39. $b + +64 = -3$ 40. $r - +53 = -58$

Lesson 17.8

Solve.

1. $-6n = -36$ 2. $\frac{x}{-4} = +7$ 3. $+3b = -39$ 4. $\frac{r}{+8} = -10$

5. $\frac{c}{+2} = +34$ 6. $\frac{d}{-12} = -5$ 7. $-5s = -65$ 8. $-7w = +105$

9. $-9t = -90$ 10. $\frac{b}{+5} = -20$ 11. $\frac{m}{+8} = +9$ 12. $-6b = +96$

13. $-4r = +120$ 14. $\frac{n}{-12} = -12$ 15. $-9e = -81$ 16. $\frac{y}{+2} = -38$

17. $\frac{w}{-30} = +5$ 18. $-14f = +56$ 19. $\frac{d}{-15} = -6$ 20. $-4s = -60$

21. $+11b = +121$ 22. $+20c = -180$ 23. $\frac{s}{+70} = -7$ 24. $\frac{w}{-8} = -100$

25. $-33a = +99$ 26. $\frac{t}{-60} = -2$ 27. $\frac{r}{+80} = -10$ 28. $-6b = +90$

29. $\frac{x}{+13} = +13$ 30. $\frac{y}{-25} = +8$ 31. $-19c = -95$ 32. $-41a = -123$

33. $+50c = -600$ 34. $-18d = -54$ 35. $\frac{e}{-75} = +4$ 36. $\frac{f}{+12} = +9$

37. $\frac{t}{-11} = -11$ 38. $\frac{w}{+2} = -44$ 39. $-4s = -48$ 40. $-8n = -32$

Solve.

1. $\frac{r}{+3} - {}^-8 = {}^+18$

2. ${}^-8s + {}^+9 = {}^+73$

3. ${}^+12y - {}^+7 = {}^-79$

4. ${}^-5w + {}^-3 = {}^-33$

5. $\frac{c}{-4} + {}^+6 = {}^+26$

6. $\frac{d}{-10} - {}^-15 = {}^+8$

7. ${}^-6t - {}^-13 = {}^-5$

8. ${}^-9y + {}^-25 = {}^-7$

9. $\frac{a}{-2} - {}^+25 = {}^-75$

10. $\frac{b}{+8} + {}^-15 = {}^-22$

11. $\frac{f}{-15} - {}^-30 = {}^+27$

12. ${}^-13r + {}^+14 = {}^-12$

13. ${}^-10a - {}^-27 = {}^+67$

14. ${}^+12t + {}^-18 = {}^-78$

15. $\frac{x}{-50} + {}^+6 = {}^-4$

16. ${}^+4c - {}^+3 = {}^-51$

17. ${}^-8b + {}^-4 = {}^-28$

18. $\frac{w}{-42} - {}^-15 = {}^+13$

19. $\frac{e}{-75} - {}^+9 = {}^-11$

20. ${}^-3d + {}^-66 = {}^-33$

21. ${}^-8x - {}^-1 = {}^-63$

22. $\frac{m}{-4} + {}^-15 = {}^-20$

23. $\frac{r}{+6} - {}^+9 = {}^+1$

24. ${}^-5w + {}^+25 = 0$

25. ${}^+9s - {}^-50 = {}^-22$

26. $\frac{y}{-10} + {}^-27 = {}^-18$

27. ${}^+14t - {}^+4 = {}^-32$

28. ${}^-3x + {}^+24 = {}^-12$

29. ${}^+7a - {}^-5 = {}^+54$

30. $\frac{b}{-6} - {}^+11 = {}^-16$

Appendix

Calculator Applications

▶ Basic Calculator Operations

All calculators have standard keys that allow you to perform the basic mathematical operations of addition ⊞, subtraction ⊟, multiplication ⊠, and division ⊡. Just enter the problem in the order you would solve it using a pencil and paper, and then press ⊨ to get the answer.

Whenever you begin a problem, clear the calculator. Not all calculators clear in the same way, but two common clear keys are Ⓒ and ⒶⒸ. If you make a mistake in the middle of a problem, you can correct the last step by pushing the clear error key, either ⒸⒺ or Ⓒ.

One Operation		Two Operations		Correcting An Error	
275 + 319 + 481		57.1 + 473.9 ÷ 0.5		149 × 3.5	
Enter	**Display**	**Enter**	**Display**	**Enter**	**Display**
Ⓒ or ⒶⒸ	0.	Ⓒ or ⒶⒸ	0.	Ⓒ or ⒶⒸ	0.
275	275.	473.9	473.9	149	149.
⊞	275.	⊡	473.9	⊠	149.
319	319.	0.5	0.5	2.5	2.5
⊞	594.	⊞	947.8	ⒸⒺ or Ⓒ	0.
481	481.	57.1	57.1	3.5	3.5
⊨	1075.	⊨	1004.9	⊨	521.5

$$275 + 319 + 481 = 1075$$

$$57.1 + 473.9 \div 0.5 = 1004.9$$

$$149 \times 3.5 = 521.5$$

Estimation is a valuable skill in calculator use, since it is easy to make mistakes when entering numbers on the calculator. Always check the answers the calculator gives to make sure that they are reasonable.

For each of the following problems, estimate the answer mentally. Then use your calculator to solve. If an estimated answer and a calculated answer are very different, recompute. Remember to use order of operations. Round each answer to the nearest hundredth.

1. $6791 + 22{,}372 + 47$
2. $23.9 \times 73.8 - 467.5$
3. $0.11336 \div 0.11336$
4. $8899 \div 47.3 \times 2.005$
5. $7.136 \times 35 - 329.4$
6. $5872 + 2.7 \div 3.234$
7. $7 \times 37{,}037 \times 3$
8. $201 - 47.3 \times 2.01$
9. $55 - 5.13 \times 6.7 + 1.2$
10. $374 + 8194 - 92 \times 5$
11. $4592 \div 61 - 43$
12. $1.993 + 4.505 \times 1.993$

▶ Interpreting Quotients

When you divide to solve a problem and the answer is not a whole number, you have to interpret the quotient. Read the Example and Questions A and B below.

Example. 194 students are going on a field trip. They will travel on buses that hold 40 passengers.

A How many buses can be completely filled?
B How many buses will be needed to hold all the students?

For both questions, enter 194 ÷ 40 = on your calculator. The answer should be 4.85.

Answer to Question A: The whole-number part of 4.85 shows that 4 buses can be completely filled.

Answer to Question B: The decimal part of 4.85 shows that 1 bus will be partially filled. Thus, 5 buses will be needed in all.

Use your calculator to solve each problem below.

1. 1 case holds 24 bottles; you have 844 bottles. How many full cases will you have?

2. 1 pencil box holds 12 pencils; you have 72 pencils. How many pencil boxes will you need?

3. Every classroom has seats for 35 students; there are 187 freshmen. How many freshman homerooms are needed?

4. The population of the Boston metropolitan area is 2,805,911. Fenway Park, home of the Boston Red Sox, has a seating capacity of 33,465. How many times must Fenway Park be filled to hold the entire Boston area population?

5. 1 bus token costs $0.35. Anne has $8.92. How many bus tokens can she buy? How much money will she have left over?

6. Maxwell builds toolboxes. It takes him 45 minutes to finish one box. How many boxes does he complete in an 8-hour day?

7. Mar Vista Cleaners charges $1.45 for each 8-pound load of laundry. Gregor has 37 pounds of laundry. How much will it cost him to have his laundry done?

8. Tamara must cut as many 5-inch strips as she can out of $6\frac{1}{2}$ yards of fabric. How many strips will she end up with?

▶ Using Percents

The calculator's %️ key, in combination with the operation keys, can make some mathematical tasks easier. Even if your calculator doesn't have a %️ key, you can still calculate percents. Simply change the percents to decimals.

Example A. Calculate a 15% tip on a meal tab of $18. Without a %️ key, enter as follows: 18 ✕️ 0.15 =️. With a %️ key, enter as follows: 18 ✕️ 15 %️. The answer is $2.70.

Find the tip, at a rate of 15%, on the following amounts. Round to the nearest ten cents.

1. $26.89   2. $46.27   3. $8.12   4. $3.50   5. $19.75

Since not all calculators use the same method to calculate percents, four methods are presented below for finding a total price with tax. Method 1 shows how to calculate a total price without a %️ key. Methods 2, 3, and 4 are for calculators with a %️ key. Try all that your calculator allows. Then use the quickest one that gives the correct answer.

Example B. Find the total price of a $17.99 item with $6\frac{1}{2}\%$ tax.

Method 1		Method 2		Method 3		Method 4	
Enter	Display	Enter	Display	Enter	Display	Enter	Display
17.99	17.99	17.99	17.99	17.99	17.99	17.99	17.99
✕️	17.99	✕️	17.99	+️	17.99	+️	17.99
0.065	0.065	6.5	6.5	6.5	6.5	6.5	6.5
+️	1.16935	%️	1.16935	%️	1.16935	%️	19.15935
17.99	17.99	+️	1.16935	=️	19.15935	**Round to $19.16**	
=️	19.15935	17.99	17.99	**Round to $19.16**			
Round to $19.16		=️	19.15935				
		Round to $19.16					

Use your calculator to find the total price of each item.

6. A $24.99 sweater; sales tax is 5%

7. A $6.99 record and a $5.79 cassette; sales tax is 7%

8. 3 pounds of cashews at $5.89 a pound; sales tax is 4%

9. A $120 coat marked down 50%; sales tax is $6\frac{1}{2}\%$

10. A $33 bookbag selling for $\frac{1}{3}$ off; sales tax is $4\frac{1}{2}\%$

▶ Pythagorean Theorem

When a variable n is multiplied by itself we refer to it as n^2 (stated as "n squared" or "n to the second power").

$$n^2 = n \times n$$

For example, $3^2 = 3 \times 3 = 9$, and
$$12^2 = 12 \times 12 = 144.$$
To find 12^2 using a calculator, press 12 ☒ 12 ☐ or 12 ☒☐.

All right triangles obey a special relationship called the *Pythagorean Theorem*. The theorem is represented by the equation $a^2 + b^2 = c^2$.

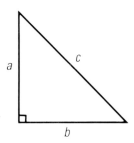

You can determine if a triangle is a right triangle by substituting the lengths of the sides given for a, b, and c in the equation. The longest measure must be substituted for c.

Use the Pythagorean Theorem to determine if the given sides form right triangles.

1. (3, 4, 5)   2. (6, 7, 8)   3. (7, 24, 25)   4. (6, 8, 10)

5. (16, 63, 65)   6. (5, 10, 15)   7. (30, 40, 50)   8. (48, 55, 73)

To find the longest side of a right triangle when the other two sides are given, find $a^2 + b^2$. Then press ☑ (called the "square root key").

Find the longest side of each right triangle with the known given sides.

9. (5, 12, ___)   10. (9, 12, ___)   11. (20, 21, ___)   12. (8, 15, ___)

13. (12, 35, ___)   14. (33, 56, ___)   15. (39, 80, ___)   16. (65, 72, ___)

To find a third side when the longest side is given, find $c^2 - a^2$ (or $c^2 - b^2$). Then press ☑.

Find the unknown side for each right triangle.

17. (___ , 16, 20)   18. (20, ___ , 101)   19. (28, ___ , 53)   20. (___ , 77, 85)

Consumer Applications

▶ Paychecks

Often when you receive a paycheck, you will find that **deductions** have been taken by your employer. These deductions cover such obligatory payments as federal income tax (FIT), state income tax, and Social Security tax (or FICA, Federal Insurance Contributions Act). In addition, voluntary payments into a health insurance plan or charities may be deducted from your paycheck.

The amount of money that you actually receive is called your **net pay**. The amount of money that you earn (hours worked × hourly wage) is called your **gross pay**.

Net pay = Gross pay − Deductions

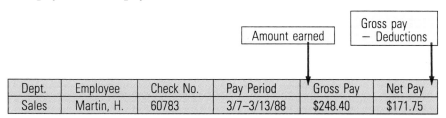

Dept.	Employee	Check No.	Pay Period	Gross Pay	Net Pay
Sales	Martin, H.	60783	3/7–3/13/88	$248.40	$171.75

Amount earned → Gross Pay
Gross pay − Deductions → Net Pay

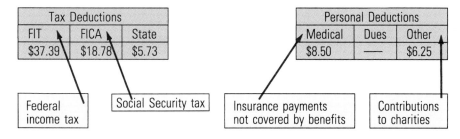

Tax Deductions		
FIT	FICA	State
$37.39	$18.78	$5.73

Federal income tax → FIT
Social Security tax → FICA

Personal Deductions		
Medical	Dues	Other
$8.50	—	$6.25

Insurance payments not covered by benefits → Medical
Contributions to charities → Other

Use the pay statement above to answer the following questions.

1. How long is the pay period covered by this check?

2. Henry Martin earns $5.40 an hour for a regular 40-hour week and time and a half for overtime. Use Henry's gross pay to determine how many hours he worked in the pay period covered by this check.

3. Calculate the total taxes paid.

4. Calculate the total deductions for health insurance ("Medical") and charities ("Other").

5. What are the total deductions listed on this pay statement?

6. Show how Henry's net pay was determined.

▶ Sales Receipts

When you go shopping, you usually receive a sales receipt as proof of purchase. The receipt may be a cash register tape listing amounts only, a computer slip listing both items purchased and amounts, or a sales slip such as the one shown below. A receipt indicates the price of each item you bought, the sales tax (if any), and the total amount you paid.

Total purchase price = Sum of items bought + Sales tax

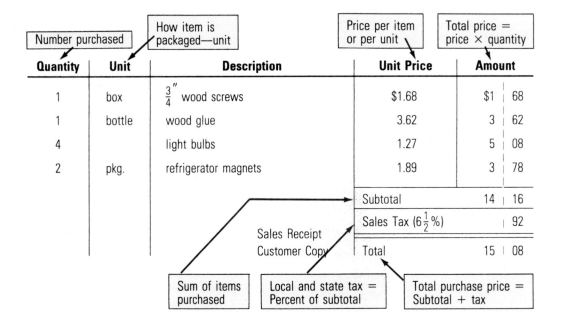

Use the sales receipt above to answer the following questions.

1. What is the sales tax rate (percent) on the items purchased?

2. Calculate the amount of sales tax on the original order at a rate of $4\frac{1}{2}\%$. What would the total purchase price be?

3. How much does 1 light bulb cost? What would the total purchase price be if 6 light bulbs were bought, instead of 4?

4. What would the sales tax on the original order be if, in addition, 2 extension cords costing $2.98 each were purchased?

5. Suppose that, instead of $0.92, the sales tax on the original order is $0.57. What is the tax rate?

▶ Checking Accounts

When you write checks, make deposits into your checking account, or use an automatic teller machine to withdraw cash from your account, it is important that you write down the amount of your transaction. Check registers are designed to help you keep track of your account. One column is for withdrawals, and another column is for deposits. The **balance** column shows how much money you have in your account.

New balance = Previous balance − Check amount
New balance = Previous balance + Deposit amount

Balance brought forward from previous page

Check No.	Date	Description	Amount of Check	Amount of Deposit	Balance
					$429.02
	3/14	Paycheck deposit		$125.53	554.55
583	3/17	Fox's Sporting Goods	$73.92		480.63
584	3/19	Marcy's Clothing Store	59.45		421.18
	3/20	Deposit of birthday money		25.00	446.18

Previous balance + Deposit amount

Previous balance − Check amount

Use the check register above for the following problems.

1. Write the calculation that gives the second balance.

2. Write the calculation that gives the third balance.

3. Write the calculation that gives the fourth balance.

4. What will the new balance be if a check is written for $27.92?

Complete the balance column in the check register below.

	Check No.	Date	Description	Amount of Check	Amount of Deposit	Balance
						$261.39
5.	954	7/1	Lions baseball tickets	$23.75		?
6.		7/5	Refund		$ 19.99	?
7.	955	7/5	Eatery Grocery Store	42.81		?
8.		7/7	Paycheck deposit		$172.64	?

Geometric Shapes and Formulas

▶ Abbreviations

General Abbreviations

l = length
w = width
s = length, for a regular polygon
n = number of sides of a regular
    polygon
b = base (for trapezoids, the bases
    are labeled a and b)
a = altitude (for cones)
h = height
d = diameter
r = radius

2-Dimensional Figures

A = area
P = perimeter
C = circumference

3-Dimensional Figures

A = surface area
V = volume
B = area of base of a pyramid,
    prism, or cylinder

Additional Facts

$\pi \approx 3.14$ or $\frac{22}{7}$

 symbol for a right
angle

▶ 2-Dimensional Figures

Polygons

3-sided Polygons

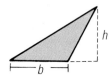 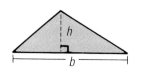

Right triangle

P = sum of sides
$A = \frac{1}{2}bh$

4-sided Polygons

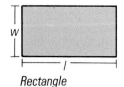

Rectangle

$P = 2l + 2w$
$A = lw$

Square

$P = 4 \times s$
$A = s \times s$

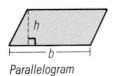

Parallelogram

P = sum of sides
$A = bh$

Trapezoid

P = sum of sides
$A = \frac{1}{2} \times (a + b) \times h$

Circles

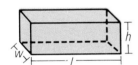

$C = \pi \times d$

$C = 2 \times \pi \times r$
$A = \pi \times r \times r$

▶ 3-Dimensional Figures

Prisms

Rectangular Prism

$A = 2(lw) + 2(wh)$
$V = Bh = lwh$

Cube

$A = 6(s \times s)$
$V = s \times s \times s$

Pyramids

$A = (s \times s) + 4(\frac{1}{2}sh)$

Cones

$V = \frac{1}{3}Bh$

Cylinders

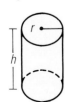

$A = 2(\pi \times r \times r) + (Ch)$
$V = Bh = (\pi \times r \times r) \times h$

Table of Measures

Time

60 seconds (sec.) = 1 minute (min.)
60 minutes (min.) = 1 hour (hr.)
24 hours = 1 day
7 days = 1 week

365 days
12 months (mo.) } = 1 year (yr.)
366 days = 1 leap year
10 years = 1 decade
100 years = 1 century

Metric System of Measurement

Length
10 millimeters (mm) = 1 centimeter (cm)
1000 millimeters
100 centimeters } = 1 meter (m)
1000 meters = 1 kilometer (km)

Capacity
1000 milliliters (mL) = 1 liter (L)
1000 liters = 1 kiloliter (kL)

Mass
1000 milligrams (mg) = 1 gram (g)
1000 grams = 1 kilogram (kg)
1000 kilograms = 1 metric ton (T)

Temperature
Water freezes at 0 degrees Celsius (0° C).
Water boils at 100 degrees Celsius (100° C).

Customary System of Measurement

Length
12 inches (in.) = 1 foot (ft.)
36 inches
3 feet } = 1 yard (yd.)
1760 yards
5280 feet } = 1 mile (mi.)

Capacity
8 fluid ounces (fl. oz.) = 1 cup (c.)
16 fluid ounces
2 cups } = 1 pint (pt.)
32 fluid ounces
2 pints } = 1 quart (qt.)
4 quarts = 1 gallon (gal.)

Mass
16 ounces (oz.) = 1 pound (lb.)
2000 pounds = 1 ton (T.)

Temperature
Water freezes at 32 degrees Fahrenheit (32° F).
Water boils at 212 degrees Fahrenheit (212° F).

Glossary

Numbers in parentheses refer to chapter and lesson numbers.

altitude (AL-tuh-tood)   In geometry, the shortest segment from a vertex to the opposite base. (15.1)

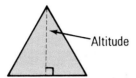
Altitude

area   The measure of the region inside a polygon or other figure. Area is measured in square units. (14.5)

average   Mean. (10.1)

balance column   In a check register, the column that shows how much money is left in an account. (Appendix)

bar graph   A diagram in which parallel bars represent quantities. (9.1)

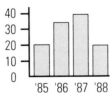

capacity (kuh-PAS-uh-tee)   The amount of space that can be filled. (8.1)

chance   The likelihood that a particular event will occur; also called *probability*. (13.1)

circle   A figure in which each point is an equal distance from its center. (14.1)

circle graph   A circular diagram divided into regions that represent data. (9.1)

circumference (sir-KUM-fer-ence)   The distance around a circle. (14.4)

cone   A three-dimensional figure, as pictured below. (15.7)

cube   A three-dimensional figure with square surfaces. (15.3)

customary measurement   The system used in the United States for nonscientific measurement. Examples: feet, yards, quarts. (8.1)

cylinder (SIL-ihn-dur)   A three-dimensional figure with equal circular bases that are parallel. (15.4)

data   Facts and figures. (10.1)

decimal (DES-uh-mul)   A number in which quantities less than 1 are expressed by place values based on 10. Examples: 1.13, 9.7, 0.004. (1.2)

deduction   An amount subtracted from a larger amount. (Appendix)

denominator (dih-NOM-uh-nay-ter) In the fraction $\frac{3}{4}$, the denominator is 4. (5.1)

diameter (dy-AM-ih-tur)   A line segment passing through the center of a circle, with endpoints on the circle. (14.4)

Diameter

difference   The result of subtraction. (1.10)

dividend (DIV-uh-dend)   In $600 \div 15 = 40$, 600 is the dividend. (3.1)

divisor (dih-VY-zur)   In $600 \div 15 = 40$, 15 is the divisor. (3.1)

equally likely outcomes   Results with the same chances of occurring. (13.2)

equation (ee-KWAY-zhun)   A statement that two expressions are equal. Example: $2n + 5 = 15$. (16.1)

equivalent fractions   Fractions that represent the same number. Example: $\frac{2}{3}$ and $\frac{4}{6}$. (5.3)

equivalent ratio   A ratio that represents the same comparison as a given ratio. Example: $1:2$ and $2:4$. (11.1)

event   In probability, an outcome or a group of outcomes. (13.1)

expression (ex-PRESH-uhn)   A mathematical phrase. Examples: x, $4x$, $n + 12$. (16.1)

factors   Numbers being multiplied; also called *multipliers*. (2.1)

favorable outcome   A desired result. (13.1)

Fermi (FER-mee) **problem**   A problem in which you must estimate an answer, using only common sense and commonly known information. (17.9)

formula (FORM-yuh-luh)   A statement of a mathematical rule. Example: The formula for the area of a rectangle is Area = length × width, or $A = l \times w$. (14.2)

fraction   A number that names part of a whole or part of a group. (5.1)

frequency (FREE-kwen-see)   The number of times an event occurs. (10.1)

frequency table   A table that tells how many times a group of events occurs. (10.2)

gram (g)   In the metric system, the basic unit used to measure mass. (4.1)

graph   A drawing or diagram that provides information. (9.1)

greatest common factor (GCF) The greatest factor shared by each number in a given set. Example: 4 is the GCF of 12 and 16. (5.5)

gross pay   The amount of pay earned before deductions are subtracted. (Appendix)

height   The length of an altitude. (14.1)

hexagon (HEX-uh-gon)   A polygon with six sides. (14.1)

histogram (HIS-tuh-gram)   A bar graph that shows frequency. (10.6)

horizontal scale   The line running across the bottom of a graph, giving information about what is represented on the graph. (9.1)

hypotenuse   The side opposite the right angle in a right triangle (Appendix).

identity property   Rule stating that any number multiplied by 1 equals that number. (2.1)

improper fraction   A fraction with a numerator that is greater than or equal to the denominator. Examples: $\frac{3}{2}$; $\frac{6}{6}$. (5.1)

integer (IN-tih-jur)   A positive or negative whole number or zero. (17.1)

line graph   A graph that uses an unbroken line to show trends or changes over time. (9.1)

liter (L)   In the metric system, the basic unit used to measure volume, or capacity. (4.1)

lowest common denominator (LCD) The smallest number that can be divided by each denominator in a given set of fractions. Example: 20 is the lowest common denominator of $\frac{3}{5}$ and $\frac{1}{10}$. (7.2)

lowest terms   A fraction is in lowest terms when 1 is the GCF of the numerator and the denominator. Examples: $\frac{1}{2}$, $\frac{7}{13}$, and $\frac{15}{4}$ are in lowest terms. (5.6)

mean   The value obtained by dividing the sum of a set of numbers by the number of items in the set. Example: the mean of 3, 4, and 8 is $15 \div 3 = 5$. (10.1)

median   The middle number in an ordered set of numbers. (10.3)

```
11
14
17  ◄── Median
19
20
```

meter (m)   In the metric system, the basic unit used to measure length or distance. (4.1)

mixed number   A number that consists of a whole number and a fraction. Example: $3\frac{5}{8}$. (5.1)

mode   The number that occurs most frequently in a set of numbers. (10.3)

```
7
7
9
9  ◄── Mode
9
11
13
```

multiple   A multiple of a given number is obtained by multiplying that number by any whole number. Example: Multiples of 4 are 4, 8, 12, etc. (2.3)

multiple event   In probability, a situation involving more than one outcome. Example: A coin tossed twice. (13.5)

multipliers   Numbers being multiplied; also called *factors*. (2.1)

negative integer   A whole number less than zero. (17.1)

net pay   The amount of pay earned after deductions for items such as income tax. (Appendix)

numerator (NOO-mer-ay-tur) In the fraction $\frac{3}{4}$, the numerator is 3. (5.1)

odds   A ratio used in mathematics to express the chance of an outcome occurring. (13.3)

opposite integers   Two whole numbers the same distance from zero on the number line. Example: +5 and −5. (17.1)

ordered pair   A pair of numbers that corresponds to a location on a grid. (17.4)

origin (OR-ih-jun)   In mathematics, the point where the horizontal and vertical number lines meet on a grid. (17.4)

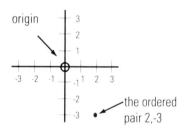

outcome   In probability, the result of an experiment. (13.1)

parallel (PAR-uh-lel) **lines**   Two lines that are the same distance apart and that never meet. (14.1)

parallelogram   A four-sided polygon whose opposite sides are parallel. (14.5)

pentagon (PEN-tuh-gon)   A polygon with five sides. (14.1)

percent   Per hundred. Example: 12% means 0.12, or $\frac{12}{100}$. (12.1)

perimeter (puh-RIM-uh-tur)   The distance around a shape or a figure. (14.1)

pi (py)   The ratio of the circumference of a circle to its diameter, written as π. π is $\approx$3.14, or $\frac{22}{7}$. (14.4)

pictograph   A graph in which pictures represent quantities. (9.1)

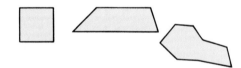

Pictograph
Test Score Distribution

= 2 Students

polygon (POL-ee-gon)   A two-dimensional figure such as a square, triangle, or octagon. (14.1)

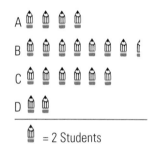

polyhedron (pol-ee-HEE-drun)   A three-dimensional figure with faces that are polygons. (15.1)

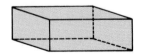

positive integer   A whole number greater than zero. (17.1)

power A number obtained by multiplying a given number by itself. Examples: Powers of 10 are 100 (10 × 10), 1000 (10 × 10 × 10), etc. (2.3)

prism (PRIZ-uhm) A three-dimensional figure with two parallel, identical faces. (15.3)

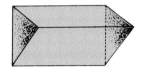

probability (prob-uh-BIL-uh-tee) The likelihood that a particular event will occur; also called *chance*. (13.1)

product The result of multiplication. In 6 × 7 = 42, the product is 42. (2.1)

proper fraction A fraction with a numerator that is less than the denominator. Example: $\frac{3}{5}$. (5.1)

property A characteristic or rule. (2.1)

proportion A mathematical statement that two ratios are equal. Example: $\frac{3}{6} = \frac{1}{2}$. (11.2)

pyramid (PIR-uh-mid) A three-dimensional figure with a base that is a polygon and triangular sides. (15.4)

Pythagorean (puh-thag-uh-REE-uhn) **Theorem** A rule stating that the square of the hypotenuse of a right triangle is equal to the sum of the squares of the two other sides. (Appendix)

quotient (KWOH-shunt) The result of division. In 600 ÷ 15 = 40, 40 is the quotient. (3.1)

radius (RAY-dee-uhs) A line segment with endpoints at the center of a circle and at a point on the circle. (14.4)

range The difference between the largest and the smallest number in a set of numbers. (10.3)

ratio A comparison of two numbers by division. Example: A ratio of 4 : 2 would represent 4 books for 2 people. (11.1)

reciprocal (ruh-SIP-ruh-cuhl) A number formed by exchanging a fraction's numerator and denominator. When reciprocals are multiplied, the product is 1. Example: $\frac{3}{2}$ is the reciprocal of $\frac{2}{3}$; $\frac{3}{2} \times \frac{2}{3} = 1$. (6.6)

rectangle (REK-tan-guhl) A four-sided polygon with four right angles. (14.5)

rectangular prism A three-dimensional figure with rectangular faces. (15.3)

regular polygon In a regular polygon, all sides have equal length. (14.1)

remainder   When 14 is divided by 3, 2 is left and is the remainder. (3.2)

right angle   An angle as pictured below. (14.5)

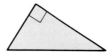

right triangle   A triangle containing a right angle. (Appendix)

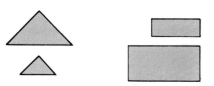

scale drawing   A drawing of an object with dimensions proportional to those of the actual object. (11.5)

similar figures   Figures that have the same shape and are proportional in size. (11.7)

square   A polygon with four right angles and four sides of equal length. (14.1)

square root   The square root of a given number is a number which, when multiplied by itself, gives the original number. Example: The square root of 9 is 3 because $3 \times 3 = 9$. (Appendix)

statistics   Numerical data organized and analyzed to give information. (10.1)

subtotal   A partial sum. (12.11)

sum   The result of addition. In $12 + 24 = 36$, 36 is the sum. (1.9)

surface area   The total area of all the faces of a three-dimensional figure. (15.1)

times   Multiplied by. (2.1)

trapezoid (TRAP-uh-zoid)   A four-sided polygon that has exactly one pair of parallel sides. (14.7)

tree diagram   A diagram showing all possible outcomes of a series of events. (13.5)

triangle   A three-sided polygon. (14.1)

unit price   The cost of one unit of a product. (11.4)

unknown   In algebra, a symbol, such as x, used to represent a quantity; also called *variable*. (16.1)

variable (VAR-ee-uh-buhl)   In algebra, a symbol, such as x, used to represent a quantity; also called *unknown*. (16.1)

vertical scale   The line running up and down at the left of a graph, giving information about what is represented on the graph. (9.1)

volume   The amount of space inside a three-dimensional figure. (15.1)

whole number   Any counting number or zero. Examples: 7, 99, 1376. (1.1)

zero property   Rule stating that any number multiplied by 0 equals 0. (2.1)

Index

mean, 230–231
range, mode, and median,
228–229
Subtracting
with customary units, 186–187
fractions with like denominators,
150–151
fractions with unlike
denominators, 158–159
integers, 376–377
mixed numbers, 166–167
mixed numbers from whole
numbers, 162–163
solving equations using, 362–363,
384–385
whole numbers and decimals,
20–21
whole numbers and decimals
with zeros, 22–23
Sums, estimating, 24–25
Surface area, 333
of cubes and prisms, 336–337
of pyramids and cylinders,
338–339

T

Table
frequency, 226–227
making a, 100–101, 145
reading a, 98–99, 144–145,
182–183
Tax, sales, 462
Temperature
line graph of, 210–211
measurement of, 188–189, 466
Terminating decimals, 125
Three-dimensional figures,
332–333, 464–465
Time, elapsed, 190–191, 466
Time cards, 192–193
Times, 30
Tons, 180
metric, 94
Trapezoid, 322–323, 464–465
Tree diagram, 296–297
Triangle, 464–465
area of, 320–321
perimeter of, 312
Pythagorean theorem for, 460
right, 460, 464–465
Two-dimensional figures,
310–311, 464–465

U

Unit pricing, 248–249
Unknown, 352

V

Value, of expression, 356–357
Variable, 352
isolation of, 364
Vertical scale, 203
Vertices, of polyhedron, 332
Volumes, 333
of cubes, prisms, and pyramids,
340–341
of cylinders and cones, 344–345
liquid. *See* Liquid volume

W

Weight
customary units of, 180–181, 466
Whole numbers
adding, 18–19
comparing and ordering, 10–11
dividing by decimals, 80–81
dividing decimals by, 74–75
multiplying, 44–45
reading and writing, 2–3
rounding, 14–15
subtracting, 20–21
subtracting mixed numbers from,
162–163
subtracting with zeros, 22–23
**Withdrawals, checking
account**, 371, 463

Y

Yards, 176

Z

Zero property, 31
**Zeros, subtracting whole
numbers and decimals with**,
22–23

Selected Answers

Pages 2–3   **5.** Hundreds   **7.** Ten millions   **9.** Millions   **11.** Tens   **13.** 15,312,000   **15.** 15,000,312,000
17. 402,000,017   **19.** 6,000,421,092   **21.** Twenty-three thousand, one   **23.** Six hundred thirty-seven million
25. Eleven billion, three hundred million   **27.** Three million, four hundred thirty-three thousand, twelve
29. 63,040   **31.** Seventeen million, forty-nine thousand, one   **33.** Nine hundred ninety-nine dollars

Pages 4–5   **7.** Tenths   **9.** Tenths   **11.** 927.3   **13.** 0.616   **15.** Four and three hundredths
17. Twenty-seven and two hundred four thousandths   **19.** Seven hundred twenty-nine and seven hundredths
21. 2   **23.** Five hundred and twenty-five thousandths   **25.** One hundred sixty-nine dollars and sixty-five cents
27. Tens   **29.** Hundreds   **31.** Tens   **33.** Millions   **35.** Hundred thousands

Pages 8–9   **1.** Gasoline   **3.** To use up   **5.** How much fuel is used up in 8 hours; how much

Pages 10–11   **5.** >   **7.** <   **9.** >   **11.** <   **13.** >   **15.** >   **17.** 75,851 > 75,581 > 57,851
19. 882,313 > 828,331 > 282,313   **21.** 490,230 > 409,320 > 409,230 > 49,320   **23.** <   **25.** >
27. 1,437,912,100 > 1,437,911,200 > 1,437,901,200   **29.** 355,997 > 355,797 > 355,779 >
353,779 > 335,797   **31.** Maravich, Selvy, Carr, Williams, McGill   **33.** Twenty eight thousand, six hundred
seven   **35.** Seven and fifty-two thousandths   **37.** Nine hundred seventy-four thousandths

Pages 12–13   **5.** <   **7.** <   **9.** <   **11.** >   **13.** 0.0601 < 0.1006 < 0.1060 < 0.6001
15. 3.6911329 < 36.091132 < 36.901132 < 36.991132 < 369.01132   **17.** <   **19.** <   **21.** =
23. 0.098 < 0.89 < 0.908 < 0.98 < 9.09   **25.** 5.35 < 5.37 < 5.57 < 5.73 < 5.75   **27.** Satherson, W.
29. <   **31.** 74,337 < 79,344 < 79,433   **33.** 998 < 8009 < 9008 < 19,090

Pages 14–15   **5.** 130   **7.** 380   **9.** 800   **11.** 8490   **13.** 17,010   **15.** 1000   **17.** 4100   **19.** 13,200
21. 2900   **23.** 8300   **25.** 17,900   **27.** 9000   **29.** 5000   **31.** 12,000   **33.** 18,000   **35.** 68,000
37. 40,000   **39.** $64,510   **41.** $65,000   **43.** 72,000   **45.** 25,020   **47.** 25,000   **49.** 16,500
51. 4.602 < 46.02 < 46.2 < 462.2

Pages 16–17   **7.** 10   **9.** $14   **11.** $219   **13.** $100   **15.** 8   **17.** 3.5   **19.** $41.50   **21.** 40.0
23. 18.2   **25.** 9.5   **27.** 1.43   **29.** 6.91   **31.** 0.09   **33.** 0.25   **35.** 8.91   **37.** 473   **39.** $840
41. $837.50   **43.** 12,900   **45.** 12,933   **47.** 446.5 cm   **49.** =   **51.** <   **53.** >

Pages 18–19   **7.** 8089   **9.** 16,433   **11.** 0.96   **13.** 1.039   **15.** 7.34   **17.** 314.123   **19.** 782.35
21. 1.31   **23.** 33.18   **25.** $76.22   **27.** 8400   **29.** 16,100   **31.** 35,600   **33.** 166.9   **35.** 474.9
37. 4.0

Pages 20–21   **7.** 2278   **9.** 4601   **11.** 8612   **13.** 6753   **15.** 1806   **17.** 1.81   **19.** 1.38   **21.** 4.388
23. 3.349   **25.** 0.632   **27.** 1.67   **29.** 10.783   **31.** 32,314   **33.** 5399   **35.** 523.84   **37.** 7.5560
39. 44,260

Pages 22–23   **5.** 1599   **7.** 1082   **9.** 33,856   **11.** 390.2   **13.** 8.7   **15.** 68.62   **17.** 422.26
19. 0.1128   **21.** 0.3868   **23.** 15,599   **25.** 47,088   **27.** 3754   **29.** 7.5   **31.** 32.1   **33.** 75
35. 2.0191 < 2.091 < 2.919   **37.** 76,536 < 78,536 < 87,536

Pages 24–25   **7.** ≈33 or ≈40   **9.** ≈70,000 or ≈71,000   **11.** ≈$100 or ≈$103   **13.** ≈53,000
15. ≈800 or ≈803   **17.** ≈450 or ≈447   **19.** ≈1400 or ≈1000   **21.** ≈10 or ≈16   **23.** ≈20
25. ≈3200 or ≈3000   **27.** ≈$162, ≈$160, or ≈$100   **29.** ≈1200 or ≈1170   **31.** ≈400   **33.** ≈$1
35. 980   **37.** 51.19   **39.** 279   **41.** 3.78

Page 26   **1.** Thousandths   **3.** 7   **5.** Eight million, seventy-five thousand, six and three tenths   **7.** 1.13
9. 0.508   **11.** 2.694 < 2.9 < 6.03 < 6.294 < 6.87   **13.** <   **15.** <   **17.** 1000   **19.** 20,000   **21.** 5
23. 25   **25.** 78   **27.** 78.84   **29.** $5

Pages 30–31   **7.** Times   **9.** Factors; multipliers   **11.** Identity property

Pages 32–33   **5.** 219   **7.** 330   **9.** 188   **11.** 552   **13.** 265   **15.** 553   **17.** 5538   **19.** 45,591
21. 18,072   **23.** 1052   **25.** 50,064   **27.** 19,220   **29.** 392   **31.** 27024   **33.** 1556   **35.** 5775
37. 220 miles   **39.** ≈900 or ≈940   **41.** ≈170 or ≈200   **43.** ≈360 or ≈400

Pages 34–35   **7.** 892,000   **9.** 7200   **11.** 520,000   **13.** 80   **15.** 4,700,000   **17.** 43,000   **19.** 48,000
21. 42,000   **23.** 24,000   **25.** 63,000   **27.** 2600   **29.** 1370   **31.** 1,600,000   **33.** 350,000   **35.** 750
37. 920,000   **39.** 3,600,000   **41.** $340   **43.** 4.368   **45.** 148,859   **47.** 15,908

Pages 36–37   **7.** 2686   **9.** 24,318   **11.** 54,270   **13.** 3084   **15.** 9240   **17.** 12,784   **19.** 6396
21. 1,176,416   **23.** 1,211,821   **25.** 134,250   **27.** 7,120,582   **29.** 2,868,544   **31.** 21,960   **33.** 548,422
35. 243,405   **37.** 67,165   **39.** 23,625   **41.** 525,375   **43.** 250,176   **45.** 10,400   **47.** 150,500
49. $11,220   **51.** 117,078   **53.** 10,253,404

Pages 40–41   **5.** ≈160   **7.** ≈350   **9.** ≈320   **11.** ≈200 or ≈150   **13.** ≈1200   **15.** ≈300 or ≈250
17. ≈3000   **19.** ≈15,000   **21.** ≈32,000   **23.** ≈1600   **25.** ≈270   **27.** ≈14,000   **29.** ≈80
31. ≈2800   **33.** ≈120   **35.** ≈1200 miles   **37.** >   **39.** <   **41.** >

Pages 42–43   **5.** 12.9   **7.** 379.5   **9.** 351.1   **11.** 93   **13.** 90   **15.** 60   **17.** 9700   **19.** 301,000
21. 380   **23.** 5420   **25.** 1000   **27.** $320   **29.** $850   **31.** 321,100   **33.** $600   **35.** $20.70   **37.** 9305
39. $10,000   **41.** $42.90   **43.** $120   **45.** One and five thousandths   **47.** Twenty-six and nine tenths
49. Seven hundred thirty one   **51.** Fifty-seven thousand, sixty   **53.** Two hundred nine thousandths
55. Twenty-five and twenty-five hundredths   **57.** Four hundred sixty-seven and nine hundredths   **59.** Four
thousand six hundred fifty-two   **61.** Fifty-six and three hundred forty-three thousandths

Pages 44–45   **5.** 20.8   **7.** 58.8   **9.** 19.4   **11.** 1.2   **13.** 29.4   **15.** 112.5   **17.** 18   **19.** 79.2
21. 5.95   **23.** 54.25   **25.** 99.99   **27.** 12.5   **29.** 64.61   **31.** 112.5   **33.** 4.545   **35.** 48.918
37. 399.600   **39.** 203.125   **41.** 28.125   **43.** 10.756   **45.** 7.72   **47.** 3.645   **49.** 20.979   **51.** 3.92
53. 504.495   **55.** 0.261   **57.** 0.48 in.   **59.** 6.6; 6.58   **61.** 9.9; 9.88   **63.** 0.4; 0.37

Pages 46–47   **1.** 4; 4; 2   **3.** 3   **5.** 15; 15; 15   **7.** 3   **9.** 1 + 5 + 9   **11.** 5 because it is a part of
four sums   **15.** 252   **17.** 21,216   **19.** 14,499

Pages 48–49   **5.** 20.65   **7.** 64.9072   **9.** 29.152   **11.** 181.24   **13.** 4.25   **15.** 131.215   **17.** 1421.625
19. 0.3696   **21.** 0.223729   **23.** 0.0365   **25.** 0.0588   **27.** 3.9528   **29.** 0.121947   **31.** 1.76
33. 10.9989   **35.** 30.351   **37.** 1.40466   **39.** 0.2565   **41.** 2.17045   **43.** 5.435073   **45.** $0.96
47. 9500; 9500   **49.** 59,500; 59,530   **51.** 729,700; 729,720   **53.** 400; 440   **55.** 19,700; 19,710

Pages 50–51   **1.** $2400   **3.** In Question 1, you could multiply 4 × $2 × 3 mentally and add two zeros to
the product. Question 2 could be answered the same way.   **5.** $862.50   **7.** $900   **9.** $2670.75; paper-and-
pencil or calculator   **11.** You must know how much Jacklyn paid the driver, how many miles she lives from the
airport, the cost of the taxi per mile, and the size of the tip.   **13.** How many weeks it will take Paul to earn
$79.95; how much Paul earns an hour or how many hours a week he works

Page 52   **1.** Product   **3.** 35,000   **5.** 519,000   **7.** 3800   **9.** 20,000   **11.** 232   **13.** 1375   **15.** 28,750
17. 1690   **19.** 154   **21.** $264   **23.** ≈200,000   **25.** ≈2,400,000   **27.** 42,890   **29.** 630   **31.** 18
33. 945.5   **35.** 145.8   **37.** 441.6   **39.** 2474.5   **41.** 10.08   **43.** 39.676   **45.** 0.01248

Pages 56–57   **9.** 38.88; 4.8; 8.1   **11.** 63; 7; 9   **13.** 5; 5   **15.** 9; 9

Pages 58–59   **7.** 12   **9.** 102   **11.** 209   **13.** 1057   **15.** 480   **17.** 19   **19.** 30 R2   **21.** 10 R4
23. 7 R3   **25.** 16 R2   **27.** 197 R3   **29.** 485 R6   **31.** 19   **33.** 255   **35.** 237   **37.** 55 R6   **39.** 15 R3
41. 474   **43.** 606 R4   **45.** 3216   **47.** $75   **49.** 344.58   **51.** 9.923   **53.** 10.71

Pages 60–61   **7.** 106   **9.** 5   **11.** 39   **13.** 0.7   **15.** 118.5   **17.** 0.011   **19.** 4.95   **21.** 5   **23.** 19
25. 12,000   **27.** 16.1   **29.** 60   **31.** 15.25   **33.** 0.095   **35.** 210   **37.** $0.04; $0.043; $0.04; the largest
or the smallest box   **39.** $5.50 per hour   **41.** 7.4698   **43.** 0.000054

Pages 62–63   **5.** 5   **7.** 4   **9.** 5   **11.** 15   **13.** 15   **15.** 20   **17.** 50 or 55   **19.** 60 or 64   **21.** 50
23. 20   **25.** 80   **27.** 4 or 5   **29.** 700   **31.** 6   **33.** 400   **35.** 9 or 10   **37.** 15   **39.** 6000   **41.** 30
students   **43.** $0.12; $0.22; 1st pad   **45.** 600,012   **47.** 5000.5

Pages 64–65   **5.** 19 R37   **7.** 14 R49   **9.** 11 R16   **11.** 7 R30   **13.** 65   **15.** 23   **17.** 93 R4
19. 591 R8   **21.** 55 R47   **23.** 31 R83   **25.** 72 R5   **27.** 66   **29.** 555   **31.** 22 R5   **33.** 333 R9
35. 10 R9   **37.** 41   **39.** 99 R49   **41.** 101   **43.** 40 R40   **45.** $60   **47.** 7.37   **49.** 4.96   **51.** 6.82
53. 2.07   **55.** 0.191

Pages 66–67   **5.** 6 R7   **7.** 14 R442   **9.** 6 R359   **11.** 7 R483   **13.** 733 R101   **15.** 150 R155
17. 870 R391   **19.** 914 R285   **21.** 6   **23.** 1386 R153   **25.** 169 R134   **27.** 235 R137   **29.** ≈3000 tickets
31. 54,000   **33.** 250,000

Pages 70–71   **1.** Yes   **3.** Division   **5.** 10   **7.** Multiplication   **9.** Yes; yes   **11.** Subtraction; 900
13. Division or multiplication; 24 quarters   **15.** Subtraction; $6.50   **17.** $21.60

Pages 72–73   **5.** 16   **7.** 1   **9.** 5   **11.** 2   **13.** 20   **15.** 5   **17.** 1   **19.** 7   **21.** 13   **23.** 14
25. 12   **27.** (6 × 15) + (10 × 14)   **29.** 12   **31.** 3085 R2

Pages 74–75   **5.** 9.8   **7.** 10.8   **9.** 0.66   **11.** 0.048   **13.** 0.145   **15.** 0.009   **17.** 0.79   **19.** 0.06
21. 1.24   **23.** 0.149   **25.** 0.84   **27.** 0.036   **29.** 10.9 mi.   **31.** $3.75   **33.** 65,000   **35.** 492,000

Pages 76–77   **7.** 13.5   **9.** 8.0   **11.** 0.3   **13.** 1.8   **15.** 2.2   **17.** 0.6   **19.** 1.9   **21.** 1.3
23. $14.05   **25.** $45.90   **27.** $22.31   **29.** $3.44   **31.** $2.15   **33.** $3.17   **35.** $0.25   **37.** $95.88
39. $7.33   **41.** $15.88   **43.** 34.6   **45.** $16.50   **47.** 0.1   **49.** 0.8   **51.** 0.7   **53.** 0.1   **55.** $20 per
month   **57.** About $400   **59.** 1034   **61.** 2740

Pages 78–79   **5.** 5.8   **7.** 3.3   **9.** 29   **11.** 4.8   **13.** 0.2   **15.** 9.1   **17.** 0.2   **19.** 1.7   **21.** 6.3
23. 16   **25.** 21   **27.** 3.6   **29.** 6.3   **31.** 8.2   **33.** 140 albums   **35.** 436 pages   **37.** 5.789
39. 20   **41.** 900   **43.** 0.037

Pages 80–81   **7.** 15   **9.** 2.5   **11.** 400   **13.** 950   **15.** 62.5   **17.** 2.5   **19.** 46.7   **21.** 200.3
23. 855.8   **25.** 136.2   **27.** 17.3   **29.** 200.5   **31.** 228.6   **33.** 122.2   **35.** 25.2   **37.** 6237.5   **39.** 92.4
41. 151.9   **43.** 1700   **45.** 90.7   **47.** 109,140   **49.** 120.6   **51.** 32 lockers   **53.** 71   **55.** 28   **57.** 11

Page 82   **1.** F; dividend   **3.** 16   **5.** 4   **7.** 90   **9.** 11 R5   **11.** 4   **13.** 40   **15.** 5   **17.** 7
19. 13 R16   **21.** 77 R79   **23.** 1 R339   **25.** 12 R3   **27.** Division; 64 packages   **29.** Subtraction; 408 years
31. 51   **33.** 137   **35.** 0.6   **37.** 1.3

Pages 86–87   **1.** mL, mg, mm   **3.** kilobyte—1000 bytes (a basic unit of digits operated on by a computer);
milligram—$\frac{1}{1000}$ of a gram (a measure of mass); millivolt—$\frac{1}{1000}$ of a volt (a measure of
electricity); centimeter—$\frac{1}{100}$ of a meter (a measure of length)   **5.** "I earn $25 a week."   **7.** $18; divide; 100
9. 1.2   **11.** 5900   **13.** 0.00083   **15.** 0.00072

Pages 88–89 **7.** 5.3 cm **9.** 1.34 m **11.** 6.384 km **13.** 2900 m **15.** 4200 cm **17.** 84 mm **19.** 0.009 m **21.** 0.05364 km **23.** 5.2 mm **25.** 3.1 km **27.** 580 cm **29.** 1900 mm **31.** 3200 m **33.** 0.00629 km **35.** 2200 m **37.** 5.0 **39.** 171.1 **41.** 0.6 **43.** 3.0

Pages 90–91 **5.** Answers may range from 20 to 30 cm and 200 to 300 mm. **7.** m **9.** m **11.** 4 cm; 39 mm **13.** cm **15.** Answers will vary. **17.** 10.17 **19.** 3.326 **21.** 7593 **23.** 1.077

Pages 92–93 **1.** The students could begin by dropping the marble from the roof of a building, timing the drop, and reducing the height of the drop until they reach 1 second. **3.** No **5.** 2 seconds **7.** 30,000 cm; 0.3 km **9.** 3 seconds **11.** About 3 seconds **13.** 7000 **15.** 15,000 **17.** 5400 **19.** 160,000 **21.** 720,000

Pages 94–95 **5.** 3.7 g **7.** 0.9 g **9.** 750,000 g **11.** 8.725 kg **13.** 45,000 g **15.** kg **17.** 0.500 kg **19.** 0.00004 g **21.** 9.257 g **23.** kg **25.** 2.3 T **27.** 76.14 **29.** 0.000032

Pages 96–97 **(Lesson 4.6)** **7.** 43 L **9.** 0.2 L **11.** 3.75 L **13.** L **15.** mL **17.** 45,000 mL **19.** 9300 mL **21.** 800 mL **23.** mL **25.** L **27.** 96 mL **29.** 4.056 4.065 4.506 4.605 4.650

Pages 98–99 **1.** 3772 km **3.** Adelaide; Melbourne **5.** 1 hour and 30 minutes or 1 hour and 40 minutes; it is not always the same **7.** 4 hours and 15 minutes; about 118 km/hr. **9.** 188.8 km; 890.5 − 188.8 = 701.7 km

Pages 100–101 **1.** $490; $490; $300; $475 **3.** Deluxe minijeep **5.** Landrover **7.** Plan C **9.** Table should reflect information in the following order: Plan A—$20, 0, $8, $12, $8, $48; Plan B—$16, $6, $8, $12, $8, $50; Plan C—$35, $6, $8, 0, 0, $49 **11.** $240; calculator or paper-and-pencil **13.** 200; any choice

Page 102 **1.** 10 **3.** 240 mm **5.** 4800 m **7.** 0.363 **9.** 54 cm **11.** 36,000 cm **13.** 2 cm; 19 mm **15.** g **17.** 0.039 L **19.** 0.00976 kg **21.** 0.00421 L **23.** 0.00041 kg **25.** 0.004751 kg **27.** New Zealand; Japan

Pages 110–111 **1.** $\frac{4}{5}$ **3.** $\frac{1}{3}$ **5.** proper fraction **7.** proper fraction **9.** improper fraction **11.** proper fraction **13.** A 6-pack of soda; an inch on a ruler divided into sixteenths; a pack of 100 sheets **15.** 1, 2, 4, 7, 14, 28

Pages 112–113 **5.** $\frac{2}{4}, \frac{3}{6}, \frac{4}{8}$, etc. **7.** $\frac{10}{8}, \frac{15}{12}, \frac{20}{16}$, etc. **9.** $\frac{12}{14}, \frac{18}{21}, \frac{24}{28}$, etc. **11.** $\frac{2}{12}, \frac{3}{18}, \frac{4}{24}$, etc. **13.** $\frac{6}{10}, \frac{9}{15}, \frac{12}{20}$, etc. **15.** $\frac{2}{16}, \frac{3}{24}, \frac{4}{32}$, etc. **17.** $\frac{1}{2}$ **19.** $\frac{6}{8}$ or $\frac{3}{4}$ **21.** $\frac{2}{9}$ **23.** $\frac{2}{5}$ **25.** $\frac{5}{4}$ **27.** $\frac{30}{36}, \frac{15}{18}$, or $\frac{5}{6}$ **29.** $\frac{2}{6}, \frac{3}{9}, \frac{4}{12}$ **31.** $\frac{6}{4}$ or $\frac{9}{6}$ **33.** $\frac{2}{20}, \frac{3}{30}, \frac{4}{40}$, etc. **35.** $\frac{3}{4}$ **37.** 24 in. **39.** 615 **41.** 838

Pages 114–115 **5.** 50 **7.** 48 **9.** $\frac{29}{6}$ **11.** $\frac{23}{8}$ **13.** $\frac{30}{9}$ **15.** $3\frac{6}{9}$ **17.** $13\frac{3}{4}$ **19.** $7\frac{2}{10}$ **21.** 5 **23.** $3\frac{5}{10}$ **25.** $\frac{31}{7}$ **27.** $\frac{3}{1}, \frac{6}{2}, \frac{9}{3}$ **29.** $\frac{31}{16}$ **31.** 8 trips **33.** 890 cm

Pages 116–117 **5.** < **7.** < **9.** < **11.** = **13.** > **15.** > **17.** = **19.** = **21.** < **23.** < **25.** > **27.** < **29.** > **31.** < **33.** = **35.** = **37.** < **39.** Greg **41.** cm **43.** kg

Pages 118–119 **5.** 1, 2, 3, 6 **7.** 1, 2, 7, 14 **9.** 1, 2, 3, 6, 9, 18 **11.** 1, 2, 3, 4, 6, 8, 12, 24 **13.** 4 **15.** 3 **17.** 20 **19.** 9 **21.** 1, 2, 4, 8 **23.** 1, 13 **25.** 1, 3, 11, 33 **27.** 14 **29.** 1 **31.** 4 **33.** 6 **35.** 1 in., 2 in., 4 in., 8 in. **37.** 66, 803 **39.** 2,000,090

Pages 120–121 **7.** $\frac{2}{3}$ **9.** $\frac{3}{5}$ **11.** $\frac{2}{3}$ **13.** $\frac{4}{7}$ **15.** $\frac{3}{10}$ **17.** $\frac{1}{3}$ **19.** $\frac{1}{3}$ **21.** $\frac{9}{40}$ **23.** $\frac{5}{6}$ **25.** $\frac{4}{3}$ **27.** $\frac{2}{5}$ **29.** $\frac{8}{5}$ **31.** 2 : 1, 2; 5 : 1, 5; $\frac{2}{5}$ **33.** 13 : 1, 13; 14 : 1, 2, 7, 14; $\frac{13}{14}$ **35.** 15 : 1, 3, 5, 15; 16 : 1, 2, 4, 8,

16; $\frac{15}{16}$ **37.** $\frac{3}{4}$ **39.** $\frac{4}{5}$ **41.** $\frac{3}{8}$ **43.** $\frac{5}{9}$ **45.** $\frac{11}{13}$ **47.** $\frac{3}{4}$ **49.** $\frac{7}{8}$ **51.** $\frac{5}{4}$ **53.** $4\frac{1}{2}$ **55.** $\frac{1}{3}$ hour **57.** $\frac{2}{3}$
59. ≈ 3000 **61.** ≈ 7000

Pages 122–123 **5.** 0.5 **7.** 0.4 **9.** 0.6 **11.** 0.5625 **13.** $0.1\overline{6}$ **15.** $0.\overline{45}$ **17.** $0.\overline{81}$ **19.** $4\frac{3}{4}$
21. $\frac{201}{1000}$ **23.** $\frac{1}{1000}$ **25.** 3.2 **27.** $4.58\overline{3}$ **29.** 11.1 **31.** $2.\overline{3}$ **33.** $\frac{19}{1000}$ **35.** $\frac{7}{10}$ **37.** $1\frac{16}{25}$
39. 5.56 **41.** $0.2\overline{6}$ **43.** $0.48\overline{3}$ **45.** $\frac{1}{5}$ lb. **47.** 5.62 **49.** 16.28

Pages 124–125 **1.** $0.\overline{18}$, $0.\overline{142857}$, $0.4\overline{71}$ **3.** The arrows from each fraction point to the first digit that begins a repeating set of decimals for that fraction. **5.** $0.\overline{0001}$. A calculator might not display the pattern correctly. **7.** 0.5, $0.\overline{3}$, 0.25, 0.2, $0.1\overline{6}$, $0.\overline{142857}$, 0.125, $0.\overline{1}$, 0.1, $0.\overline{09}$, $0.08\overline{3}$, $0.\overline{076923}$, $0.0\overline{714285}$, $0.0\overline{6}$, 0.0625, $0.\overline{0588235294117647}$, $0.\overline{05}$, $0.\overline{052631578947368421}$, 0.05, $0.0\overline{47619}$, 0.045, $0.\overline{0434782608695652173913}$, $0.041\overline{6}$, 0.04 **9.** It neither terminates nor repeats; the calculator does not answer the question because it doesn't show whether the pattern would repeat if division were carried more places; long division shows this.

Page 126 **1.** Numerator **3.** $\frac{1}{3}$ **5.** $\frac{1}{2}$ or $\frac{2}{4}$ **7.** $\frac{14}{20}, \frac{21}{30}, \frac{28}{40}$ **9.** $\frac{8}{18}, \frac{12}{27}, \frac{16}{36}$ **11.** $\frac{2}{11}$ **13.** $\frac{1}{2}$ **15.** $\frac{9}{16}$
17. $\frac{6}{1}$ **19.** $\frac{79}{10}$ **21.** $=$ **23.** $>$ **25.** 9 **27.** 4 **29.** 4 **31.** $\frac{14}{27}$ **33.** $\frac{2}{1}$ **35.** $\frac{5}{8}$ **37.** 0.25
39. $9\frac{3}{10}$ **41.** $\frac{1}{8}$

Pages 130–131 **7.** $\frac{1}{15}$ **9.** $\frac{9}{20}$ **11.** $\frac{8}{45}$ **13.** $\frac{6}{25}$ **15.** $\frac{2}{45}$ **17.** $\frac{5}{12}$ **19.** $\frac{5}{8}$ **21.** $\frac{1}{6}$ **23.** $\frac{20}{27}$ **25.** $\frac{1}{4}$
27. $\frac{1}{5}$ **29.** $\frac{2}{5}$ **31.** 3 **33.** $\frac{6}{7}$ **35.** $3\frac{8}{9}$ **37.** $1\frac{3}{8}$ **39.** $\frac{8}{11}$ **41.** $1\frac{2}{7}$ **43.** $2\frac{2}{3}$ **45.** $1\frac{2}{5}$ **47.** $6\frac{2}{9}$ **49.** $\frac{3}{4}$
51. $\frac{3}{8}$ **53.** $\frac{3}{64}$ **55.** $\frac{1}{18}$ **57.** $\frac{3}{4}$ **59.** $2\frac{1}{2}$ **61.** $18 **63.** 6 **65.** 1 **67.** 5 **69.** 2 **71.** 17 **73.** 4
75. 3 **77.** 2

Pages 132–133 **5.** $\frac{5}{21}$ **7.** $\frac{5}{36}$ **9.** $\frac{3}{28}$ **11.** $\frac{1}{6}$ **13.** $\frac{2}{15}$ **15.** $\frac{9}{44}$ **17.** $\frac{1}{9}$ **19.** $\frac{26}{75}$ **21.** $\frac{2}{7}$ **23.** $\frac{4}{33}$
25. $\frac{3}{5}$ **27.** $\frac{24}{365}$ **29.** $\frac{13}{34}$ **31.** $\frac{1}{2}$ **33.** $\frac{1}{8}$ **35.** $\frac{14}{33}$ **37.** $\frac{25}{34}$ **39.** $\frac{5}{12}$ **41.** $\frac{9}{20}$ **43.** $\frac{3}{20}$ **45.** $\frac{11}{18}$ **47.** 4
49. $\frac{35}{72}$ **51.** 1 qt. **53.** $\frac{59}{8}$ **55.** $\frac{9}{1}$ **57.** $\frac{39}{5}$ **59.** $\frac{4}{1}$ **61.** $\frac{53}{6}$ **63.** $\frac{20}{1}$ **65.** $\frac{43}{5}$

Pages 134–135 **1.** She was awake less time than she was asleep. Therefore, she was awake less than half the time the movie ran. **3.** The difference between the movie time (109 minutes) and the sum that results from a guess of 40 minutes (105) is closer than the sum that results from a guess of 50 minutes (125). **5.** Too high. José spent $16.50 and Cecile spent $11.50. **7.** $0.6\overline{1}$ **9.** $0.1\overline{36}$

Pages 136–137 **7.** 14 **9.** $34\frac{2}{3}$ **11.** $24\frac{7}{10}$ **13.** $13\frac{31}{45}$ **15.** $\frac{8}{15}$ **17.** $2\frac{5}{16}$ **19.** $3\frac{1}{6}$ **21.** $2\frac{34}{35}$ **23.** $\frac{80}{99}$
25. 79 **27.** $47\frac{1}{2}$ **29.** 57 **31.** $3\frac{3}{4}$ **33.** $3\frac{2}{3}$ **35.** $73\frac{1}{3}$ **37.** $1\frac{1}{10}$ **39.** $9\frac{1}{2}$ **41.** $1\frac{2}{3}$ cups oats; 5 cups water **43.** 0.00036 **45.** 0.6572 **47.** 17.01 **49.** 26,839.8 **51.** 6318

Pages 138–139 **1.** Yes, because it is $\frac{1}{8}$ less than 1; 0.875 **3.** Close to 1: $\frac{11}{10}, \frac{21}{20}, \frac{15}{16}$; close to $\frac{1}{2}$: $\frac{5}{9}, \frac{9}{20}, \frac{100}{205}$; close to neither: $\frac{7}{29}, \frac{1}{20}$ **5.** She knew that the answer would be somewhere between $8 \times 9 = 72$ and $9 \times 10 = 90$. Answers may include numbers between 72 and 90. **7.–33.** Answers may include the following but any answers you can justify are acceptable: **7.** $\approx \frac{1}{4}$ **9.** $\approx 4\frac{1}{2}$ **11.** ≈ 5 **13.** ≈ 3 or 6 **15.** ≈ 18
17. ≈ 6 **19.** $\approx \frac{1}{2}$ **21.** ≈ 4 **23.** about 5 mi. **25.** 11 R2 **27.** 7 **29.** 21 R1 **31.** 22 R5

Pages 140–141   **7.** $\frac{3}{2}$ **9.** $\frac{1}{4}$ **11.** $\frac{16}{5}$ **13.** 5 or $\frac{5}{1}$ **15.** 2   **17.** $1\frac{1}{2}$ **19.** $\frac{84}{121}$ **21.** $1\frac{5}{11}$ **23.** $3\frac{1}{3}$
25. $\frac{1}{4}$ **27.** $\frac{1}{10}$ **29.** $\frac{1}{12}$ **31.** $\frac{2}{7}$ **33.** $\frac{12}{125}$ **35.** $\frac{2}{9}$ **37.** $\frac{3}{32}$ **39.** $1\frac{13}{32}$ **41.** $\frac{1}{16}$ **43.** $\frac{2}{5}$ **45.** $\frac{1}{4}$ yd.
47. $\approx$40,000   **49.** $\approx$100,000   **51.** $\approx$360,000

Pages 142–143   **7.** $2\frac{18}{35}$ **9.** $\frac{16}{27}$ **11.** $1\frac{1}{3}$ **13.** $2\frac{43}{100}$ **15.** $\frac{1}{3}$ **17.** $\frac{7}{90}$ **19.** $\frac{1}{5}$ **21.** $\frac{5}{52}$ **23.** $\frac{5}{27}$
25. $\frac{8}{91}$ **27.** $11\frac{1}{2}$ **29.** $5\frac{7}{16}$ **31.** $1\frac{37}{40}$ **33.** $8\frac{5}{8}$ **35.** $10\frac{14}{25}$ **37.** 3   **39.** $1\frac{11}{16}$ **41.** 5   **43.** $\frac{1}{12}$ **45.** $\frac{4}{85}$
47. $1\frac{1}{3}$ **49.** $3\frac{6}{7}$ **51.** 16 panels   **53.** 48,000   **55.** 85.7   **57.** 4000   **59.** 5400   **61.** 600   **63.** 554

Pages 144–145   **15.** 19.32   **17.** 410.625   **19.** 0.7425   **21.** 9.1

Page 146   **1.** $\frac{3}{8}$ **3.** $\frac{8}{15}$ **5.** $\frac{1}{6}$ **7.** $7\frac{7}{8}$ **9.** $\frac{1}{4}$ **11.** $\frac{2}{3}$ **13.** 5 video; 8 audio   **15.** 58   **17.** $16\frac{1}{2}$
19. $\approx$1   **21.** $\approx$35   **23.** $\approx$12   **25.** $\frac{16}{13}$ **27.** $\frac{22}{9}$ **29.** $\frac{27}{16}$ **31.** $1\frac{1}{2}$ **33.** 8   **35.** $2\frac{7}{10}$ **37.** $2\frac{1}{3}$ **39.** $\frac{3}{28}$
41. 5 posters

Pages 150–151   **5.** $\frac{10}{13}$ **7.** 1   **9.** $1\frac{1}{15}$ **11.** $1\frac{2}{3}$ **13.** $1\frac{11}{16}$ **15.** $\frac{2}{5}$ **17.** $\frac{2}{3}$ **19.** $\frac{1}{4}$ **21.** $\frac{1}{3}$ **23.** $\frac{3}{5}$
25. $\frac{18}{25}$ **27.** $1\frac{1}{2}$ **29.** $\frac{1}{5}$ **31.** $1\frac{3}{4}$ **33.** $\frac{5}{9}$ **35.** $\frac{8}{25}$ **37.** $1\frac{3}{4}$ **39.** $1\frac{4}{9}$ **41.** $\frac{1}{3}$ **43.** 3.10   **45.** 0.93
47. 1.84   **49.** 7.5   **51.** 4.4   **53.** 4.7

Pages 152–153   **7.** 15, 30, 45, 60, 75   **9.** 22, 44, 66, 88, 110   **11.** 18, 36, 54, 72, 90   **13.** 9   **15.** 20
17. 12   **19.** 60   **21.** 35   **23.** 11, 22, 33, 44, 55   **25.** 30, 60, 90, 120, 150   **27.** 32, 64, 96, 128, 160
29. 18   **31.** 27   **33.** 30   **35.** 70   **37.** 77   **39.** 1 touchdown, 2 field goals, and 1 safety; 2 touchdowns and 2 extra points; 4 fields goals and 1 safety; 7 safeties; 1 touchdown and 4 safeties; 1 touchdown, 1 extra point, 1 field goal, and 2 safeties; 2 touchdowns and 1 safety; 2 field goals and 4 safeties   **41.** 7990
43. 176,270

Pages 154–155   **7.** $\frac{7}{10}$ **9.** $\frac{11}{16}$ **11.** $\frac{11}{14}$ **13.** $\frac{7}{9}$ **15.** $\frac{19}{24}$ **17.** $1\frac{1}{4}$.   **19.** $1\frac{3}{20}$ **21.** $1\frac{7}{36}$ **23.** $1\frac{19}{30}$
25. $1\frac{1}{32}$ **27.** $\frac{5}{8}$ **29.** $1\frac{13}{20}$ **31.** $1\frac{5}{48}$ **33.** $\frac{41}{44}$ **35.** $1\frac{5}{8}$ **37.** Yes   **39.** 726 mm   **41.** 28,100 g

Pages 156–157   **5.** $5\frac{3}{5}$ **7.** $9\frac{3}{4}$ **9.** $6\frac{2}{7}$ **11.** $7\frac{1}{2}$ **13.** $16\frac{1}{2}$ **15.** $12\frac{7}{10}$ **17.** $16\frac{13}{20}$ **19.** $11\frac{11}{24}$
21. $11\frac{1}{12}$ **23.** $4\frac{3}{14}$ **25.** $14\frac{5}{12}$ **27.** $10\frac{1}{3}$ **29.** $12\frac{5}{6}$ **31.** $4\frac{4}{5}$ **33.** $10\frac{1}{3}$ **35.** 15   **37.** $12\frac{1}{10}$ **39.** $17\frac{19}{28}$
41. $1053\frac{1}{3}$ ft.   **43.** 0.7   **45.** 0.875   **47.** $0.8\overline{3}$

Pages 158–159   **5.** $\frac{1}{8}$ **7.** $\frac{1}{4}$ **9.** $\frac{1}{9}$ **11.** $\frac{1}{5}$ **13.** $\frac{1}{8}$ **15.** $\frac{1}{9}$ **17.** $\frac{3}{10}$ **19.** $\frac{1}{20}$ **21.** $\frac{4}{15}$ **23.** $\frac{1}{12}$
25. $\frac{11}{36}$ **27.** $\frac{1}{8}$ **29.** $\frac{3}{8}$ **31.** $\frac{13}{28}$ **33.** $\frac{10}{21}$ **35.** $\frac{5}{36}$ **37.** $\frac{9}{35}$ **39.** $\frac{1}{2}$ **41.** $\frac{3}{10}$ **43.** $\frac{3}{40}$ **45.** Corn chips;
$\frac{3}{16}$ oz.   **47.** 6   **49.** 70   **51.** 33,000   **53.** 670,000

Pages 160–161   **5.** $\frac{7}{40}$ **7.** $1\frac{17}{70}$ **9.** $\frac{37}{84}$ **11.** 1   **13.** $\frac{31}{35}$ **15.** $1\frac{13}{28}$ **17.** $\frac{1}{18}$ **19.** $\frac{51}{52}$ **21.** $\frac{2}{7}$ **23.** $\frac{1}{2}$
25. $1\frac{46}{105}$ **27.** $\frac{31}{42}$ **29.** $1\frac{3}{8}$ **31.** $\frac{7}{8}$ **33.** $5\frac{1}{3}$ **35.** $1\frac{7}{8}$ **37.** $\frac{15}{32}$ **39.** $2\frac{2}{3}$

Pages 162–163   **5.** 4   **7.** 8   **9.** 7   **11.** $1\frac{4}{5}$ **13.** $8\frac{3}{10}$ **15.** $12\frac{3}{8}$ **17.** $3\frac{6}{11}$ **19.** $25\frac{3}{20}$ **21.** $3\frac{5}{8}$
23. $6\frac{7}{10}$ **25.** $25\frac{19}{25}$ **27.** $14\frac{17}{30}$ **29.** $18\frac{13}{16}$ **31.** $4\frac{1}{5}$ **33.** $13\frac{17}{21}$ **35.** $30\frac{5}{14}$ **37.** $24\frac{63}{100}$ **39.** $1\frac{3}{4}$ oz.
41. Eight thousand thirteen and six hundredths   **43.** Three hundred ninety-five thousandths   **45.** Thousands
47. Hundred thousands   **49.** Millions

Pages 164–165 **1.** Multiplication **3.** $85\frac{4}{5}$ calories **5.** 4 calories; 36 minutes **7.** 7 hours; 1 hour; 1 hour **9.** 9 hours **11.** Division; 7 doors **13.** Addition; $31\frac{1}{2}$ in. **15.** Subtraction; $585\frac{1}{5}$ mi. **17.** 10 **19.** 55

Pages 166–167 **5.** 4 **7.** 17 **9.** $4\frac{5}{24}$ **11.** $3\frac{7}{20}$ **13.** $2\frac{23}{36}$ **15.** $2\frac{2}{3}$ **17.** $4\frac{5}{8}$ **19.** $2\frac{37}{40}$ **21.** $3\frac{3}{4}$ **23.** $5\frac{7}{12}$ **25.** $1\frac{9}{20}$ **27.** $\frac{4}{5}$ **29.** $2\frac{20}{21}$ **31.** $3\frac{81}{100}$ sec. **33.** $1\frac{2}{25}$ **35.** $4\frac{2}{3}$

Pages 168–169 **5.** 2 **7.** 12 **9.** 4 **11.** 2 **13.** 1 **15.** $9\frac{1}{2}$ **17.** 18 **19.** 2 **21.** 3 **23.** 2 **25.** 2; $1\frac{1}{2}$ **27.** 2; $2\frac{1}{2}$ **29.** 5; 5 **31.** 4; $3\frac{1}{2}$ **33.** 2; 2 **35.** 3 in. **37.** km **39.** cm

Page 170 **1.** $1\frac{2}{3}$ **3.** $\frac{1}{2}$ **5.** 10, 20, 30, 40, 50 **7.** 9, 18, 27, 36, 45 **9.** 15, 30, 45, 60, 75 **11.** 8 **13.** 20 **15.** 24 **17.** $\frac{23}{24}$ **19.** $16\frac{19}{24}$ **21.** $10\frac{4}{15}$ **23.** $2\frac{1}{2}$ mi. **25.** 6 **27.** 7 **29.** $\frac{1}{2}$ **31.** $5\frac{3}{4}$ **33.** $2\frac{2}{5}$ **35.** $3\frac{7}{10}$ **37.** $6\frac{13}{20}$ **39.** $\frac{13}{15}$ **41.** 13 **43.** 8

Pages 174–175 **1.** Feet, inches, pounds, years, months, days, degrees Fahrenheit, quarts, miles, hours, dollars **3.** all of them **5.** time, temperature **7.** capacity—the ability to receive, hold, or absorb; volume **9.** temperature—the degree of hotness or coldness of a body or environment **11.** distance—inch, foot, yard, mile **13.** weight—ounce, pound, ton **15.** capacity—cup, pint, quart, gallon **17.** d **19.** e **21.** a **23.** 240 **25.** $0.284\overline{09}$ **27.** $4.73\overline{48}$ **29.** $7.\overline{6}$

Pages 176–177 **7.** 48 in. **9.** 10,560 ft. **11.** 240 in. **13.** 360 in. **15.** 187 yd. **17.** 3 mi. **19.** 4 mi. **21.** 48 in. **23.** 99 yd. **25.** 150 ft. **27.** 46,145 yd. **29.** 1 **31.** $\frac{11}{20}$ **33.** 9

Pages 178–179 **5.** 2 in.; 2 in.; $1\frac{3}{4}$ in. **7.** 1 yd. **9.** Answers will vary. **11.** 60 in. **13.** $7\frac{3}{4}$ in. ; $9\frac{1}{2}$ in. **15.** $2\frac{2}{3}$ **17.** $5\frac{7}{12}$

Pages 180–181 **5.** 304 oz. **7.** 352,000 oz. **9.** 344 oz. **11.** 3 lb. **13.** 6 lb. 2 oz. **15.** 4 T. **17.** lb. **19.** 46 oz. **21.** 2 lb. 15 oz. **23.** 8 T. 1860 lb. **25.** lb. **27.** 6756 bags **29.** > **31.** > **33.** <

Pages 182–183 **1.** 100 lb. **3.** 200 lb. **5.** 0.38 lb. **7.** 74 lb. **9.** 263 lb. **11.** 394.5 lb. **13.** Jupiter, 100 lb.; Earth, 39 lb.; Moon, 6 lb.; Mercury, 11 lb.; Venus, 33 lb.; Mars, 15 lb.; Saturn, 42 lb.; Uranus, 31 lb.; Neptune, 46 lb.; Pluto, 0.38 lb. **15.** Subtract; 167 lb.

Pages 184–185 **5.** 14 pt. **7.** 12 gal. **9.** 9 pt. 1 c. **11.** 26 c. **13.** 76 qt. **15.** 38 qt. **17.** fl. oz. **19.** 80 fl. oz. **21.** 17 c. **23.** 36 pt. **25.** gal. **27.** 6 pack **29.** 40,000 **31.** 5750

Pages 186–187 **5.** 3 hr. 51 min. **7.** 32 gal. 3 qt. **9.** 20 hr. 11 min. **11.** 1 hr. 9 min. **13.** 15 lb. 13 oz. **15.** 7 hr. 54 min. **17.** 39 min. 44 sec. **19.** 5 gal. 2 qt. **21.** 13 yd. 7 in. **23.** 5 hr. 3 min. 11 sec. **25.** 4 min. 15 sec. **27.** 126 **29.** 13 **31.** 21

Pages 188–189 **5.** 5°F **7.** 98.6°F **9.** 20°C **11.** 37°C **13.** 90°F; 32°C **15.** 125°F; 52°C **17.** 79°F **19.** $\frac{14}{25}$ **21.** 24 **23.** $4\frac{4}{7}$ **25.** $1\frac{1}{2}$

Pages 190–191   **1.** A possible answer would be to figure out the elapsed time between 8:10 and 2:10 and then add the 15 minutes from 2:10 to 2:25.   **3.** 8 hours, 15 minutes   **5.** 12 hours, 6 minutes   **7.** Yes; by changing 1:15 p.m. to 13:15 and then subtracting it from 8:40 to get 4 hours, 35 minutes; this is similar to decimals since you would regroup as you do with decimals; this is different from decimals since you would have to regroup based on 60 instead of 10 or powers of 10   **9.** 5:45 p.m.; one possible method is to subtract 2 hours and 15 minutes from 8:00   **11.** Yes; one possible method is to add 4 hours and 30 minutes to 7:00 (11:30) and then add 20 minutes (11:50)   **13.** 9:13 p.m.   **15.** 7:19 a.m.   **17.** $0.8\overline{3}$ **19.** $0.\overline{6}$ **21.** $0.\overline{27}$

Pages 192–193   **5.** 37.5 hours   **7.** $163.13; $97.88   **9.** 8 hours   **11.** $7\frac{3}{4}$ hours   **13.** $4\frac{1}{2}$ hours   **15.** $41\frac{1}{2}$ hours   **17.** 400,000   **19.** 30,000   **21.** 600,000   **23.** 420,000

Page 194   **1.** Inch; foot; yard, mile   **3.** 84 in.   **5.** 8 ft. 3 in.   **7.** 2 lb. 3 oz.   **9.** 3 mi.   **11.** oz.   **13.** 2 in.; $2\frac{1}{2}$ in.; $2\frac{1}{4}$ in.   **15.** $2.50   **17.** 17 yd. 2 ft. 5 in.   **19.** 14 lb. 6 oz.

Pages 202–203   **1.** The price of farmland per acre.   **3.** The price of farmland rose, then fell over the period 1978–1987, with the prices highest in 1982.   **5.** 1 year   **7.** $25   **9.** $750   **11.** $1800

Pages 204–205   **5.** East Germany   **7.** Soviet Union   **9.** 16   **11.** Finland   **13.** Answers will vary.   **15.** 45,000   **17.** 7000   **19.** 600   **21.** 100,000   **23.** 13,000

Pages 206–207   **5.** San Francisco   **7.** ≈$35,000   **9.** Most national parks, monuments and recreational areas are found west of the Mississippi. Most national historic sites are found east of the Mississippi.   **11.** 2700   **13.** 0.22   **15.** 0.006211   **17.** 88,720.00

Pages 208–209   **1.** Yes. One scale can have activities; the other can have calories.   **5.** The value of the longest bar must be included on the scale.   **13.** She could have organized the information in order of calories expended, from greatest to least. This would give a graph with bars of steadily decreasing size (or vice versa).   **15.** 0.099 m   **17.** 480 g   **19.** 54.3 cm   **21.** 43,000 mL   **23.** 0.157 kg   **25.** 600 mL

Pages 210–211   **7.** January   **9.** 62 days   **11.** Honolulu's average temperature remains above 70°F all year round. The Minneapolis temperature ranges from 10°F to just under 80°F.   **13.** 0.8046   **15.** 0.378   **17.** 68.932   **19.** 31,990   **21.** 1.47

Pages 212–213   **11.** The increase seems sharper on the first graph because the intervals differ. The graphs give the same information but the intervals on the first graph are closer together.   **13.** 9600   **15.** 33,000   **17.** 57,200

Pages 214–215   **9.** 95 million; 165 million   **11.** The urban population grew significantly, whereas the rural population remained fairly static; the 1990 rural population will probably be about 60 million; the 1990 urban population will probably be about 175 million   **13.** 11.993   **15.** 13.2   **17.** 74,161

Pages 216–217   **9.** School   **11.** $\frac{1}{4}$ **13.** $\frac{1}{4}$ **15.** $1141.7 billion   **17.** $\frac{1}{2}$ **19.** $\frac{1}{4}$; 294 billion   **21.** 11 R56   **23.** 12   **25.** 390   **27.** 2830   **29.** 6041   **31.** 0.024

Pages 218–219   **9.** $451.51

Page 220   **1.** Australia   **3.** 16 or 17   **5.** 22 million   **9.** $\frac{1}{4}$

Pages 224–225   **1.** average   **3.** To mean "normal" or "not out of the ordinary"   **7.** 6, 12, 17, 19, 20, 21, 53, 85, 206   **9.** 7, 9, 30, 54, 98, 706, 800, 1200   **11.** $2500, $10,600, $18,500, $22,000, $56,000   **13.** 14.1   **15.** 2.8   **17.** 207.0

Pages 226–227 **5.** 22 pairs **7.** 5 pairs **9.** one 8, three 9s, five 10s, one 11, eight 12s, one 13, four 14s, one 15 **11.** 12 **13.** Answers will vary. **15.** 32.24 **17.** 2.608 **19.** 88.628 **21.** 877

Pages 228–229 **5.** 3.8 **7.** $1.25 **9.** $14,700 **11.** 5.5 **13.** 13 **15.** 77 **17.** $58; 21 oz.; $27; 5 oz. **19.** 30 **21.** 205

Pages 230–231 **7.** 13.4 **9.** 473.25 **11.** $18,430 **13.** 80.3 **15.** 184.83 **17.** 8.54 mi. **19.** $\frac{7}{8}$ **21.** $1\frac{1}{2}$ **23.** $2\frac{2}{3}$ **25.** $1\frac{19}{30}$ **27.** 10

Pages 232–233 **1.** Median: $13,875; mean: $14,112.50; the managers' salaries are higher than the clerks', so calculating with the managers' salaries pushes the mean up **3.** Mean: $25,956.67; median: $14,090; mode: $13,600. Mode and median stay the same; the mean becomes greater since the new highest salary is so much higher than any of the others. **5.** Median income. High salaries tend to raise the mean to levels that do not represent the "average" American. The median is less influenced by unusually high salaries. **7.** Median

Pages 234–235
9. $1\frac{5}{12}$ **11.** $\frac{5}{9}$ **13.** $\frac{1}{4}$ **15.** $\frac{1}{8}$ **17.** $2\frac{2}{5}$

Pages 236–237
11. The answer should be a bar graph showing the type of job on one axis and the number of employed students on the other. The freshmen, sophomores, juniors, and seniors in each category can be either added together or kept separate by including four bars for each category. **13.** Multiplication; $13.23

Page 238 **1.** Average **3.** Range **5.** Range: 17; mean: 81.8; median: 79; modes: 74, 91 **7.** $231,250 **11.** Answers will vary. A possible answer is to show the profit made by a business over several years.

Pages 242–243 **5.** 55 to 1; $\frac{55}{1}$; 55 : 1 **7.** 10 to 5; $\frac{10}{5}$; 10 : 5 **9.** $1.56 to 4; $\frac{\$1.56}{4}$; $1.56 : 4 **11.** $\frac{2}{1}$; $\frac{4}{2}$; $\frac{6}{3}$ **13.** $\frac{6}{8}$; $\frac{9}{12}$; $\frac{12}{16}$ **15.** 7 to 2; $\frac{7}{2}$; 7 : 2; $\frac{14}{4}$; $\frac{21}{6}$; $\frac{28}{8}$ **17.** $10 to 8; $\frac{\$10}{8}$; $10 : 8; $\frac{\$5}{4}$; $\frac{\$20}{16}$; $\frac{\$40}{32}$ **19.** 7 to 9; $\frac{7}{9}$; 7 : 9; $\frac{14}{18}$; $\frac{21}{27}$; $\frac{28}{36}$ **21.** 200 to 10; $\frac{200}{10}$; 200 : 10; $\frac{100}{5}$; $\frac{20}{1}$; $\frac{400}{20}$ **23.** $\frac{3}{5}$; $\frac{6}{10}$; $\frac{9}{15}$; $\frac{12}{20}$; 18 hits **25.** 17.1 **27.** 122,174 **29.** 8.40 **31.** 44.625 **33.** 324,425 **35.** 82.838

Pages 244–245 **7.** = **9.** ≠ **11.** ≠ **13.** ≠ **15.** = **17.** ≠ **19.** ≠ **21.** ≠ **23.** ≠ **25.** ≠ **27.** = **29.** = **31.** ≠ **33.** = **35.** = **37.** ≠ **39.** ≠ **41.** = **43.** Yes **45.** 9 **47.** $\frac{24}{13}$ **49.** $\frac{3}{1}$

Pages 246–247 **9.** 8 **11.** 8 **13.** $13\frac{1}{2}$ **15.** 52 **17.** $16\frac{4}{5}$ **19.** 5 **21.** 105 **23.** 32 **25.** $85\frac{1}{2}$ **27.** 4 **29.** 33 **31.** $93\frac{1}{3}$ **33.** 14 **35.** 3 **37.** 84 **39.** 84 points **41.** 0.1 **43.** 20.0 **45.** 1900 **47.** 44,300 **49.** 40,000

Pages 248–249 **7.** $0.10 **9.** $0.05 **11.** $0.0075 **13.** $7.12 for 8 batteries **15.** $4 for 6 pairs of socks **17.** $1.08 for a dozen eggs **19.** 3-pack of 24 prints **21.** 42,000 **23.** 810,000 **25.** 5.3 **27.** 448.6

Pages 250–251 **7.** 60 ft. **9.** 3 ft. **11.** 5 in. **13.** $3\frac{1}{4}$ in. **15.** $7\frac{1}{2}$ in. **17.** 90 ft. **19.** 5 in. **21.** Yes **23.** 107

Pages 252–253 **1.** 10 ft. **3.** $27\frac{1}{2}$ ft. **5.** $32\frac{1}{2}$ ft. **7.** Yes **9.** sofa—$\frac{3}{4}$ in. × $\frac{1}{4}$ in.; chairs—$\frac{1}{4}$ in. × $\frac{1}{4}$ in.; table—$\frac{1}{3}$ in. × $\frac{1}{10}$ in.; recliner—$\frac{1}{3}$ in. × $\frac{1}{3}$ in.; cabinet—$\frac{1}{5}$ in. × $\frac{1}{10}$ in.; t.v.—$\frac{1}{3}$ in. × $\frac{1}{5}$ in. **11.** 6:25 p.m.

Pages 254–255   **5.** 24 cm; 7.5 cm   **7.** $4\frac{1}{2}$ in.; $7\frac{1}{2}$ in.; $10\frac{1}{2}$ in.; 6 in.; 9 in.   **9.** $l = 9$ cm   **11.** 15 cm; 19.2 cm   **13.** 20 cm; 15 cm; 2.8 cm; 3.2 cm; 3.6 cm   **15.** $w = 26$ cm   **17.** No; a proportion is not formed   **19.** 27   **21.** 88

Pages 256–257   **1.** 1000 mi.   **3.** 2000 mi.   **5.** $\frac{\text{\tiny▨}}{900} = \frac{25}{3100}$; $\frac{\text{\tiny▨}}{1750} = \frac{25}{3100}$; 7 ft.; 14 ft.   **7.** $\frac{400}{250} = \frac{650}{\text{\tiny▨}}$; 400 frankfurters   **9.** $\frac{900}{1\frac{1}{2}} = \frac{\text{\tiny▨}}{5}$; 3000 mi.   **11.** Yes; $\frac{160}{3} = \frac{\text{\tiny▨}}{7}$; $373\frac{1}{3}$ mi. or 370 mi.   **13.** 0.95   **15.** 700   **17.** 80   **19.** 0.0313

Page 258   **1.** 60 to 7; $\frac{60}{7}$; 60 : 7   **3.** 63 to 18; $\frac{63}{18}$; 63 : 18   **5.** $\frac{\$2.50}{6}$, $\frac{\$3.75}{9}$, $\frac{\$5}{12}$ **7.** $\frac{\$50}{4}$, $\frac{\$75}{6}$, $\frac{\$100}{8}$ **9.** $\neq$ **11.** $\neq$ **13.** 2   **15.** 12   **17.** $3.99   **19.** $34.50 for 6 oz. of cologne   **21.** $\frac{1}{64} = \frac{4.5}{\text{\tiny▨}}$; 288 mi.   **23.** $\frac{1}{1\frac{1}{2}} = \frac{9}{\text{\tiny▨}}$; $\frac{1}{1\frac{1}{2}} = \frac{7}{\text{\tiny▨}}$; $13\frac{1}{2}$ ft. by $10\frac{1}{2}$ ft.

Pages 262–263   **1.** Less; you can find out how much less by multiplying $60 by 0.3 or $\frac{3}{10}$ **3.** Examples might include bank interest, sales tax, results of opinion polls, sales commission.   **5.** Equals   **7.** 3.7   **9.** 0.472   **11.** 0.87   **13.** 3.43

Pages 264–265   **7.** 65%   **9.** 50%   **11.** 100%   **13.** 21.5%   **15.** 7.5%   **17.** 0.08   **19.** 0.114   **21.** 0.135   **23.** 0.445   **25.** 0.665   **27.** 20%   **29.** 0.75   **31.** 0.165   **33.** 13.5%   **35.** 50.5%   **37.** 1%   **39.** 0.78   **41.** 2.1%   **43.** 0.078   **45.** 0.5   **47.** 0.75   **49.** 0.875   **51.** $\frac{33}{100}$ **53.** $\frac{7}{125}$ **55.** $\frac{1}{8}$

Pages 266–267   **5.** 20%   **7.** 75%   **9.** 68%   **11.** 3%   **13.** 20%   **15.** 15%   **17.** $33\frac{1}{3}$%   **19.** $83\frac{1}{3}$%   **21.** $12\frac{1}{2}$%   **23.** $68\frac{3}{4}$%   **25.** $33\frac{1}{3}$%   **27.** $6\frac{2}{3}$%   **29.** $37\frac{1}{2}$%   **31.** 40%   **33.** $6\frac{1}{4}$%   **35.** 75%   **37.** $16\frac{2}{3}$%   **39.** $93\frac{3}{4}$%   **41.** 70%   **43.** $63\frac{7}{11}$%   **45.** 25%; $16\frac{2}{3}$%; 15%; 10%; $8\frac{1}{3}$%; 20%; 5%   **47.** $\frac{3}{50}$ **49.** $\frac{19}{100}$ **51.** $\frac{31}{200}$

Pages 268–269   **7.** $\frac{3}{4}$ **9.** $\frac{99}{100}$ **11.** $\frac{1}{25}$ **13.** $\frac{6}{25}$ **15.** $\frac{73}{100}$ **17.** $\frac{1}{10}$ **19.** $\frac{11}{20}$ **21.** $\frac{16}{25}$ **23.** $\frac{1}{4}$ **25.** $\frac{3}{8}$ **27.** $\frac{5}{12}$ **29.** $\frac{151}{300}$ **31.** $\frac{39}{200}$ **33.** $\frac{1}{8}$ **35.** $\frac{39}{400}$ **37.** $\frac{1}{3}$ **39.** $\frac{11}{25}$ **41.** $\frac{23}{40}$ **43.** $\frac{7}{50}$ **45.** $\frac{1}{6}$ **47.** $\frac{7}{8}$ **49.** $\frac{27}{50}$ **51.** $\frac{13}{200}$ **53.** $\frac{31}{50}$ **55.** $\frac{1}{20}$ **57.** $\frac{1}{2}$

Pages 270–271   **1.** 30%; 20%   **3.** 30%; $\frac{3}{10}$; 30% or $\frac{3}{10}$ **5.** 100%   **7.** $\frac{2}{5}$; 40%   **9.**

Vehicle	Under 21	21–50	Over 50
Sports Car	$\frac{2}{5}$; 40%	$\frac{1}{10}$; 10%	$\frac{1}{8}$; $12\frac{1}{2}$%
Van	$\frac{1}{4}$; 25%	$\frac{1}{10}$; 10%	$\frac{1}{10}$; 10%
Economy	$\frac{1}{5}$; 20%	$\frac{1}{2}$; 50%	$\frac{1}{4}$; 25%
Full-size	$\frac{1}{10}$; 10%	$\frac{1}{4}$; 25%	$\frac{1}{2}$; 50%
Other	$\frac{1}{20}$; 5%	$\frac{1}{20}$; 5%	$\frac{1}{40}$; $2\frac{1}{2}$%

11. ≈800 or ≈810   **13.** ≈8000 or ≈8300

Pages 272–273   **7.** 170%   **9.** 400%   **11.** 70%   **13.** 0.1%   **15.** 0.75%   **17.** 0.5%   **19.** 0.7%   **21.** 425%   **23.** 50,000%   **25.** 22,500%   **27.** $\frac{1}{400}$; 0.0025   **29.** $\frac{1}{160}$; 0.00625   **31.** $\frac{81}{25}$; 3.24   **33.** $\frac{441}{200}$; 2.205   **35.** $\frac{1}{320}$; 0.003125   **37.** 0.75%   **39.** 425%   **41.** 0.4%   **43.** 0.3%   **45.** $387\frac{1}{2}$%   **47.** 708%   **49.** 625%   **51.** $741\frac{2}{3}$%   **53.** 0.03%   **55.** 1000%   **57.** 0.20%   **59.** 5.6   **61.** 99.11

Pages 274–275   **9.** $35.20   **11.** 20   **13.** 0.78   **15.** 154   **17.** 24.4   **19.** 60   **21.** 15   **23.** 437.5
25. 35.5   **27.** 18   **29.** 21   **31.** $7   **33.** $3.75   **35.** 0.21   **37.** 1065.9   **39.** 0.425   **41.** $1237.50
43. 7.2   **45.** 99.9   **47.** 2.2   **49.** 0.87   **51.** 1.96   **53.** 8.34

Pages 276–277   **7.** 75%   **9.** 50%   **11.** 43%   **13.** 42.5%   **15.** 12.5%   **17.** $1\frac{1}{3}$%   **19.** $33\frac{1}{3}$%
21. 35%   **23.** 2.5%   **25.** 43.75%   **27.** 100%   **29.** 150%   **31.** 0.0987   **33.** 30   **35.** 540,000

Pages 278–279   **5.** 55   **7.** 45   **9.** 515   **11.** 400   **13.** 72   **15.** 1000   **17.** 270   **19.** 51   **21.** 25
23. 2000   **25.** 18 mg   **27.** 2.5 in.

Pages 280–281   **7.** 25%   **9.** 50%   **11.** 150%   **13.** 200%   **15.** 40%   **17.** $12\frac{36}{47}$%   **19.** $8\frac{1}{3}$%
21. 50%   **23.** 14% increase   **25.** 40% decrease   **27.** 0.1% increase   **29.** $16\frac{2}{3}$% decrease   **31.** 30%
decrease   **33.** 2% increase   **35.** 10%   **37.** 10.13% decrease   **39.** ≈$110 or ≈$106   **41.** ≈$900 or
≈$899   **43.** ≈300 or ≈302

Pages 282–283   **7.** $5.40   **9.** $80.00   **11.** $64.00   **13.** $17.91   **15.** $36.00   **17.** $30.60   **19.** $4.44
21. $5.00   **23.** $4.75   **25.** $7.63   **27.** $154.73   **29.** $\frac{15}{14}$; $\frac{30}{28}$; $\frac{45}{42}$ **31.** $\frac{13}{1}$; $\frac{26}{2}$; $\frac{39}{3}$

Page 284   **1.** Hundred   **3.** 17.5%   **5.** 40%   **7.** 87.5%   **9.** 62.5%   **11.** 5%   **13.** $\frac{12}{25}$ **15.** $\frac{3}{8}$
17. $\frac{5}{12}$ **19.** 500%   **21.** 0.7%   **23.** $\frac{3}{400}$; 0.0075   **25.** $\frac{7}{800}$; 0.00875   **27.** $\frac{3}{500}$; 0.006   **29.** $1.12
31. 0.22   **33.** 5%   **35.** 800%   **37.** 900   **39.** 12   **41.** 20% decrease   **43.** $180.70

Pages 288–289   **1.** Favorable outcome   **3.** Event   **5.** $\frac{4}{4}$ **7.** $\frac{20}{20}$ **9.** $\frac{1}{5}$ **11.** 26 choices   **13.** 41

Pages 290–291   **5.** $\frac{5}{8}$ **7.** $\frac{1}{2}$ **9.** $\frac{1}{10}$ **11.** $\frac{1}{10}$ **13.** $\frac{3}{5}$ **15.** $\frac{2}{5}$ **17.** 0   **19.** $\frac{2}{5}$ **21.** 5 to 8; 5 : 8; $\frac{5}{8}$

Pages 292–293   **7.** 3 to 4   **9.** 2 to 5   **11.** 3 to 2   **13.** 2 to 3   **15.** 1 to 19   **17.** 1 to 1   **19.** 9 to 1
21. $\frac{1}{3}$ **23.** $\frac{7}{12}$ **25.** $\frac{28}{45}$ **27.** $\frac{3}{5}$

Pages 294–295   **1.** They are alike because they all have 1 as the numerator or denominator. They are different
because the ratios are all different values.   **3.** $\frac{2}{4}$ or $\frac{1}{2}$ **5.** odds of Cleaver winning   **7.** $\frac{1}{101}$ **9.** Chance
means probability and a $\frac{50}{50}$ probability would mean that she would definitely win. Madelyn's odds of winning are
$\frac{50}{50}$ (that is, she has a 50% chance of winning).   **11.** A probability can never be greater than 1. The statement
should be "The odds of winning are 2 to 1" (or "the probability of winning is $\frac{2}{3}$").   **13.** ≈4   **15.** ≈12

Pages 296–297
5.

```
           H — H H H
      H <
           T — H H T
 H <
           H — H T H
      T <
           T — H T T

           H — T H H
      H <
           T — T H T
 T <
           H — T T H
      T <
           T — T T T
```

7.

```
        David —— Ann, David
Ann <   Ed —————— Ann, Ed
        Fred ———— Ann, Fred

        David —— Betty, David
Betty < Ed —————— Betty, Ed
        Fred ———— Betty, Fred

        David —— Cathy, David
Cathy < Ed —————— Cathy, Ed
        Fred ———— Cathy, Fred
```

9. $\frac{1}{28}$

11.

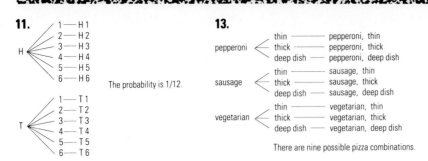

The probability is 1/12.

13.

pepperoni ← thin —— pepperoni, thin
thick —— pepperoni, thick
deep dish —— pepperoni, deep dish

sausage ← thin —— sausage, thin
thick —— sausage, thick
deep dish —— sausage, deep dish

vegetarian ← thin —— vegetarian, thin
thick —— vegetarian, thick
deep dish —— vegetarian, deep dish

There are nine possible pizza combinations.

15. $\frac{3}{10}$ **17.** $\frac{3}{4}$ **19.** $\frac{2}{7}$

Pages 298–299 **5.** 8 outcomes **7.** 24 outcomes **9.** $\frac{1}{36}$ **11.** $\frac{1}{12}$ **13.** 100; $\frac{1}{100}$ **15.** 15 pairs

17. $1\frac{1}{6}$ **19.** $\frac{1}{32}$ **21.** $\frac{15}{56}$

Pages 300–301 **1.** Disagree because the probability is still $\frac{1}{2}$. **3.** Heads and tails are still equally likely.
He can call either. **7.** $\frac{1}{40}$ **9.** If the card is replaced, the chance of winning both prizes is $\frac{1}{40} \times \frac{1}{40}$, or $\frac{1}{1600}$.
If the card is not replaced, the chance of winning both prizes is 0. **11.** The probability is $\frac{1}{2}$, but again other
factors might enter into the situation, such as the team might feel that they will lose because it is raining.
13. $0.\overline{09}$ **15.** $0.\overline{7}$ **17.** $0.91\overline{6}$ **19.** $7.\overline{6}$ **21.** $9.\overline{36}$ **23.** 2.19 **25.** $9\frac{1}{25}$ **27.** $8\frac{1}{8}$ **29.** $1\frac{111}{200}$

Page 302 **1.** $\frac{1}{15}$ **3.** $\frac{8}{15}$ **5.** 1, 2 **7.** 5 to 7; 7 to 5

9.

black ← white —— black shoes, white socks
blue —— black shoes, blue socks
green —— black shoes, green socks

brown ← white —— brown shoes, white socks
blue —— brown shoes, blue socks
green —— brown shoes, green socks

red ← white —— red shoes, white socks
blue —— red shoes, blue socks
green —— red shoes, green socks

11. $\frac{1}{3}$ **13.** 24 **15.** $\frac{1}{4}$

17. $\frac{1}{45}$ **19.** $\frac{1}{12}$

Pages 310–311
1.

length
width

3.

length
width

5. octa- **7.** deca- **9.** quad- **11.** Not regular **13.** Not regular **15.** 100 **17.** 567
19. 41,929.14 **21.** lw **23.** 2l

Pages 312–313 **7.** 18 in. **9.** 34 in. **11.** 126 ft. **13.** 50 cm **15.** 58 cm **17.** 18 cm **19.** 19 m
21. 22.4 cm **23.** 35.8 m **25.** 36 ft. **27.** 9 cm **29.** 294 ft. **31.** ≠ **33.** ≠

Pages 314–315 **7.** ≈5 cm **9.** ≈5 cm **11.** ≈6 cm **13.** ≈8 cm **15.** Answers will vary. A possible
answer is 20 cm. **17.** m **19.** cm

Pages 316–317 **9.** ≈25.1 in. **11.** ≈$14\frac{1}{7}$ or ≈14.1 ft. **13.** ≈33 in. **15.** ≈12.56 in. **17.** ≈5.0 km

19. ≈31.4 yd. **21.** ≈3.14 m **23.** ≈11.9 m **25.** ≈81.64 in. **27.** $12\frac{1}{4}$

Pages 318–319 **5.** 264 square inches **7.** 5 square meters **9.** 6500 square meters **11.** 324 square
yards **13.** 468 square meters **15.** 2500 square yards **17.** 84 square meters **19.** 180 square centimeters

21. 25 square meters **23.** 10.2 square meters **25.** 3.2 square centimeters **27.** 420 square inches
29. $93\frac{1}{2}$ square inches **31.** 63 **33.** 300

Pages 320–321 **5.** 23.22 square centimeters **7.** 90.25 square meters **9.** 13.44 square meters **11.** 2.66 square centimeters **13.** 700 square feet **15.** 2.24 square centimeters **17.** 387.5 square meters **19.** 180 square feet **21.** 19.6 square centimeters **23.** 12.6 square meters **25.** 182 square inches **27.** $127.52 **29.** 6.09

Pages 322–323 **7.** 45 square centimeters **9.** 3.08 square meters **11.** 153 square centimeters **13.** 9.75 square meters **15.** 81 square inches **17.** 21.21 square centimeters **19.** 2 gallons **21.** $1\frac{9}{92}$ **23.** 5

Pages 324–325 **7.** 452.16 sq. ft. **9.** $213\frac{51}{56}$ sq. in. **11.** 132.7 sq m **13.** 1384.7 sq m **15.** 19.6 sq. yd. **17.** 116.8 sq m **19.** 314 sq. yd. **21.** 706.5 sq. in. **23.** 12,167.7 sq. in. **25.** 3.1 sq. mi. **27.** 283.4 sq. mi. **29.** 167.3 sq mm **31.** 9498.5 sq. mi. **33.** 57 **35.** 94

Pages 326–327 **1.** 30 yd. **3.** 7200 square yards **5.** 20 ft. **7.** 8 ft. **9.** 204 square feet **11.** 519.25 square feet **13.** 120,000 square feet **15.** 30 ft. **17.** 15 ft.

Page 328 **1.** regular **3.** 54 in. **5.** 15.12 m **7.** 235.2 cm **9.** 25.12 m; 50.24 square meters **11.** 81.64 m; 530.7 square meters **13.** 104.88 square meters **15.** 98 sq cm

Pages 332–333 **1.** Polygons make up the second figure; circles make up part of the first and third figures. **3.** a square and 4 triangles **5.** 31.36 square centimeters **9.** 5 faces; 8 edges; 5 vertices

Pages 334–335 **1.** 6 faces; 12 edges; 8 vertices **3.** Forms cube **5.** Does not form cube **7.** Drawings will vary. **9.** A square **11.** 5 faces; 8 edges; 5 vertices **13.** 5 faces; 8 edges; 5 vertices **15.** $13\frac{1}{8}$ square inches **17.** $35\frac{1}{16}$ square inches **19.** 257.92 square centimeters **21.** $162\frac{9}{16}$ square inches

Pages 336–337 **5.** 864 square centimeters **7.** 40.56 square centimeters **9.** 1160 square centimeters **11.** 232.12 square centimeters **13.** 1858.56 square centimeters **15.** $277\frac{1}{8}$ square inches **17.** $121\frac{1}{2}$ square inches **19.** $198\frac{3}{4}$ square inches **21.** No; it is 4 times as large **23.** 108 square inches **25.** 45.5 square inches

Pages 338–339 **5.** 56 square inches **7.** 280 square inches **9.** $\approx$314 square inches **11.** $\approx$1406.72 or $\approx$1408 square centimeters **13.** $\approx$3428.88 or $\approx$3432 square inches **15.** $\approx$414$\frac{6}{7}$ or $\approx$414.48 square inches **17.** 180 square feet **19.** 4.8 square meters **21.** $66\frac{1}{2}$ square feet

Pages 340–341 **5.** 1000 cubic centimeters **7.** 4096 cubic centimeters **9.** 630 cubic meters **11.** 3040 cubic feet **13.** 60 cubic inches **15.** 288 cubic inches **17.** 200 cubic inches **19.** 204.8 cubic centimeters **21.** 91,636,272 cubic feet **23.** = **25.** =

Pages 342–343 **1.** 10 cubic inches **3.** 720 boxes **5.** 105,120 marbles **7.** Multiply 250 by 90 **9.** About 2925 cubic feet **11.** Find the weight of 100 nickels and multiply by 20. **13.** Estimate the number of people served by a gallon. Estimate by dividing the number attending by the first part. **15.** Estimate gallons used in a week and multiply by 52. **17.** Multiply or divide; $6.23

Pages 344–345 **5.** $\approx$226.08 cubic inches **7.** $\approx$836.5 cubic meters **9.** $\approx$113.04 cubic centimeters **11.** $\approx$180.06 cubic meters **13.** $\approx$5652 cubic feet **15.** $\approx$12.95 cubic meters **17.** $\approx$192.33 cubic inches **19.** $\approx$184.6 or $184\frac{4}{5}$ cubic feet **21.** 226 cubic inches **23.** 98 **25.** 52 **27.** 39

Pages 346–347   1. 18.84 cubic yards   **3.** $V = 67\frac{1}{2}$ cubic inches; $A = 112\frac{7}{8}$ square inches   **5.** 516 square yards; 5160 square yards   **7.** 3870 square yards   **9.** $3.95; paper-and-pencil or calculator
11. $\approx$36.8 miles per gallon; calculator

Page 348   1. cylinder   **3.** pyramids   **5.** 82 square meters   **7.** 160 square centimeters   **9.** 8.64 square centimeters   **11.** 432 square inches   **13.** 280 cubic feet   **15.** $\approx$552.64 cubic centimeters   **17.** $\approx$923.16 or $\approx$924 cubic inches   **19.** $\approx$2486.88 cubic centimeters

Pages 352–353   1. Unknown; variable   **3.** Equation   **5.** How much does Nora earn after taxes?
7. How much profit did the Morgans make when they sold their house?   **9.** How much will the costume cost?
11. 27   **13.** 13   **15.** 15   **17.** 18   **19.** 30   **21.** 15

Pages 354–355   7. $72 - 12$ **9.** $40 + 18$ **11.** $99\,n$ **13.** $5 + n$ **15.** $16 - 2n$ **17.** $5n - 5$
19. Some number divided by three   **21.** Thirty-six times some number plus seven   **23.** $36 \div 3$ or $\frac{36}{3}$
25. $7 - n$ **27.** $11n$ **29.** $33 - 22$ **31.** 1   **33.** 6   **35.** $6\frac{1}{3}$ **37.** 4   **39.** 3   **41.** 9

Pages 356–357   5. 16   **7.** 6   **9.** 8   **11.** 126   **13.** 13   **15.** 122   **17.** 18   **19.** 2   **21.** 14
23. 10   **25.** $s + 5$; 14   **27.** $\frac{6}{8}; \frac{9}{12}; \frac{12}{16}$ **29.** $\frac{2}{4}; \frac{3}{6}; \frac{4}{8}$ **31.** $\frac{2}{24}; \frac{3}{36}; \frac{4}{48}$ **33.** $\frac{4}{5}$ **35.** $\frac{2}{3}$ **37.** $\frac{3}{4}$

Pages 358–359   5. $n + 8 = 13$ **7.** $\frac{12}{n} = 4$ **9.** $2n - 1 = 37$ **11.** $\frac{4n}{2} + 2 = 10$ **13.** $\frac{12}{n} + 2 = 3$
15. $3n + 8 = 16$ **17.** $\frac{n}{2} = 28$ **19.** $2s + 450,000 = 1,850,000$ **21.** 23.14   **23.** 111   **25.** 12,073
27. 2143

Pages 360–361   7. $s = 26$ **9.** $v = 30$ **11.** 34   **13.** $b = 100$ **15.** $d = 2$ **17.** $a = 2\frac{1}{2}$
19. $t = 13$ **21.** $z = 10$ **23.** $z = 33$ **25.** $2 + n = 22$; $n = 20$ **27.** $n - 5 = 13$; $n = 18$
29. $5n + 6 = 31$; $n = 5$ **31.** $n - 7 = 41$; $n = 48$ **33.** Yes; a possible answer is to do an opposite operation   **35.** $10; mental math

Pages 362–363   5. $n = 6$ **7.** $d = 17$ **9.** $c = 7$ **11.** $a = 27$ **13.** $n = 23$ **15.** $f = 15$
17. $n = 23$ **19.** $c = 7$ **21.** $a = 26$ **23.** $b = 40$ **25.** $d = 7$ **27.** $p = 1$ **29.** $n = 7$
31. $f = 20$ **33.** $e + 80,773 = 88,700$; $e = 7927$ **35.** 17   **37.** 95   **39.** 4361   **41.** 65   **43.** 5
45. 16

Pages 364–365   5. $d = 7$ **7.** $b = 11$ **9.** $r = 7$ **11.** $h = 13$ **13.** $k = 6$ **15.** $c = 13$
17. $s = 56$ **19.** $n = 40$ **21.** $t = 25$ **23.** $r = 88$ **25.** $k = 196$ **27.** $y = 105$ **29.** $k = 12$
31. $y = 132$ **33.** $f = 6$ **35.** $m = 85$ **37.** $k = 100$ **39.** $w = 108$ **41.** $4n = 72$; $n = 18$ **43.** $\frac{9}{32}$
45. $\frac{3}{5}$ **47.** $2\frac{1}{3}$

Pages 366–367   5. $y = 4$ **7.** $f = 7$ **9.** $n = 2$ **11.** $q = 8$ **13.** $r = 190$ **15.** $b = 34$
17. $d = 256$ **19.** $c = 60$ **21.** $h = 99$ **23.** $k = 3$ **27.** $24 + 0.08m = 72$; $m = 600$

Page 368   1. Expression; variable or unknown   **3.** $5 + n$ **5.** 5×4 **7.** Five plus eighteen times some number   **9.** Eight more than four times some number   **11.** Forty-two minus two times some number   **13.** 12
15. $\frac{n}{9} = 10$ **17.** $d = 40$; $40 - 8 = 32$ **19.** $a = 57$; $57 + 33 = 90$ **21.** $e = 44$; $44 + 12 = 56$
23. $h = 51$; $51 - 30 = 21$ **25.** $c = 22$; $4 \times 22 = 88$ **27.** $e = 6$; $12 \times 6 = 72$ **29.** $a = 9$;
$12 \times 9 = 108$ **31.** $n = 729$; $\frac{729}{9} = 81$ **33.** $a = 25$ **35.** $c = 6$ **37.** $a = 5$ **39.** $w = 639$
41. $b = 7$

Pages 372–373 **1.** Possible answers include positive self-image, positive that you are going to do something; negative attitude, negative self-esteem; across or opposite someone, opposite opinion. **3.** 14,776 ft. **5.** 75 **7.** 4887 **9.** 5781 **11.** 11,467 **13.** 537 **15.** 753 **17.** 17 **19.** 8 **21.** 317 **23.** 2566 **25.** 32 **27.** 97 **29.** $5344 - 4538 = n$; 806 ft.; Mississippi

Pages 374–375 **5.** +8 **7.** +13 **9.** +15 **11.** +9 **13.** +8 **15.** −4 **17.** −8 **19.** −12 **21.** −5 **23.** −9 **25.** +1 **27.** −1 **29.** +4 **31.** −4 **33.** +1 **35.** −6 **37.** −1 **39.** −11 **41.** −9 **43.** −2 **45.** Loss of 5 yd. **47.** 120 **49.** 36

Pages 376–377 **5.** +5 **7.** +7 **9.** +9 **11.** +16 **13.** +7 **15.** −3 **17.** −5 **19.** +7 **21.** −2 **23.** −6 **25.** −4 **27.** +1 **29.** +4 **31.** 0 **33.** +1 **35.** +6 **37.** −3 **39.** +24 **41.** −3 **43.** −8 **45.** −4 **47.** −15 **49.** +4 **51.** 84 °F **53.** 942 **55.** 380

Pages 378–379 **7.** Arch & 12th **9.** Race & 11th **11.** Afro-American History Museum **13.** U.S. Mint **15.** Answers will vary. **17.** 50% **19.** 10% **21.** 20% **23.** $81\frac{1}{4}$%

Pages 380–381 **7.** +20 **9.** +10 **11.** +56 **13.** +72 **15.** +200 **17.** +240 **19.** +18 **21.** +99 **23.** +28 **25.** +48 **27.** +81 **29.** −24 **31.** −24 **33.** −63 **35.** −84 **37.** +15 **39.** +9 **41.** +105 **43.** +80 **45.** −90 **47.** +25 **49.** −56 **51.** +80 **53.** −14 points **55.** 50 **57.** 328 **59.** 12

Pages 382–383 **9.** +5 **11.** +5 **13.** +7 **15.** +12 **17.** +8 **19.** +9 **21.** +2 **23.** +1 **25.** +11 **27.** +9 **29.** −5 **31.** −4 **33.** −10 **35.** −9 **37.** −7 **39.** −5 **41.** +24 **43.** −3 **45.** +13 **47.** +6 **49.** +14 **51.** −12 **53.** $0.09 **55.** $a = 86$ **57.** $x = 4.4$ **59.** $b = 82$ **61.** $r = 45$

Pages 384–385 **7.** $x = -4$ **9.** $z = -1$ **11.** $e = +16$ **13.** $b = +16$ **15.** $x = -5$ **17.** $z = -12$ **19.** $a = 0$ **21.** $c = +5$ **23.** $e = 0$ **25.** $g = -1$ **27.** $i = +3$ **29.** $k = 0$ **31.** $m = -10$ **33.** $p = +2$ **35.** $r = -4$ **37.** $v = +3$ **39.** $n - 12 = -3$; $n = +9$ **41.** $n = 112$ **43.** $s = 12$

Pages 386–387 **5.** $p = -3$ **7.** $n = 0$ **9.** $p = -8$ **11.** $n = -18$ **13.** $p = +1$ **15.** $n = 0$ **17.** $a = -9$ **19.** $c = -1$ **21.** $e = -15$ **23.** $q = -2$ **25.** 3 below par **27.** $a = 8$ **29.** $b = 15$

Pages 388–389 **1.** Any reasonable estimate is acceptable. **3.** Answers may range from 1 to 3 words. **5.** Answers may range from 2700 to 8100 words. **7.** Any reasonable estimate is acceptable. Steps may include determining the number of minutes in a day and multiplying by the number of breaths in a minute, and so on. **9.** Any reasonable answer is acceptable. Steps may include counting the number of teenagers in the classroom, estimating the total number in the school, the community, the state, and so on. **11.** Any reasonable answer is acceptable. Steps may include estimating the number of dentists needed for 100 people, estimating the population of California, and so on. **13.** $n \div 13 = 4$; $n = 52$ **15.** $20 \times n = 60$; $n = 3$ **17.** $18 + 2 = n$; $n = 20$ **19.** $1 + 1 \times n = 6$; $n = 5$

Pages 390–391 **5.** $k = +4$ **7.** $n = -2$ **9.** $p = -12$ **11.** $k = -7$ **13.** $n = +3$ **15.** $r = +32$ **17.** $w = +4$ **19.** $j = +3$ **21.** 12 bird houses **23.** 5 **25.** 100 **27.** 132

Pages 392–393 **1.** 7 games; 3 games; 10 games **3.** 2 games **5.** 1 game **7.** question 4 **9.** 0.554 **11.** 0.504 **13.** 1 **15.** 0.446 **17.** 12.5 **19.** 1.74 **21.** 2.49 **23.** 7:07 a.m.

Page 394 **1.** False **3.** True **5.** −20 **7.** −6 **9.** +12 **11.** 0 **13.** +7 **15.** −8 **17.** +10 **19.** +10 **21.** −72 **23.** −50 **25.** +26 **27.** +49 **29.** +9 **31.** −5 **33.** −8 **35.** −4 **37.** $a = -18$ **39.** $e = +30$ **41.** $n = -4$ **43.** $z = -81$ **45.** $n = +8$ **47.** $r = -36$

Photo/Art Credits

Cover John Payne
Title Page John Payne

Assignment Photography

Ralph Brunke: 1, *r* 2, 9, 10, 11, 12, 14, 15, 16, 18, 19, 20, 21, 23, 25, 29, *t* 30, 31, 32, 33, *r* 34, 36, 40, 44, 45, 46, 48, 49, *b* 50, 58, 59, 62, 64, *r* 76, 77, 78, 79, 80, 85, *b* 88, *b* 92, *l* 94, 95, 96, 97, *l* 99, *t* 100, 109, *l* 112, 113, 114, 116, 117, 129, 130, *r* 132, 133, *l* 144, 145, 149, 150, 151, 160, 175, *l* 218, 223, *tr,b* 226, 228, 229, 241, 243, 250, 251, 252, 253, 255, *r* 266, 270, 272, 273, *r,l* 274, 280, 282, 287, *t* 288, *r* 290, 291, 293, 297, 302, 303, 309, 314, 315, 316, 317, 331, 338, 342, 347, 351, *l* 352, 354, 356, 357, 365, *l* 366, 371, *l* 374, *l* 378, 380, *b* 386, 389, 391, *b* 392; Charles Shotwell: facing 1, *l* 2, 4, 28, *b* 30, *l* 34, *t* 50, 56, 60, 66, 72, *l* 76, 84, *t* 88, 90, *br* 92, *r* 94, 108, *r* 112, 120, 122, 128, *l* 132, 134, 136, 138, 140, 142, *r* 144, 148, 156, 164, 174, *r* 218, 222, *tl* 226, 230, 232, 240, *in* 242, 244, 248, 260, *l* 266, *c* 274, 276, 286, *l* 290, 292, 296, 298, 300, 308, 310, *in* 317, 322, 330, 344, 350, *r* 352, 360, *l* 364, *r* 366, 370, *r* 374, 376, *r* 378, *l* 382, *t* 386; 22 Robert Beck, Focus West; 91 Stephen Earley; 92 *t* Bettmann Archive; 93 Michael Melford, Wheeler Pictures; 98 Georgine Knoll; 99 Fritz Bronzel, Peter Arnold; 100 *b* Peter Hendrie, Image Bank; 101 Robert Frerck, Click/Chicago; 166 Michael Melford, Image Bank; 201 Frithfoto, Bruce Coleman; 202 Steve Satushek, Image Bank; 212 Bettmann Archive; 226 Frans Lanting; 233 NASA; 242 H. Armstrong Roberts; 254 Suzanne Murphy, Click/Chicago; 257 *l* FPG; *r* FPG; 264 Focus on Sports; 270 *t* Frank Loose; *b* Thomas Craig, FPG; 288 *bl* William Warren, Click/Chicago; *br* Steve Meltzer, West Stock; 301 Marc Pokempner, Click/Chicago; 309 Aram Gesar, Image Bank; 326 Chuck O'Rear, West Light; 346 *l* Robert Perron; *c* Robert Perron; *r* Robert Perron; 358 Terry Murphy, Animals, Animals; 363 JPL; 364 *r* Walter Chandoha; 372 *b* Stock Photos, Image Bank; *t* T. Mareschal, Image Bank; 382 Frank Loose; 388 *t* Eric Meola, Image Bank; *b* Fermilab; 392 Robert Soltis, Nawrocki Stock Photo.

Illustrators

Martin Austin: 92, 130, 139, 157, 165, 169, 177, 189, 194, 261, 310, 353, 358, 377, 383; Steven Boswick: 29, 40, 46, 51, 66, 70, 71, 72, 90, 98, 100, 113, 134, 185, 190, 194, 223, 234, 253, 334, 340, 373, 386, 399; David Lee Csicsko: 2, 24, 176, 287, 360; Steve George: 12; Linda Kelen: 56, 57, 110, 188, 294, 351; Ligature: 3, 4, 6, 8, 13, 20, 30, 37, 38, 45, 47, 48, 55, 63, 68, 85, 88, 90, 98, 100, 109, 110, 111, 124, 125, 126, 127, 129, 137, 138, 139, 149, 165, 173, 174, 178, 187, 192, 202, 203, 204, 205, 206, 207, 210, 211, 212, 213, 214, 215, 217, 220, 221, 225, 226, 229, 230, 231, 234, 249, 250, 252, 255, 263, 264, 265, 268, 270, 271, 272, 275, 277, 281, 289, 291, 307, 309, 311, 312, 313, 315, 316, 317, 318, 319, 320, 321, 322, 323, 324, 325, 326, 327, 328, 329, 332, 333, 334, 336, 337, 339, 340, 341, 342, 344, 346, 347, 348, 349, 360, 372, 378, 379, 381, 392, 399; Tim McWilliams: 6, 60, 74, 158, 199, 224, 258, 262, 263, 290, 291, 343, 384, 390; Precision Graphics: 9, 56, 98, 119, 140, 154, 155, 168, 182, 210, 236, 237, 312, 332, 343, 344, 353, 356, 372; William Rieser: 124, 250, 292, 293, 346, 362; Slug Signorino: 118, 162, 163, 324; Dave Stuckey: 216, 218, 246, 300; Paul Vaccarello: 107, 109, 129, 140, 149, 159, 224, 232, 248, 278, 279, 280, 296, 336, 376.